AF769118

Early Life Origins of Health and Disease

ADVANCES IN EXPERIMENTAL MEDICINE AND BIOLOGY

Recent Volumes in this Series

Volume 564
GLYCOBIOLOGY AND MEDICINE: PROCEEDINGS OF THE 7TH JENNER GLYCOBIOLOGY AND MEDICINE SYMPOSIUM
Edited by John S. Axford

Volume 565
SLIDING FILAMENT MECHANISM IN MUSCLE CONTRACTION: FIFTY YEARS OF RESEARCH
Edited by Haruo Sugi

Volume 566
OXYGEN TRANSPORT TO TISSUE XXVI
Edited by Paul Okunieff, Jacqueline Williams, and Yuhchyau Chen

Volume 567
THE GROWTH HORMONE-INSULIN-LIKE GROWTH FACTOR AXIS DURING DEVELOPMENT
Edited by Isabel Varela-Nieto and Julie A. Chowen

Volume 568
HOT TOPICS IN INFECTION AND IMMUNITY IN CHILDREN II
Edited by Andrew J. Pollard and Adam Finn

Volume 569
EARLY NUTRITION AND ITS LATER CONSEQUENCES: NEW OPPORTUNITIES
Edited by Berthold Koletzko, Peter Dodds, Hans Akerbloom, and Margaret Ashwell

Volume 570
GENOME INSTABILITY IN CANCER DEVELOPMENT
Edited by Erich A. Nigg

Volume 571
ADVANCES IN MYCOLOGY
Edited by J.I. Pitts, A.D. Hocking, and U. Thrane

Volume 572
RETINAL DEGENERATIVE DISEASES
Edited by Joe Hollyfield, Robert Anderson, and Matthew LaVail

A Continuation Order Plan is available for this series. A continuation order will bring delivery of each new volume immediately upon publication. Volumes are billed only upon actual shipment. For further information please contact the publisher.

Early Life Origins of Health and Disease

Edited by

E. Marelyn Wintour

Department of Physiology, Monash University, Clayton, Victoria, Australia

Julie A. Owens

Department of Obstetrics and Gynaecology, University of Adelaide, Adelaide, South Australia

Springer Science+Business Media

Landes Bioscience / Eurekah.com

Springer Science+Business Media
Eurekah.com / Landes Bioscience

Springer Science+Business Media, 233 Spring Street, New York, New York 10013, U.S.A.

Please address all inquiries to the Publishers:
Eurekah.com / Landes Bioscience, 810 South Church Street, Georgetown, Texas, U.S.A. 78626
Phone: 512/ 863 7762; FAX: 512/ 863 0081
http://www.eurekah.com
http://www.landesbioscience.com

Early Life Origins of Health and Disease edited by E. Marelyn Wintour and Julie A. Owens, Landes Bioscience / Springer Science+Business Media dual imprint / Springer series: Advances in Experimental Medicine and Biology

ISBN 978-1-4419-3950-0 eISBN 978-0-387-32632-0

While the authors, editors and publisher believe that drug selection and dosage and the specifications and usage of equipment and devices, as set forth in this book, are in accord with current recommendations and practice at the time of publication, they make no warranty, expressed or implied, with respect to material described in this book. In view of the ongoing research, equipment development, changes in governmental regulations and the rapid accumulation of information relating to the biomedical sciences, the reader is urged to carefully review and evaluate the information provided herein.

Library of Congress Cataloging-in-Publication Data

Early life origins of health and disease / edited by E. Marelyn Wintour, Julie A. Owens.
p. ; cm. -- (Advances in experimental medicine and biology ; v. 573)
Includes bibliographical references and index.

1. Prenatal influences. 2. Diseases--Causes and theories of causation. 3. Diseases--Susceptibility. I. Wintour, E. Marelyn. II. Owens, Julie A. III. Title. IV. Series.
[DNLM: 1. Embryonic Development. 2. Fetal Development. 3. Disease--etiology. 4. Disease Susceptibility. 5. Maternal Exposure. 6. Risk Factors. WQ 210.5 E125 2006]
RJ91.E27 2006
618.2'4--dc22

2005023215

PREFACE

When we agreed to edit this book we established some guidelines for the authors who were asked to contribute. The authors were asked to include all relevant material currently available, but to be critical of the methodology used to obtain the various types of data. In addition to evaluating the methodology used in each quoted study, they were asked to suggest the optimal state of the art methodology which should be used for each type of investigation. Thus Denton et al (Ch. 9) considered the optimal way (conscious and undisturbed) and length of time for which blood pressure should be measured, for the results to be most meaningful. They also considered all the components of the cardiovascular system which required examination if the full impact of some programming stimulus were to be investigated fully. This is continued in the examination of other aspects of cardiovascular dysfunction (Poston et al, Ch. 10) The same criteria were applied to assessment of nephron number (Moritz and Cullen-McEwen, Ch.11) and optimal methodology (unbiased stereology) suggested. All authors paid attention to the type of statistics to be used and stressed where appropriate the importance of studying both sexes of offspring. Equal rigor was used in assessing metabolic changes induced by pre/perinatal conditions (Gatford et al, Ch.13). Simon Langley-Evans (Ch. 8) was asked to make sure that readers would appreciate that not all low-protein diets are equivalent. Ruth Morley (Ch. 3) was commissioned to make it clear that not all monozygotic twins share one placenta and to give a critical evaluation of what can or cannot be learned from a study of twins. In short, any investigator planning a study of the early life origins of health and disease, having read this book, should be able to devise the best possible experiment, using optimal methodology, to give the utmost reliable outcomes.

The book covers data relevant to humans (Chs. 1-5), and various animal models (Chs. 8-14). After the whole background to the concept is set by the current president and secretary of the international society devoted to this area (DOHaD), an expert in epidemiology (Fall, Ch. 2) gives a masterly summation of the past findings. In a timely reminder the peri-implantation embryo is considered as a vulnerable stage (Thompson et al, Ch. 5). In addition, the potential mechanisms by which such programming might occur are covered in the two chapters on epigenetics (Chs. 6, 7).

Finally there are four chapters which cover emerging areas of great potential interest (Chs. 15-18). In all these areas (vitamin D deficiency, hypoxia, alcohol exposure, adult mental health) there are limited data which suggest that an influence exerted during development might have long-term consequences for adult offspring, but much more investigation is required.

This should be a most valuable resource book for all those currently engaged in the study of the influences of the prenatal environment on future health, as well as for those who are just contemplating beginning work in this area.

E. Marelyn Wintour and Julie A. Owens

PARTICIPANTS

James A. Armitage
Maternal and Fetal Research Unit
Division of Reproductive Health
Endocrinology and Development
King s College
London
UK
james.armitage@kcl.ac.uk

Leanne Bellinger
Centre for Reproduction and Early Life
School of Biosciences
University of Nottingham
Sutton Bonington, Loughborough
Leicestershire
UK
sbxlbl@gwmail.Nottingham.ac.uk

Laura Bennet
Fetal Physiology and Neuroscience Group
Department of Physiology
University of Auckland
Auckland
New Zealand
l.bennet@auckland.ac.nz

Bernhard H. Breier
Liggins Institute for Medical Research
Faculty of Medical and Health Sciences
University of Auckland
Auckland
New Zealand
b.breier@auckland.ac.nz

Luke C. Carey
Department of Obstetrics and Gynecology
Wake Forest University School of Medicine
Winston-Salem, North Carolina
USA
lcarey@wfubmc.edu

John B. Carlin
Department of Paediatrics
University of Melbourne
and
Murdoch Childrens Research Institute
Royal Children s Hospital
Parkville, Victoria
Australia
jbcarlin@unimelb.edu.au

Luise A. Cullen-McEwen
Department of Anatomy and Cell Biology
Monash University
Clayton, Victoria
Australia
luise.cullen-mcewen@monash.edu.au

Miles J. De Blasio
Department of Obstetrics and Gynaecology
Monash University
Victoria
Australia
miles.deblasio@adelaide.edu.au

Kate M. Denton
Department of Physiology
Monash University
Victoria
Australia
kate.denton@med.monash.edu.au

Miodrag Dodic
Department of Physiology
Monash University
Victoria
Australia
miodrag.dodic@med.monash.edu.au

Terence Dwyer
Murdoch Childrens Research Institute
Royal Children s Hospital
Parkville, Victoria
Australia
terry.dwyer@mcri.edu.au

Caroline H.D. Fall
Reader in Epidemiology and Consultant in Child Health
MRC Environmental Epidemiology Unit
University of Southampton
Southampton
UK
chdf@mrc.soton.ac.uk

Kathryn L. Gatford
Department of Obstetrics and Gynaecology
Monash University
Victoria
Australia
kathy.gatford@adelaide.edu.au

Dino A. Giussani
Department of Physiology
University of Cambridge
Cambridge
UK
dag26@cam.ac.uk

Peter D. Gluckman
Liggins Institute
University of Auckland and National Research Centre
for Growth and Development
Auckland
New Zealand
pd.gluckman@auckland.ac.nz

Alistair J. Gunn
Fetal Physiology and Neuroscience Group
Department of Physiology
University of Auckland
Auckland
New Zealand
aj.gunn@auckland.ac.nz

Mark A. Hanson
Centre for Developmental Origins of Health and Disease
University of Southampton
Princess Anne Hospital
Southampton
UK
m.hanson@soton.ac.uk

David J. Henderson-Smart
NSW Centre for Perinatal Health Services Research
Queen Elizabeth II Research Institute
University of Sydney
Sydney, NSW
Australia
dhs@perinatal.usyd.edu.au

Dane M. Horton
Department of Physiology
University of Adelaide
Adelaide
South Australia
dane.horton@adelaide.edu.au

Michelle M. Kett
Department of Physiology
Monash University
Victoria
Australia
michelle.kett@med.monash.edu.au

Karen L. Kind
Department of Obstetrics
and Gynaecology
Monash University
Victoria
Australia
karen.kind@adelaide.edu.au

Stefan O. Krechowec
Liggins Institute for Medical Research
Faculty of Medical and Health
Sciences
University of Auckland
Auckland
New Zealand
s.krechowec@auckland.ac.nz

Michelle Lane
Department of Obstetrics
and Gynaecology
Research Centre for Reproductive
Health
University of Adelaide
Adelaide
South Australia
michelle.lane@adelaide.edu.au

Alison Langley-Evans
Centre for Reproduction and Early Life
School of Biosciences
University of Nottingham
Sutton Bonington, Loughborough
Leicestershire
UK
alison.langley-
evans@nottingham.ac.uk

Simon C. Langley-Evans
Centre for Reproduction and Early Life
School of Biosciences
University of Nottingham
Sutton Bonington, Loughborough
Leicestershire
UK
simon.langley-
evans@nottingham.ac.uk

Sarah McMullen
Centre for Reproduction and Early Life
School of Biosciences
University of Nottingham
Sutton Bonington, Loughborough
Leicestershire
UK
sarah.mcmullen@nottingham.ac.uk

Karen M. Moritz
Department of Anatomy
and Cell Biology
Monash University
Clayton, Victoria
Australia
karen.moritz@med.monash.edu.au

Ruth Morley
Department of Paediatrics
University of Melbourne
and
Murdoch Childrens Research Institute
Royal Children s Hospital
Parkville, Victoria
Australia
morleyr@unimelb.edu.au

Timothy J.M. Moss
School of Women s and Infants Health
University of Western Australia
Crawley
Australia
tmoss@cyllene.uwa.edu.au

Julie A. Owens
Department of Obstetrics and Gynaecology
University of Adelaide
Adelaide
South Australia
julie.owens@adelaide.edu.au

Helena C. Parkington
Department of Physiology
Monash University
Clayton, Victoria
Australia
helena.parkington@med.monash.edu.au

Julia B. Pitcher
Department of Obstetrics and Gynaecology
University of Adelaide
Women s and Children s Hospital
North Adelaide
South Australia
julia.pitcher@adelaide.edu.au

Lucilla Poston
Maternal and Fetal Research Unit
Division of Reproductive Health
Endocrinology and Development
King s College
London
UK
lucilla.poston@kcl.ac.uk

Sarah Robertson
Department of Obstetrics and Gynaecology
Research Centre for Reproductive Health
University of Adelaide
Adelaide
South Australia
sarah.robertson@adelaide.edu.au

Jeffrey S. Robinson
Department of Obstetrics and Gynaecology
University of Adelaide
Women s and Children s Hospital
North Adelaide
South Australia
jeffrey.robinson@adelaide.edu.au

Jeff Schwartz
Discipline of Physiology
School of Molecular and Biomedical Science
University of Adelaide
Adelaide
South Australia
jeff.schwartz@adelaide.edu.au

Dean Sculley
Centre for Reproduction and Early Life
School of Biosciences
University of Nottingham
Sutton Bonington, Loughborough
Leicestershire
UK
dean.sculley@nottingham.ac.uk

Deborah M. Sloboda
School of Women s and Infants Health
University of Western Australia
Crawley
Australia
dsloboda@obsgyn.uwa.edu.au

Marianne Tare
Department of Physiology
Monash University
Clayton, Victoria
Australia
marianne.tare@med.monash.edu.au

Paul D. Taylor
Maternal and Fetal Research Unit
Division of Reproductive Health
Endocrinology and Development
King s College
London
UK
paul.taylor@auckland.ac.uk

Jeremy Thompson
Department of Obstetrics
and Gynaecology
Research Centre for Reproductive
Health
University of Adelaide
Adelaide
South Australia
jeremy.thompson@adelaide.edu.au

Mark H. Vickers
Liggins Institute for Medical Research
Faculty of Medical and Health Sciences
University of Auckland
Auckland
New Zealand
m.vickers@auckland.ac.nz

Nicola Vickaryous
School of Molecular and Microbial
Biosciences
University of Sydney
Sydney
Australia
n.vickaryous@mmb.usyd.edu.au

Robert A. Waterland
Departments of Pediatrics
and Molecular and Human Genetics
Baylor College of Medicine
USDA Children s Nutrition Research
Center
Houston, Texas
USA
waterland@bcm.tmc.edu

Emma Whitelaw
School of Molecular and Microbial
Biosciences
University of Sydney
Sydney
Australia
e.whitelaw@mmb.usyd.edu.au

E. Marelyn Wintour
Department of Physiology
Monash University
Clayton, Victoria
Australia
mwc@med.monash.edu.au

CONTENTS

CHAPTER 1

The Developmental Origins of Health and Disease:

The Breadth and Importance of the Concept

Peter D. Gluckman* and Mark A. Hanson

Fetal Origins of Adult Disease (FOAD)

The concept of a 'fetal origins of adult disease' (FOAD) or 'fetal programming' was developed by Barker and colleagues to describe the relationship between birth size and subsequent risks of cardiovascular disease and insulin resistance/Type 2 diabetes mellitus. As a concept, FOAD was initially received with criticism. Some held the view that the answers lay within genetics and the gene; for others that the original epidemiological interpretations were flawed.

It is now nearly two decades since these landmark observations and concepts first appeared. It is apparent that those original findings have had far-reaching implications regarding human health and lifestyle choices, not only explaining the rapid societal rise in diabetes and obesity, but also covering areas as diverse as osteoporosis, depression and sedentary behaviour. With the wisdom of hindsight, we can see that some of the reluctance to accept the FOAD concept arose precisely from the problem which FOAD addressed: namely that the underlying causes of the common chronic diseases of adulthood (heart disease, diabetes, stroke) could not be explained purely in terms of genetic inheritance or lifestyle factors, such as diet or exercise. That instead, gene-environment interactions would hold the clues.

The concept of FOAD has expanded since the initial observations. The term 'fetal origins of adult disease' has now been replaced with 'developmental origins of adult disease' (DOHaD) to take into account its influence over an expanded developmental time-frame. Moreover, it has launched a new way of thinking about the evolution of human health and disease, which we refer to as the 'predictive adaptive response' and will be discussed further.

Early Clues

The linkages between early developmental events and eventual adult susceptibilities had been noted before, and in some rather unexpected ways. Kawahata and Sakamoto[1] noted that of those soldiers stationed in the tropics during WWII, those born in hotter climates had more sweat glands and were least at risk for heat stroke than soldiers born in cooler climates. Moreover, Roland[2] reported on anecdotal evidence that WWII prisoners of war who were smaller in size—presumably of smaller birth size—were more likely to survive the conditions of their captivity, such as starvation, than larger prisoners.

*Corresponding Author: Peter D. Gluckman—Liggins Institute, University of Auckland and National Research Centre for Growth and Development; 2-6 Park Avenue, Grafton, Private Bag 92019, Auckland, New Zealand. Email: pd.gluckman@auckland.ac.nz

Early Life Origins of Health and Disease, edited by E. Marelyn Wintour and Julie A. Owens.

In the 1970s, Forsdahl reported on the relationship between poor childhood living conditions in Norway and later adult heart disease.[3,4] But it wasn't until the landmark studies of Barker and colleagues[5-7] that these linkages received renewed attention. Their observations have since been supported by other large cohort studies in the USA,[8-10] Jamaica[11] and India.[12]

As a marker of early life conditions, birth size had never really been considered causal in the pathway to disease risk, an element often misunderstood by critics and supporters alike. Namely, birth size as a parameter has limitations—it is a reflection of maturation (i.e., gestational age) and growth, and both are influenced by environmental and genetic factors. Also, not all adverse events that occur in utero result in reduced birth size.[13] Nevertheless, birth size and, in particular, birth weight, remain the most accessible parameters to consider when assessing the impact of early developmental factors.

The Importance of Timing

The FOAD concept focused attention on environmental factors operating during *fetal* life. However, both experimental and clinical data suggest that adult disease risk may well be influenced by events that occur prior to conception until well after birth, and that there is an interaction between the prenatal and postnatal environments. For instance, it has been shown that those most at risk for insulin resistance and cardiovascular disease have evidence of an impaired fetal environment and put on weight rapidly in childhood,[14,15] though there are other data that suggest simply the altered pattern of infant and/or childhood growth alone is enough.[16,17] More recently, data regarding the premature infant[18] further confirm the linkages between pre and postnatal life. Not much is known about the optimal level of nutrition for these infants or the required dietary components, but it has been a long-held view that nutrient-enriched formula is beneficial in promoting growth and brain development. Alan Lucas and colleagues have shown that in adolescents born premature and fed nutrient-enriched formula, their likelihood for insulin resistance and heart disease in later life is increased than if fed a lower-nutrient diet.[19,20] Hence, metabolic regulation and nutritional compartmentalization at one stage in life can be influenced by events earlier in life.

There is also increasing focus on the consequences of the peri-conceptual status of the mother (her diet, body composition, level of physical activity, for example) for the health outcome of her offspring. In the Dutch winter famine of 1944-45, those who were conceived during the famine were not necessarily of smaller birth weight, but as adults became prone to insulin resistance and obesity.[21] Similar observations have been reported in sheep, in which periconceptual undernutrition has been shown to reset the HPA axis,[22] and in rodents where the conditions in which the preimplantation embryo develops later influence fetal growth and postnatal cardiovascular phenotype.[23] In humans, similar considerations are likely to apply as IVF is associated with a greater incidence of anomalies,[24] and in embryo donation the birthweight of the resulting offspring relates more closely to the birthweight of the recipient than the donor.[25] For these reasons, the earlier terminology 'fetal origins' has been replaced by 'developmental origins'. Moreover, it is clear that greater awareness of the importance of this time of life will not only have an impact in reducing the incidence of disease, but will provide an important platform for new measures aimed at promoting a healthy lifestyle. FOAD has therefore been replaced by DOHaD (Developmental Origins of Health and Disease) and an international society was formed in 2003 to promote this endeavour (www.dohadsoc.org).

DOHaD—Underlying Mechanisms

This book is very much concerned with mechanisms, so only general comments are appropriate here. Despite the plethora of epidemiological and experimental animal observations, it is clear that only a limited number of mechanisms can drive a long-term change in phenotype. One likely mechanism may be epigenetic regulation.

While epigenetic regulation is more generally understood in terms of parental imprinting of certain genes, it is also the basis of many other changes in gene expression. Recently, clear evidence has emerged, both with respect to parentally imprinted and nonimprinted genes, of environmentally-induced epigenetic change. In ruminant embryos subject to prolonged culture in vitro and then reimplanted, there are long-term changes in the expression of imprinted components of the IGF-2 system.[26] In the agouti mouse, the degree of imprinting can be influenced by the folate/B12 status of mother at the time of conception.[27] In the rat subject to experimental reduction of uterine blood flow, altered methylation of the p53 gene has been described in the kidney.[28] And in rat pups, Meaney and colleagues have shown that an altered neonatal behavioural environment results in altered methylation patterns in the promotor region of the glucocorticoid receptor gene.[29] If the latter could be generalized, it would suggest that there are a number of yet to be discovered mechanisms by which epigenetic change at very specific sites in the genome have evolved. Their specificity means these have not have been selected by evolution randomly, but rather because they have adaptive value. Mitochondrial DNA has been suggested as a particular target for epigenetic change[30] and this would explain the nongenomic inheritance of the phenomenon via the maternal lineage (as mitochondrial DNA is only of maternal origin); it also provides a potent way in which cellular metabolism and the response to nutritional stress can be programmed, complementing the well-known effects on growth.

At the next level, the DOHaD phenomenon involves changes in the growth and function of tissues and organs in relation to overall body size. These changes may be induced during development to reduce energy consumption by these tissues and organs during development itself, and also in expectation of a deprived postnatal environment (see Fig. 1). Much interest in the field concerns such effects on the developing kidney,[31] pancreas[32] and heart.[33] These organs are particularly interesting for several reasons. First, changes induced in them in early life, such as reduced numbers of nephrons, pancreatic beta cells or cardiomyocytes, may limit the individual's ability to respond adequately to a physiological challenge in later life, especially in the face of declining function of these organs with age, and may explain the link between developmental factors and diseases occurring after middle age, rather than in adolescence. Furthermore, cell number within these organs is set prenatally in the human, so that prenatal environment can exert permanent effects on their development.

The DOHaD phenomenon also involves changes in vascularity and, indeed, in vascular endothelial cell function,[34] with vascularity reduced in several tissues of the rat pup exposed to a low protein diet in utero.[35,36] Skeletal muscle mass may also be reduced.[37] It may be that these observations are adaptive responses to survive a poor intrauterine environment. However, it can also be viewed as a 'thrifty' adaptive response to a poor environment postnatally, staying small and trading off growth for reproduction and survival of the genotype. Because skeletal muscle is a major determinant of peripheral insulin sensitivity, these adaptive responses can also play a part in the aetiology of Type 2 diabetes and the metabolic syndrome. Endothelial function is now increasingly realized to be linked to organ and tissue growth, including that of adipocytes, and endothelial dysfunction is linked to Type 2 diabetes, hypertension and atherogenesis. A range of animal studies support the concept that this dysfunction can be induced before glucose intolerance, elevated blood pressure or obesity set in.

Lastly, there is considerable evidence both from human and from experimental animal studies that the early environment can induce permanent changes in homeostatic regulatory function. In humans born small for gestational age, there is often an altered feedback of the HPA axis, leading towards a tendency of hypercortisolemia;[38] similar changes are noted in experimental animal models, as in Meaney's behavioural model,[29] or following exposure to maternal undernutrition[39] or glucocorticoids in utero.[40] Overall, these changes are underpinned by alterations in the glucocorticoid receptors of the central nervous system via processes that are not necessarily distinct from the epigenetic processes referred to above. Other changes to homeostatic regulatory function include altered regulation of insulin release and action,[41] altered hepatic handling of glucose,[42] and altered muscular sensitivity, both to insulin and to insulin like growth factors.[43]

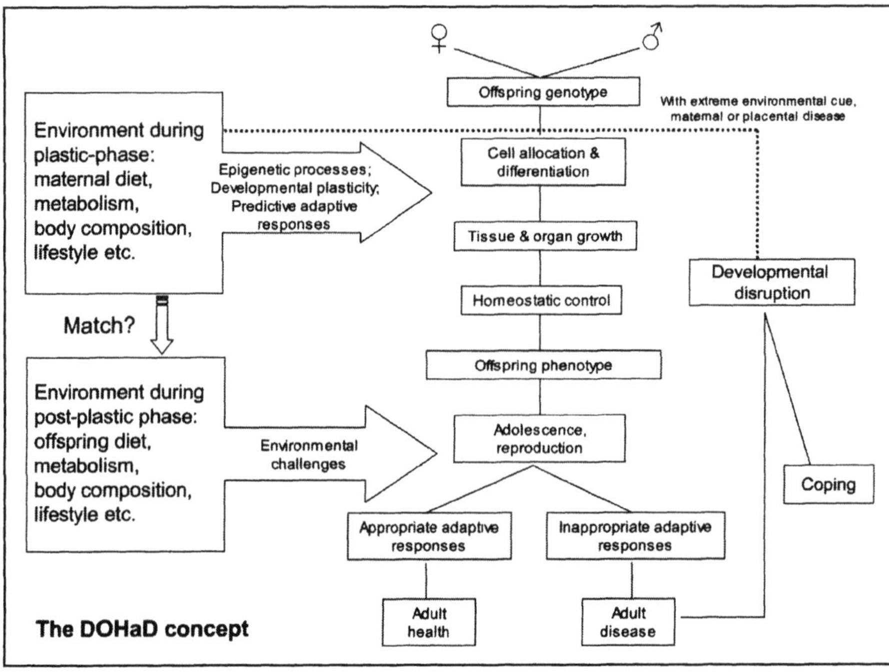

Figure 1. Diagram of the DOHaD concept. The offspring's genome inherited from both parents is subjected to a range of environmental influences in early life, which determine its phenotype. These include epigenetic processes and utilize the normal processes of developmental plasticity. Some of these effects may have predictive advantage in the postnatal environment. The processes involved can be categorized as changes in cell allocation and differentiation, tissue and organ growth and homeostatic control, and to an extent these occur sequentially as shown. If the environmental effects are severe (e.g., famine, excess, maternal or placental disease), they may induce clear disruption of development (equivalent to teratogenesis, dotted line) of reduction in developmental trajectory such that the offspring only 'copes' and is at risk both pre and postnatally. Even within the normal physiological range of adaptive responses, the risk of health vs. disease in the post-reproductive period depends on the extent of matching between the prenatal and the postnatal environments, or between the predicted and the actual environment in the post-plastic phase.

A Conceptual Framework

Some of the earliest concepts of the 'programming' of human disease have focused on the processes of mutation, genetic drift and selection, which may have created genotypes that were thrifty—the so called "thrifty genotype" hypothesis.[44] The thrifty genotype would have created a level of insulin resistance which allowed a population to cope in times of nutritional stress, but in times of plenty would have led to insulin resistance. As insulin is important to fetal growth, a genotypic change in insulin secretion or action would lead to smaller fetuses. The glucokinase mutation has been suggested as one such thrifty mutation.[45] Nevertheless, there is much that the thrifty genotype model cannot explain, such as the broader range of consequences of early life events (i.e., osteoporosis), the fact that the induction of these responses need not involve altered fetal growth, or the rapid appearance of increased risk of disease in a population subject to a brief environmental challenge (i.e., Dutch winter famine).

We think that some of these questions can be answered by a developing concept we refer to as 'predictive adaptive responses' (PARs). PARS are responses made by the developing, plastic organism, not for immediate advantage but as an adaptive strategy aimed at thriving in the environment predicted to be experienced as an adult. In other words, if the environmental stimuli are suggestive of a deprived future environment, the fetus adjusts its physiology accordingly. The biology of PARs has been the subject of several recent reviews.[46-48]

Examples of PARs in nature include the setting of the HPA axis in response to glucocorticoid exposure during development. There may be no immediate advantage to having a hyper-responsive HPA axis, but clearly if the developing organism interprets a surge in glucocorticoids as a sign of a stressful environment, there may be survival advantage in having a hyper-responsive HPA postnatally. Another clear-cut example is that of coat thickness in the meadow vole.[49] This phenomenon (reviewed in ref. 50) shows that coat thickness in the vole at birth is determined by the thermal environment anticipated some weeks later. PARs provide an evolutionary compatible explanation of why the DOHaD phenomenon has evolved and is maintained.

The fetus has a genetically determined repertoire of responses for responding to its immediate environment in order to ensure either its immediate or its future survival. If it needs to ensure its immediate survival it may select a developmental strategy which, for example, reduces growth even though this has postnatal costs which may become manifest as disease. Yet it appears increasingly that most normally grown fetuses set their development and their homeostatic control not so much for immediate advantage but rather in expectation of assisting postnatal survival to reproduction. If the fetal prediction of its future environment is correct, then the developmental path chosen in utero should lead to health in adult life. But if the fetal prediction is wrong for some reason then as an adult it will not have physiological settings appropriate for its environment and disease risk is enhanced. In other words, the risk of disease is determined by the degree of match between the environment the developing organism anticipates, on the basis of cues from its mother and placenta, and the environment it actually faces as an adult.

Such a model explains why so called life-style diseases appear in high frequency in populations undergoing rapid nutritional transition; why the relationship between birth size and the risk of adult disease is influenced by measures of the postnatal environment, such as rapid adiposity rebound; as well as why the phenomenon is not limited to those of small size, but occurs across the full range of birth sizes.

The Influence of DOHaD in Human Health

We are left with the challenge of identifying how important this phenomenon might be for human medicine. The only human estimate suggests that avoiding a developmental mismatch could reduce the incidence of heart disease and diabetes by over 50%—but caution must be applied to a single estimate.[51] However evolutionary, developmental and experimental considerations all suggest that environmental influences acting in early development can permanently alter the trajectory of development and determine the risk of later disease. Together with the human data, these approaches suggest that a mismatch between the environment as perceived during the phase of developmental plasticity and the actual environment the organism is exposed to later in life will increase the risk of disease.

Clinical practice will increasingly have to address issues of maternal health prior to and during pregnancy. The periconceptual period may be particularly critical. Sadly we still do not know the optimal nutritional recommendations for women at different stages in the reproductive cycle. Postnatal management may need to become increasingly individualized to match the postnatal environment more closely to the phenotype induced in early life. The research effort is likely to focus on a search for epigenetic processes and markers which could be prognostic, and on achieving a greater understanding of the processes of developmental plasticity and its potential reversibility.

Acknowledgements

M.A.H. is supported by the British Heart Foundation. P.D.G. is supported by the National Research Centre for Growth and Development (New Zealand). We thank Dr. C. Pinal for her assistance.

References

1. Kawahata A, Sakamoto H. Some observations on sweating of the Aino. Jap J Physiol 1951; 2:166-169.
2. Roland CG. Stripping away the veneer: P.O.W. survival in the Far East as an index of cultural atavism. J Mil Hist 1989; 53:79-94.
3. Forsdahl A. Living conditions in childhood and subsequent development of risk factors for arteriosclerotic heart disease. The cardiovascular survey in Finnmark 1974-75. J Epidemiol Community Health 1978; 32:34-37.
4. Forsdahl A. Are poor living conditions in childhood and adolescence an important risk factor for arteriosclerotic heart disease? Br J Prevent Soc Med 1977; 31:91-95.
5. Barker DJP, Osmond C, Golding J et al. Growth in utero, blood pressure in childhood and adult life, and mortality from cardiovascular disease. Br Med J 1989; 298:564-567.
6. Barker DJ, Osmond C. Low birth weight and hypertension. Br Med J 1988; 297:134-135.
7. Barker DJ, Osmond C. Infant mortality, childhood nutrition, and ischaemic heart disease in England and Wales. Lancet 1986; 1:1077-1081.
8. Rich-Edwards JW, Colditz GA, Stampfer MJ et al. Birthweight and the risk for type 2 diabetes mellitus in adult women. Ann Intern Med 1999; 130:278-284.
9. Rich-Edwards JW, Stampfer MJ, Manson JE et al. Birth weight and risk of cardiovascular disease in a cohort of women followed up since 1976. Br Med J 1997; 315:396-400.
10. Curhan GC, Willett WC, Rimm EB et al. Birth weight and adult hypertension, diabetes mellitus, and obesity in US men. Circulation 1996; 94:3246-3250.
11. Forrester TE, Wilks RJ, Bennett FI et al. Fetal growth and cardiovascular risk factors in Jamaican schoolchildren. Br Med J 1996; 312:156-160.
12. Yajnik CS, Fall CHD, Pandit AN et al. Fetal growth and glucose and insulin metabolism in four-year-old Indian children. Diabet Med 1995; 12:330-336.
13. Hanson M. Birth weight and the fetal origins of adult disease. Pediatr Res 2002; 52:473-474.
14. Eriksson JG, Forsen T, Tuomilehto J et al. Catch-up growth in childhood and death from coronary heart disease: Longitudinal study. Br Med J 1999; 318:427-431.
15. Bavdekar A, Yajnik CS, Fall CH et al. Insulin resistance syndrome in 8-year-old Indian children: Small at birth, big at 8 years, or both? Diabetes 1999; 48:2422-2429.
16. Eriksson JG, Forsen T, Tuomilehto J et al. Early growth and coronary heart disease in later life: Longitudinal study. Br Med J 2001; 322:949-953.
17. Ong KK, Ahmed ML, Emmett PM et al. Association between postnatal catch-up growth and obesity in childhood: Prospective cohort study. Br Med J 2000; 320:967-971.
18. Hofman PL, Regan F, Jackson WE et al. Premature Birth and later insulin resistance. N Engl J Med 2004; 351:2179-2186.
19. Singhal A, Cole TJ, Fewtrell M et al. Breastmilk feeding and lipoprotein profile in adolescents born preterm: Follow-up of a prospective randomised study. Lancet 2004; 363:1571-1578.
20. Singhal A, Fewtrel M, Cole TJ et al. Low nutrient intake and early growth for later insulin resistance in adolescents born preterm. Lancet 2003; 361:1089-1097.
21. Roseboom TJ, van der Meulen JH, Ravelli AC et al. Effects of prenatal exposure to the Dutch famine on adult disease in later life: An overview. Mol Cell Endocrinol 2001; 185:93-98.
22. Bloomfield FH, Oliver MH, Hawkins P et al. Periconceptional undernutrition in sheep accelerates maturation of the fetal hypothalamic-pituitary-adrenal axis in late gestation. Endocrinology 2004; 145:4278-4285.
23. Kwong WY, Wild AE, Roberts P et al. Maternal undernutrition during the preimplantation period of rat development causes blastocyst abnormalities and programming of postnatal hypertension. Development 2000; 127:4195-4202.
24. Jackson RA, Gibson KA, Wu YW et al. Perinatal outcomes in singletons following in vitro fertilization: A meta analysis. Obstet Gynecol 2004; 103:551-563.
25. Brooks AA, Johnson MR, Steer PJ et al. Birth weight: Nature or nuture? Early Hum Dev 1995; 42:29-35.
26. Young LE, Fernandes K, McEvoy TG et al. Epigenetic change in IGF2R is associated with fetal overgrowth after sheep embryo culture. Nat Genet 2001; 27:153-154.
27. Wolff GL, Kodell RL, Moore SR et al. Maternal epigenetics and methyl supplements affect agouti gene expression in A^{vy}/a mice. FASEB J 1998; 12:949-957.
28. Pham TD, MacLennan NK, Chiu CT et al. Uteroplacental insufficiency increases apoptosis and alters p53 gene methylation in the full-term IUGR rat kidney. Am J Physiol 2003; 285:R962-R970.

29. Weaver ICG, Cervoni N, Champagne FA et al. Epigenetic programming by maternal behavior. Nat Neurosci 2004; 7:847-854.
30. McConnell JL, Petrie L. Mitochondrial DNA turnover occurs during preimplantation development and can be modulated by environmental factors. Reprod Biomed Online 2004; 9:418-424.
31. Wintour EM, Moritz KM, Johnson K et al. Reduced nephron number in adult sheep, hypertensive as a result of prenatal glucocorticoid treatment. J Physiol 2003; 549(3):929-935.
32. Fowden AL, Hill DJ. Intra-uterine programming of the endocrine pancreas. Br Med Bull 2001; 60:123-142.
33. Schwartz J, Thornburg KL. The influence of various physiological challenges on permanent changes to the cardiovascular system. Arch Physiol Biochem 2003; 111:3-7.
34. Leeson CPM, Kattenhorn M, Morley R et al. Impact of low birth weight and cardiovascular risk factors on endothelial function in early adult life. Circulation 2001; 103:1264-1268.
35. Torrens C, Brawley L, Barker AC et al. Maternal protein restriction in the rat impairs resistance artery but not conduit artery function in the pregnant offspring. J Physiol 2003; 547(1):77-84.
36. Bennis-Taleb N, Remacle C, Hoet JJ et al. A low-protein isocaloric diet during gestation affects brain development and alters permanently cerebral cortex blood vessels in rat offspring. J Nutr 1999; 129:1613-1619.
37. Vickers MH, Krechowec S, Gluckman PD et al. Maternal undernutrition leads to lethargy and hyperphagia combined with increased hepatic lipid accumulation, reduced muscle mass and obesity in offspring. Pediatr Res 2003; 53:39A.
38. Phillips DI, Barker DJ, Fall CH et al. Elevated plasma cortisol concentrations: A link between low birth weight and the insulin resistance syndrome? J Clin Endocrinol Metab 1998; 83:757-760.
39. Bloomfield FH, Oliver MH, Giannoulias CD et al. Brief undernutrition in late-gestation sheep programmes the HPA axis in adult offspring. Endocrinology 2003; 144:2933-40.
40. Challis JR, Sloboda D, Matthews SG et al. The fetal placental hypothalamic-pituitary-adrenal (HPA) axis, parturition and post natal health. Mol Cell Endocrinol 2001; 185:135-144.
41. Ozanne SE, Wang CL, Coleman N et al. Altered muscle insulin sensitivity in the male offspring of protein-malnourished rats. Am J Physiol 1996b; 271(6):E1128-E1134.
42. Ozanne SE, Smith GD, Tikerpae J et al. Altered regulation of hepatic glucose output in the male offspring of protein-malnourished rat dams. Am J Physiol 1996a; 270(4 Pt 1):E559-E564.
43. Ozanne SE, Olsen GS, Hansen LL et al. Early growth restriction leads to down regulation of protein kinase C zeta and insulin resistance in skeletal muscle. J Endocrinol 2003; 177:235-241.
44. Neel JV. Diabetes mellitus: A "thrifty" genotype rendered detrimental by "progress"? Am J Hum Genet 1962; 14:353-362.
45. Spyer G, Hattersley AT, Sykes JE et al. Influence of maternal and fetal glucokinase mutations in gestational diabetes. Am J Obstet Gynecol 2001; 185:240-241.
46. Gluckman PD, Hanson MA. The developmental origins of the metabolic syndrome. Trends Endocrinol Metab 2004a; 15:183-187.
47. Gluckman PD, Hanson MA. Living with the past: Evolution, development and patterns of disease. Science 2004b; 305:1773-1776.
48. Bateson P, Barker D, Clutton-Brock T et al. Developmental plasticity and human health. Nature 2004; 430:419-421.
49. Lee TM, Zucker I. Vole infant development is influenced perinatally by maternal photoperiodic history. Am J Physiol 1988; 255:R831-R838.
50. Gluckman PD, Hanson MA. The fetal matrix: Evolution, development, and disease. Cambridge: Cambridge University Press, 2005.
51. Barker DJP, Eriksson JG, Forsén T et al. Fetal origins of adult disease: Strength of effects and biological basis. Int J Epidemiol 2002; 31:1235-1239.

Chapter 2

Developmental Origins of Cardiovascular Disease, Type 2 Diabetes and Obesity in Humans

Caroline H.D. Fall*

Abstract

Fetal growth restriction and low weight gain in infancy are associated with an increased risk of adult cardiovascular disease, type 2 diabetes and the Metabolic Syndrome. The fetal origins of adult disease hypothesis proposes that these associations reflect permanent changes in metabolism, body composition and tissue structure caused by undernutrition during critical periods of early development. An alternative hypothesis is that both small size at birth and later disease have a common genetic aetiology. These two hypotheses are not mutually exclusive. In addition to low birthweight, fetal 'overnutrition' caused by maternal obesity and gestational diabetes leads to an increased risk of later obesity and type 2 diabetes. There is consistent evidence that accelerated BMI gain during childhood, and adult obesity, are additional risk factors for cardiovascular disease and diabetes. These effects are exaggerated in people of low birthweight. Poor fetal and infant growth combined with recent increases in childhood adiposity may underlie the high rates of disease in developing countries undergoing nutritional transition. Sub-optimal maternal nutritional status is a major cause of low birthweight globally but its impact on fetal growth in 'well-nourished' western populations has been inadequately studied. In experimental animals hypertension and insulin resistance can be programmed in the offspring by restricting maternal diet in pregnancy but there are currently insufficient data to determine whether maternal nutritional status and diet programme cardiovascular disease risk in humans.

Low Birthweight and Adult Cardiovascular Disease

The concept that events in early life have long-term effects on human health life is not new. In 1934, Kermack showed that death rates from all causes in the UK and Sweden fell between 1751 and 1930 with each successive year-of-birth cohort.[1] He rejected one possible explanation, that babies were born healthier in successive generations, and concluded that it was the result of social reforms and better childhood living conditions. In 1977, Forsdahl discovered a geographical correlation in Norway between coronary heart disease (CHD) mortality in 1964-67 and infant mortality rates 70 years earlier (1896-1925).[2] He suggested that growing up in poverty caused 'permanent damage' perhaps due to a 'nutritional deficit', which resulted in 'life-long vulnerability' to an affluent adult lifestyle. Studies in the UK a decade later shifted the

*Caroline H.D. Fall—MRC Environmental Epidemiology Unit, University of Southampton, Southampton S016 6YD, U.K. Email: chdf@mrc.soton.ac.uk

Early Life Origins of Health and Disease, edited by E. Marelyn Wintour and Julie A. Owens.

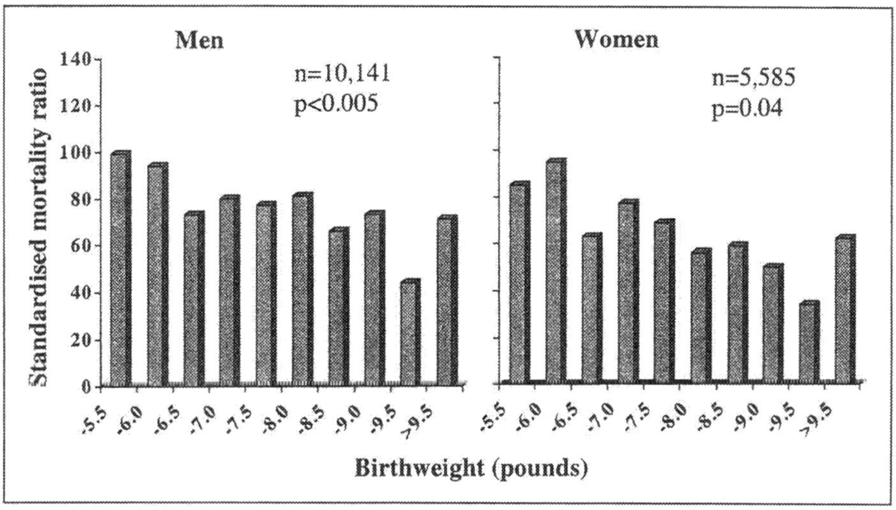

Figure 1. Standardised mortality ratios for cardiovascular disease in men and women born in Hertfordshire, UK, according to birthweight.[6]

focus back to prenatal rather than childhood events. Blood pressure was found to be inversely related to birthweight in young men and women[3] and Barker showed that regional differences in stroke and CHD mortality in the UK in 1968-78 were predicted by neonatal mortality (a marker for low birthweight) in 1921-25.[4] He went on to propose that the roots of cardiovascular disease (CVD) lay in the effects of poverty on the mother and undernutrition in fetal life and early infancy.

Using birth records dating back to 1911-1930 from one English county (Hertfordshire), Barker showed that lower birthweight and weight at one year were associated with an increased risk of death from CHD and stroke.[5,6] There was an approximate doubling of mortality from the highest to the lowest extremes of birthweight (Fig. 1), similar in men and women. Since then, studies in the UK, Europe, and the USA have confirmed these findings[7-14] and shown that it is restricted fetal growth rather than preterm delivery which carries the risk of CVD.[12] The effects are linear, graded across the whole range of birthweight (Fig. 1) and independent of adult socio-economic status.[9,10,12] Many studies were limited to birthweight as a measure of fetal growth but there is evidence that body proportions at birth show stronger associations with CVD. For example low ponderal index (weight/ length^3) predicted CHD better than birthweight in Finland,[11] and a low birthweight/head circumference ratio predicted stroke mortality in the UK.[8]

CVD Risk Factors

Subsequent work has shown that lower birthweight and other measures of small size at birth are also associated with higher levels of CVD risk factors.

Metabolic Syndrome

Hypertension, type 2 diabetes, insulin resistance, and combinations of these (the Metabolic Syndrome, Insulin Resistance Syndrome or Syndrome X) are consistently related to low birthweight in a large number of studies in different populations.[15-25] The strength of these associations led Barker and Hales to suggest that the Metabolic Syndrome should be renamed the 'small baby syndrome'.[16] It is notable that the associations are stronger for disease outcomes, such as hypertension (Fig. 2)[26] than for blood pressure measurements.

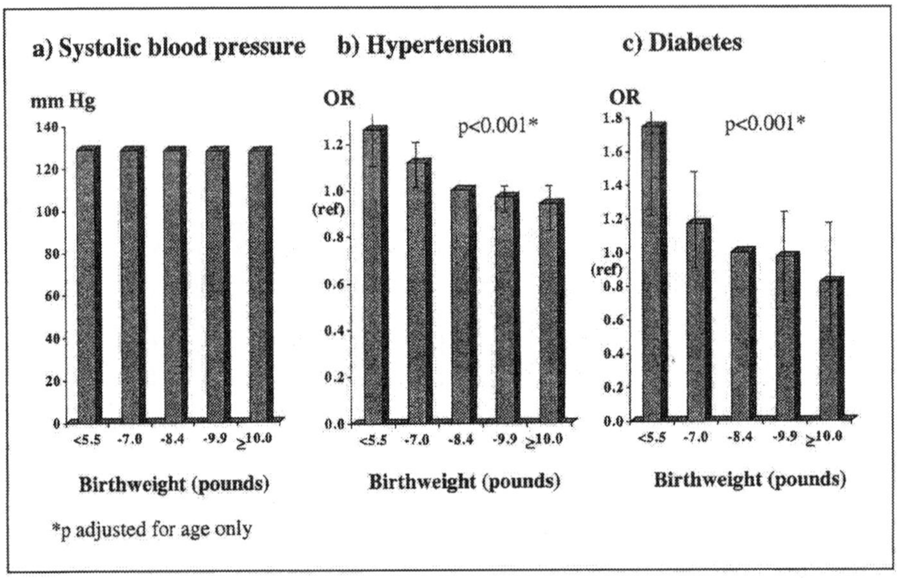

Figure 2. Mean systolic blood pressure and odds ratios for hypertension and incident diabetes in 22,846 US men aged 48-83 years.[26]

Serum Lipids and Clotting Factors

Although serum lipid concentrations show some associations with size at birth, these are less consistent. Total cholesterol shows a weak inverse association with birthweight.[27] HDL-cholesterol concentrations are positively, and triglyceride concentrations inversely, related to birthweight in some[28] but not most studies.[18,21-23,29,30] Post-prandial lipid concentrations, and Lp(a), fibrinogen and factor VII concentrations were unrelated to birthweight in Hertfordshire.[28,31,32] Total- and LDL-cholesterol, apolipoprotein B and fibrinogen concentrations were associated with smaller abdominal circumference at birth, recorded in obstetric records in Sheffield, UK.[33,34] PAI-1 was increased in low birthweight men in one published study.[22]

Measurements of Cardiovascular Structure and Function

Martyn et al showed that arterial intima media thickness (IMT) and the risk of carotid stenosis, examined using ultrasound, were increased in lower birthweight men and women.[35] In a follow up of the Newcastle (UK) 1000 Families Study, carotid IMT was increased in lower birthweight men and in women of lower socio-economic class at birth.[36] In the only study to examine peripheral vascular disease in relation to size at birth, ankle brachial pulse index, was not significantly related to birthweight.[35] Pulse wave velocity, a measure of poor arterial compliance, was increased in UK men and women with small head and abdominal circumferences at birth,[37] but showed no association with size at birth in a study in India.[38] Left ventricular mass was unrelated to birthweight in three studies.[38-40] Flow-mediated dilatation, a measure of endothelial function, and indices of microvascular function, are reduced in children and young adults of lower birthweight.[41-44]

Obesity

People of higher birthweight tend to become 'fatter' adults as measured by body mass index[22,45] (Table 1, Fig. 3). However, there is growing evidence that this reflects increased lean mass rather than adiposity.[46-51] Higher birthweight men had higher lean but not fat mass, measured using DEXA at the age of 70 years[49] (Fig. 4). Using anthropometric measurements of

Table 1. Birthweight and adult obesity: Uppsala, Sweden (men, n = 1268)[22]

	Birthweight (kg)					
	<3.25	-3.75	-4.25	⊕4.25	p	p*
50 years						
Triceps skinfold (mm) (TR)	10.3	10.9	11.2	11.3	0.02	0.2
Subscapular skinfold (mm) (SS)	17.4	16.6	16.7	16.6	0.5	0.001
SS/TR	1.78	1.60	1.57	1.54	<0.001	<0.001
70 years						
BMI (kg/m^2)	25.9	26.4	26.6	26.8	0.007	–
Waist (cm)	93.7	94.9	95.5	96.2	0.01	0.8
Hip (cm)	99.2	100.4	101.4	101.7	<0.001	0.02
Waist/Hip	0.94	0.94	0.94	0.94	0.9	0.03

p values adjusted for age; p* for age and BMI. Reproduced with permission from Byberg at al. Birthweight and the insulin resistance syndrome: Association of low birthweight with truncal obesity and raised plasminogen activator inhibitor-1 but not with abdominal obesity or plamsa lipid disturbances. Diabetalogia 43:54-60. ©2000 Springer-Verlag.

thigh diameter and skinfold thickness in young men, Kahn showed that the birthweight-BMI association was attenuated by 68% after adjustment for thigh muscle+bone area, but only by 30% after adjustment for subcutaneous fat area.[47] There is no evidence from these studies that low birthweight leads to increased adiposity, but leptin concentrations were increased in low birthweight men and women in one study.[52]

There is some evidence that small size at birth is associated with an increased risk of later central obesity (indicated by high waist circumference, waist/hip ratio and subscapular/triceps skinfold ratio). The subscapular/triceps ratio (SS/TR) is consistently higher in adults and children of lower birthweight (Table 1).[22,53-55] In contrast, although waist circumference and waist/hip ratio are inversely related to birthweight in some studies[22,56,57] this is

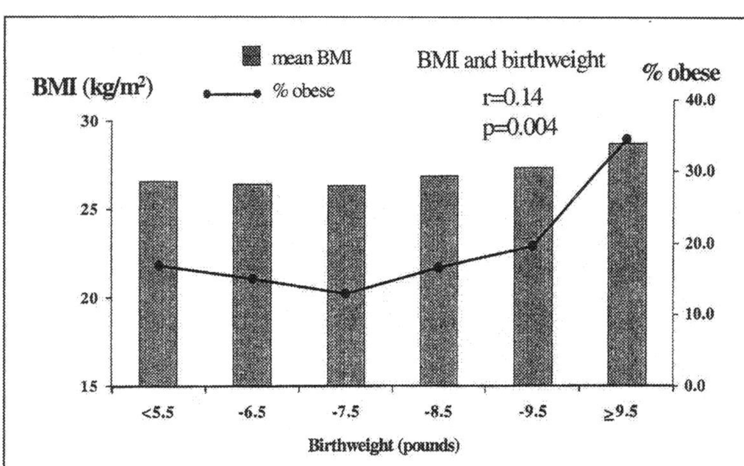

Figure 3. Adult body mass index (BMI) according to weight at birth; Hertfordshire men aged 60-70 years (n = 845).

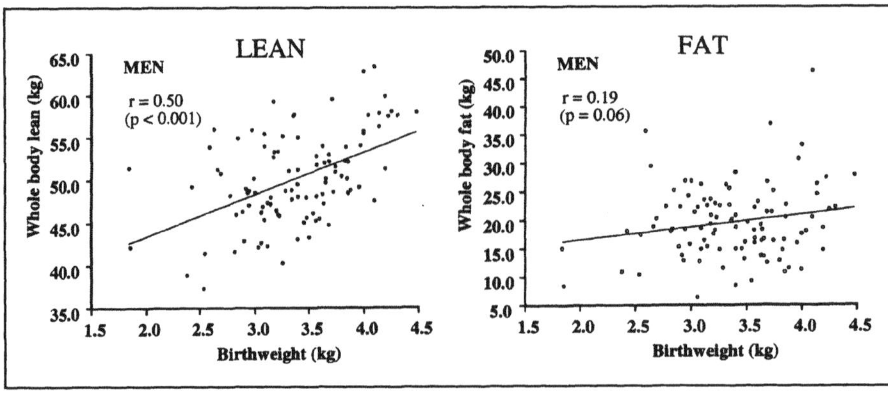

Figure 4. Lean and fat mass measured using dual energy absorption (DEXA) in men aged 70-75 years, born in Sheffield, UK (n = 102).[49] Reproduced with permission from Gale C et al. Intrauterine programming or adult body composition. J Clin Endocrinol Metab 86:267-272, ©2001, The Endocrine Society.

weaker and less consistent.[17] Where present, the association may reflect larger hip circumference (and therefore frame size and muscle mass) in higher birthweight individuals, rather than abdominal obesity in people of low birthweight (Table 1).

Post-Natal Growth and Adult Obesity

In addition to size at birth, cardiovascular disease and its risk factors also show associations with patterns of growth in infancy and childhood. In Hertfordshire, men with lower weight at the age of one year had increased cardiovascular disease mortality[5,6] (Fig. 5) and type 2 diabetes.[15] There are few adult cohorts with infant data but these findings were confirmed in men born in Helsinki, Finland and men and women born in New Delhi, India[58,59] (Fig. 6). In Hertfordshire, men who had a low weight at one year also had higher left ventricular mass,[39]

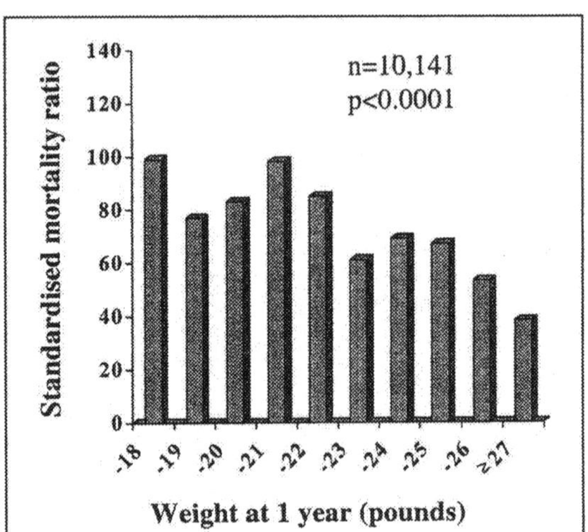

Figure 5. Standardised mortality ratios for cardiovascular disease in men born in Hertfordshire, UK, according to weight at the age of one year.[6]

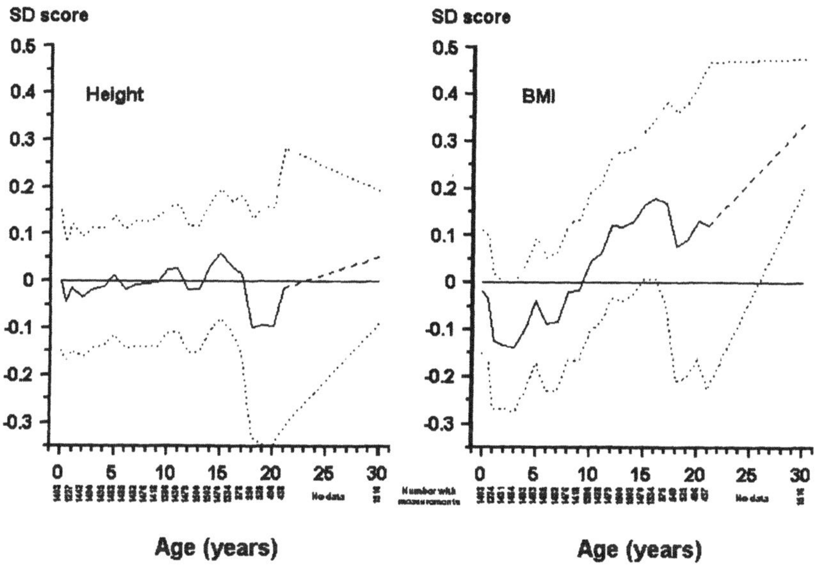

Mean SD scores are indicated by the solid line, and 95% confidence intervals by dotted lines. SD scores were derived internally; the mean SD for the whole cohort is zero.

Figure 6. Sex-specific SD scores for height (left) and BMI (right) at every age from birth to 21 years and at 30 years, for men and women in New Delhi, India who developed IGT or diabetes.[59] Reproduced with permission from Bhargava SK et al. Relation of serial changes in childhood body mass index to impaired glucose tolerance in young adulthood. N Eng J Med 350:865-875.

and fibrinogen[32] and cholesterol concentrations.[29] Adult blood pressure is not related to infant weight.[15,60]

The relationship between infant weight and weight gain and later health requires further research. It is current pediatric practice to encourage weight gain in low birthweight babies, and this is associated with increased infant survival in some populations.[61] The data described above suggest that infancy may be an accessible stage of the lifecourse in which improved nutrition could benefit adult health. This is controversial however, as studies in children show higher levels of CVD risk factors in children who gained weight rapidly in infancy.[62]

Greater weight or BMI gain in childhood (after the period of infancy) is consistently associated with an increased risk of later disease. Accelerated childhood weight gain (upward crossing of centiles) was associated with higher blood pressure in young adults in the UK.[60] Similarly, in the Finland and New Delhi birth cohorts, an increase in BMI SD score between birth and adolescence was associated with an increased risk of CHD and/or type 2 diabetes.[13,59,63-66] In both cohorts, early adiposity rebound was also associated with increased adult diabetes.[59,66] The determinants of age at adiposity rebound are unknown, but both studies showed that lower weight at one year predicted an earlier rebound.[59,66]

Increased BMI in childhood and adult obesity add to, and in some studies interact with, the effects of low birthweight and infant weight. In Finland, an increase in BMI from birth to seven years was only associated with an increased risk of adult CHD in those who were small at birth (Fig. 7).[63] The most adverse cardiovascular disease risk profile is consistently

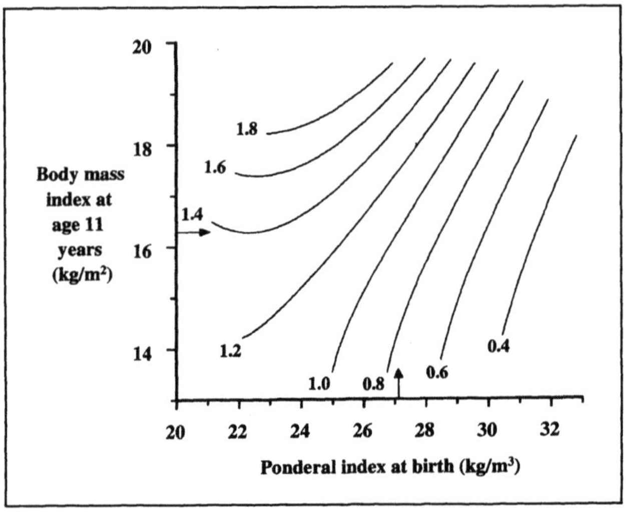

Figure 7. Hazard ratios for death from CHD for men born in Helsinki 1924-33 according to ponderal index at birth and BMI at age 11 years. Arrows indicate average values.[63] Reproduced from Eriksson JG et al. Catch-up growth in childhood and death from coronary heart disease: Longitudinal study. BMJ 1999; 318:427-431, with permission from the BMJ Publishing Group.

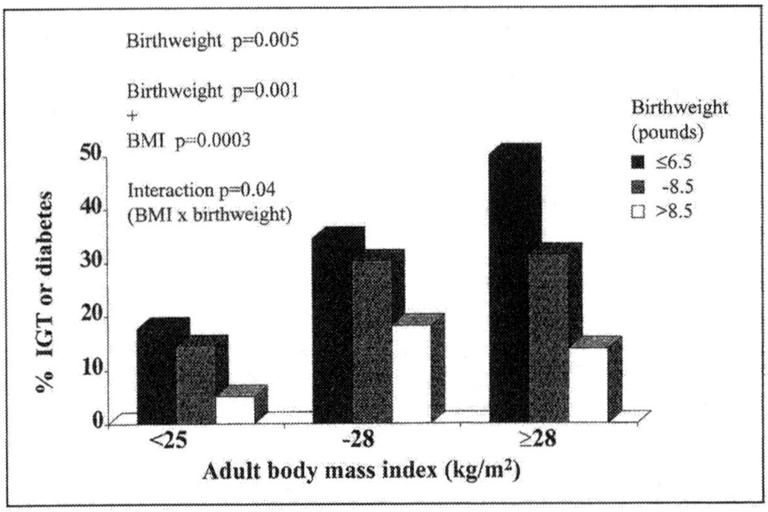

Figure 8. Prevalence of impaired glucose tolerance (IGT) and Type 2 diabetes (%) in Hertfordshire men aged 60-71 (n = 370).[15]

found in men and women who were small at birth but became obese adults (Fig. 8).[15] The intriguing corollary of this is that high birthweight may protect against the adverse effects of adult obesity. Similar additive and/or interactive effects have been described between size at birth and other aspects of adult lifestyle, for example between ponderal index at birth and adult socio-economic status on CHD[67] and between weight in infancy and the effects of smoking on fibrinogen concentrations.[31]

Variation with Sex

In general, relationships between small size at birth and adult cardiovascular disease and risk factors are similar in both sexes. However, in Finland, CHD was most strongly associated with low ponderal index at birth in men and with short birth length in women.[11,13] The strong associations of type 2 diabetes and cardiovascular disease mortality with low weight at one year in the Hertfordshire men were not seen in women.[6,15,28] Similarly, apolipoprotein B, fibrinogen and factor VII concentrations, which were inversely related to small abdominal circumference at birth[33] and low weight at one year[29] in men, showed no associations with either size at birth or infant weight in women.[28]

Variation with Ethnicity

Studies from India, Jamaica, China, and Japan, and among US Hispanics and black South Africans have shown associations between lower birthweight and higher blood pressure, glucose intolerance and insulin resistance.[68 73] Higher SS/TR ratios have been shown in Indian children[55] and white, black and hispanic US children of lower birthweight.[54] There are some inconsistencies, however. Type 2 diabetes was associated with low birthweight in young Indian adults[59,74] but with a high ponderal index at birth in older men and women.[75] A study of blood pressure in children in 5 countries showed differences in associations with birth measurements.[76] In China and Central and South America, higher blood pressure was associated with 'proportionate' smallness at birth (reductions in birthweight, length and head or chest circumferences), while in Sweden it was associated with 'asymmetrical' smallness at birth (low ponderal index). In Nigeria, blood pressure was not related to size at birth.

Studies in India have shown differences in neonatal body composition from UK babies (Fig. 9).[77,78] Indian newborns were lighter by almost 2 standard deviations (SDs) and nonfat soft tissues such as muscle (mid-upper-arm circumference) and abdominal wall and viscera (abdominal circumference) showed a similar deficit. Measurements of fat, however, were

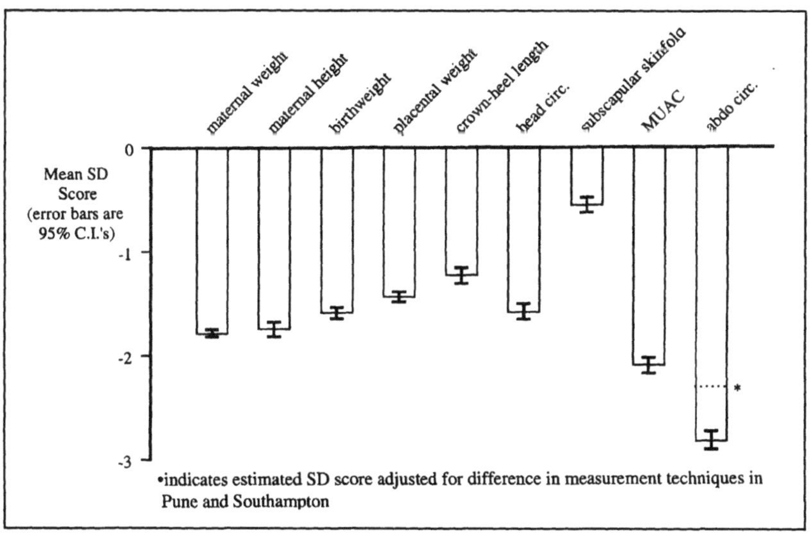

Figure 9. Mean SD scores for maternal prepregnant weight and height, and measurements of the babies, in Pune, India compared with Southampton, UK.[77] The Southampton mean is represented by 0. Reproduced from Yajnik CS et al. Neonatal anthropometry: The thin-fat Indian baby; the Pune Maternal Nutrition Study. Internat J Obes 2003; 27:173-180 with permission of Nature Publishing Group (http://www.nature.com).

substantially 'spared'; although small and thin, the Indian babies were relatively adipose. The aetiology and long-term implications of this 'muscle-thin but adipose' or 'thin-fat' phenotype are unknown. However the neonatal findings are of interest because the high rates of type 2 diabetes and CHD among South Asian populations are related to a similar body phenotype in adults.[79-82]

The 'Developmental Origins of Adult Disease' (DOHaD) Hypothesis

Barker proposed that the associations between small size at birth and cardiovascular disease reflect permanent effects of fetal undernutrition[83] (Fig. 10). The fetus is dependent on the transfer of nutrients from the mother and adapts to an inadequate nutrient supply in a number of ways: prioritisation of brain growth at the expense of other tissues such as the abdominal viscera, reduced secretion of and sensitivity to the fetal growth hormones insulin and IGF-I, and up-regulation of the hypothalamo-pituitary-adrenal (HPA) axis. Barker's 'fetal origins' hypothesis proposed that although occurring in response to a transient phenomenon (fetal undernutrition) these changes become permanent or 'programmed' because they occur during critical periods of development.

The mechanisms by which this could occur at a cellular and tissue level have been reviewed.[84,85] Programmed changes may include reduced insulin sensitivity,[86] low muscle mass,[46] pancreatic beta cell mass[87] and nephron numbers,[88] altered arterial structure[89] and increased left ventricular mass,[39] and up-regulation of the HPA axis[90] and sympathetic nervous system[91] (Fig. 10). The fetal origins hypothesis proposes that these changes not only lead directly to adult cardiovascular disease, but also make the individual more susceptible to environmental stressors such as obesity in later life. The hypothesis is supported by examples in experimental

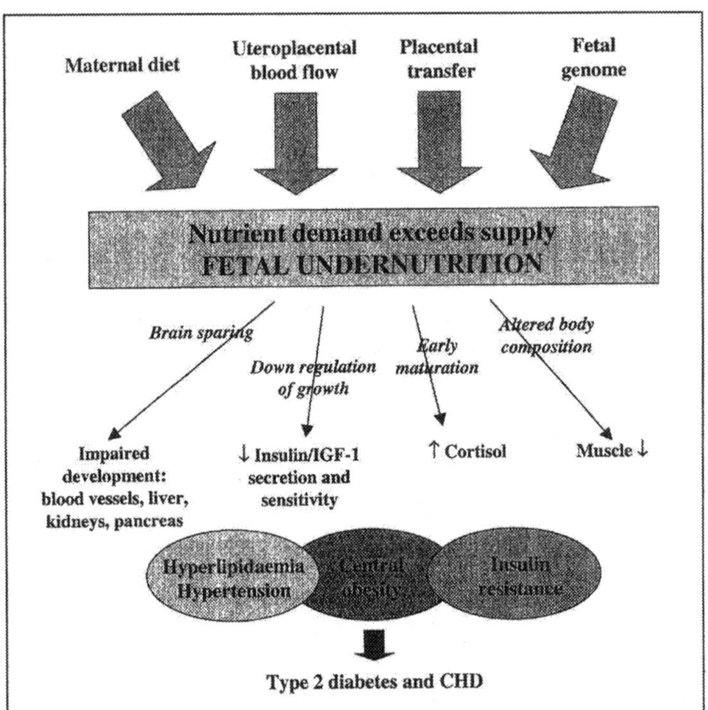

Figure 10. The fetal origins hypothesis.

animals of permanent structural, metabolic, endocrine and behavioural effects resulting from transient nutritional insults in utero, including undernutrition of the mother (reviewed extensively in other chapters in this book).

Associations of cardiovascular disease and risk factors with different body proportions at birth may reflect undernutrition during particular periods in gestation when different tissues and organ systems develop.[92] It has been suggested that male and female fetuses have different growth priorities and thus adapt differently to undernutrition, reflected in sex differences in associations with neonatal body proportions.[92] Since size at birth is a strong determinant of infant growth, the associations with low infant weight may reflect prenatal events. Alternatively, since the hyperplastic development of many tissues continues in early infancy, the infant as well as prenatal environment may have programming effects.

There are several reasons why weight gain in childhood, on a background of fetal restriction, might be associated with disease. Low birthweight babies tend to catch-up (compensatory growth), and the rapidity of post-natal growth may simply indicate the severity of fetal growth retstriction.[93] Alternatively the process of catch-up may be disadvantageous in itself.[62,94,95] There may be excess demand on other tissues which are not capable of compensatory hyperplasia, such as the pancreas.[87] Another adverse effect may be altered body composition. McCance observed excessive fat gain in pigs if they were placed on a high plane of nutrition after a period of early post-natal undernutrition. He suggested that this emphasised the development of fat which maintains the capacity for growth throughout life but could not recover muscle tissue which develops earlier and lose the capacity for cell division.[96] Another possibility is that the hormones driving catch-up growth have adverse cardiovascular and metabolic effects.[97] Low birthweight children who catch up in weight or height have higher insulin-like growth factor-1 (IGF-I) concentrations, and these in turn correlate with blood pressure.[98]

Because adult disease is linked to both post-and prenatal growth, and because a number of other health outcomes show associations with growth in early life, the term 'developmental origins of adult health and disease' (DOHaD) has supplanted 'fetal origins of adult disease' (FOAD) in recent years.

Genes 'versus' Environment

The fetal origins hypothesis has also been called the 'thrifty phenotype' hypothesis, a name coined by Hales and Barker[87] after Neel's 'thrifty genotype' hypothesis.[99] 'Thrift' can mean 'saving' (storing resources in times of plenty in preparation for leaner times) or 'economy' (making judicious use of meagre resources). Neel suggested that diabetes is caused by 'thrifty' genes that were selected for in mankind's distant past when the supply of food was precarious.[99] He proposed that they conveyed a 'fast insulin trigger' and thus the ability to store food rapidly as fat (savings) which became diabetogenic in modern times. The thrifty phenotype hypothesis, on the other hand, suggests that the undernourished fetus develops insulin resistance and other metabolic changes as a strategy for immediate survival, to down-regulate and prioritise growth (economy) for which it pays a price later in life, generally after the reproductive period. Both thrifty genes and the thrifty phenotype could become detrimental on exposure to plentiful nutrition. Variations on the thrifty genotype hypothesis are the strongest contenders to the 'fetal origins' hypothesis as an explanation for the associations between low birthweight and CVD risk.

Correlations between parent and offspring birthweights and between birthweights of half-siblings related through either the mother or the father show stronger maternal than paternal effects, which suggests that the 'maternal environment' is a more powerful influence on fetal growth than genes.[100,101] The birthweights of babies born after ovum donation are strongly related to the weight of the recipient mother but not that of the donor mother.[102] Nevertheless there are significant genetic effects on size at birth.[103] Hattersley proposed that, since insulin is a major growth hormone in fetal life, genes associated with either insulin resistance or reduced insulin secretion would lead to reduced fetal growth as well as an increased risk of adult diabetes (the 'fetal insulin hypothesis').[104] That this is biologically feasible is shown by the fact that birthweight is reduced in a number of genetic syndromes causing impaired insulin secretion or insulin resistance.[103] These are too rare to explain the observed birthweight-disease associations but more frequently occurring polymorphisms, linked both to small size at birth and adult diabetes, have recently been described.[105-107] The robustness of these associations remains to be tested.

Twin studies have classically been used to distinguish between genetic and environmental effects. Studies linking CVD risk to the difference in birthweight within twin pairs have shown inconsistent results. For example a study using the Danish twin registry showed that the smaller of monozygous twin pairs was more likely to become diabetic,[108] but this has not been confirmed elsewhere.[109] The mechanisms underlying the growth restriction of twins differ from those limiting growth in singleton fetuses. Higher disease concordance rates for monozygous than dizygous twins may reflect their shared intra-uterine environment as well as shared genes. Twin-twin interactions, for example the diffusion of steroid hormones from one twin to another, may reduce within-pair differences in programming effects. These features of the biology of fetal growth in multiple pregnancies limit the conclusions that can be drawn, in relation to the fetal origins hypothesis, from twin studies.[110]

Recent reports of associations between low offspring birthweight and an increased risk of CVD and insulin resistance in the parents (both mothers and fathers) could be evidence of common genetic factors.[111-114] However, intergenerational effects may also have environ-mental explanations. Assortive mating and shared lifestyle (socio-economic status, nutrition, smoking, and stress) could lead to both low birthweight and later CVD in both parents. Some but not all of these were taken into account in these studies. Associations shown between low offspring birthweight and type 2 diabetes in fathers but not mothers is more powerful evidence of genetic effects.[115,116] The lack of effect in mothers argues against these resulting from a shared environment. Fathers of low birthweight babies did not, however, have increased insulin resistance or diabetes in two other recent studies from India.[117,118]

Time trends in CVD and type 2 diabetes in western countries during the 20th century, and the recent rise in developing countries suggest a susceptibility to environmental changes that could have a either a genetic basis or arise from early-life programming. However these would make different predictions for the future. The former would predict continuing high levels of disease unless people reduce their lifestyle risk factors and become less obese. The thrifty phenotype hypothesis would predict a slowdown or downturn in disease as better nutrition of girls and mothers leads to improved fetal nutrition. CHD has been falling in the USA and Europe for 35 years despite increasing adult obesity, and only modest reductions in lifestyle-related risk factors. The incidence of stroke has also fallen since the early 1950s in the UK. In contrast, type 2 diabetes is increasing in all populations worldwide. However, the increase has been less marked in developed than developing countries[119] and a fall in incidence has been reported in one population.[120]

With increasing understanding of epigenetic effects and gene-environment interactions, it is no longer possible to think of diseases as being *either* 'genetic' *or* 'environmental'.[121] It was recently shown that an allele of the PPAR-γ gene is associated with increased insulin resistance, but only in men and women of low birthweight.[122] It is clearly possible to permanently alter gene expression by manipulation of intra-uterine nutrition.[123] Such epigenetic effects may persist across generations. For example, feeding 'agouti' mice with a methyl-supplemented diet during pregnancy leads to permanent and heritable effects on offspring coat colour, which is regulated by genes.[124] Imprinted genes, several of which play a role in fetal growth, are thought to be particularly susceptible to epigenetic effects. It seems likely that environmental effects, genes and interactions between the two contribute to the observed associations linking birthweight to adult disease, and that the DOHaD and fetal insulin hypotheses are not mutually exclusive.

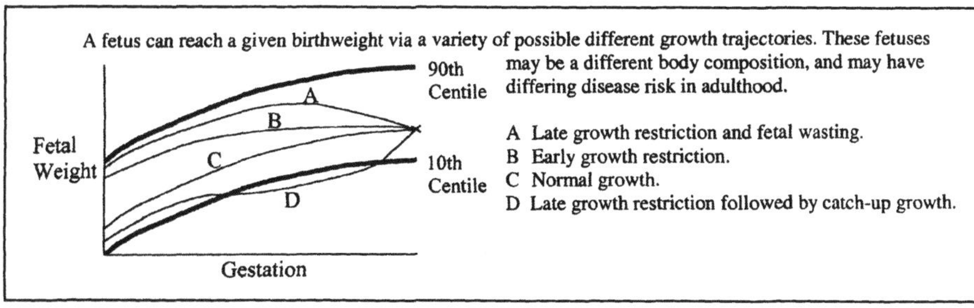

Figure 11. Different fetal growth trajectories.[128] Reproduced from: Harding JE. The nutritional basis of the fetal origins of adult disesase. Int J Epidemiol 2001; 30:15-23, by permission of Oxford University Press.

Clinical Importance of the Effects of Poor Fetal Growth

It has been argued that fetal undernutrition is of little clinical importance because statistically birthweight explains little of the variation in adult disease.[125-127] Effects are most marked at the extremes of birthweight, where there are relatively fewer individuals. Based on the Hertfordshire data Joseph estimated that 26% of CHD deaths would be averted if all babies weighed 9-9.5 pounds at birth, 9% if babies were born one birthweight category (approximately 1 pound) heavier, and 2% if more realistic increases in birthweight (100-200g) were achieved.[125]

These calculations are potentially misleading. Crude birth measurements and imprecise estimates of gestational age at birth would lead to under-estimation of birthweight effects. Birthweight is an insensitive marker of the dynamic process of fetal growth[128] (Fig. 11) and does not describe body composition or the development of specific tissues. Babies born to mothers exposed to the Dutch Famine of 1944-45 during early gestation had normal birthweights, and yet an increased risk of adult obesity and dyslipidaemia (see below). The true impact of intra-uterine undernutrition will not be known until there are better markers of programming than birthweight.

A further reason why calculations based on birthweight alone may underestimate the importance of intra-uterine growth is that, as described above, effects of size at birth are conditioned by childhood growth and adult obesity. Data from Finland, show that combinations of birth, infant and childhood measurements predict large differences in the risk of CHD, hypertension and diabetes[63,64,129] (Table 2).

The Role of Maternal Nutrition

If maternal nutritional status has important programming effects on adult disease, correlations would be expected between measures of maternal nutrition and CVD in the offspring. There are limited data to test this, especially in populations old enough to measure disease outcomes. Those that are available are limited to maternal anthropometry recorded in clinical obstetric notes (usually only weight and height), old diet and nutrition surveys, and famine studies.

Maternal Anthropometry

Several studies have shown that low maternal weight gain, BMI or skinfolds in pregnancy are associated with higher offspring blood pressure,[130,131] though not consistent in all studies.[70,132] Low maternal weight or BMI is associated with increased adult insulin resistance in the offspring.[70] and a study in India showed that CHD was associated with low maternal weight.[69] Conversely, in Finland, mortality from CHD was increased in men and women whose mothers were short and had a high BMI[11] and in India type 2 diabetes in older men and women was strongly associated with higher maternal weight and pelvic diameters.[75] It is possible that these associations with large maternal size or obesity reflect the effects of gestational diabetes, but this is speculative.

Table 2. *Odds ratios (95% CI) for type 2 diabetes and hypertension, and hazard ratios for coronary heart disease according to birthweight and body mass index at 11 years; 13,517 men and women born in Helsinki, Finland, 1924-1944*[129]

	Body Mass Index at 11 Years (kg/m^2)			
Birthweight (kg)	**-15.7**	**-16.6**	**-17.6**	**>17.6**
No. of men and women				
-3.0	991	719	581	560
-3.5	1394	1422	1264	1246
-4.0	827	984	1122	1110
>4.0	167	254	413	463
Type 2 diabetes (698 cases)†				
-3.0	1.3 (0.6-2.8)	1.3 (0.6-2.8)	1.5 (0.7-3.4)	2.5 (1.2-5.5)
-3.5	1.0 (0.5-2.1)	1.0 (0.5-2.1)	1.5 (0.7-3.2)	1.7 (0.8-3.5)
-4.0	1.0 (0.5-2.2)	0.9 (0.4-1.9)	0.9 (0.4-2.0)	1.7 (0.8-3.6)
>4.0	1.0 (ref)	1.1 (0.4-2.7)	0.7 (0.3-1.7)	1.2 (0.5-2.7)
Hypertension (2997 cases)†				
-3.0	2.0 (1.3-3.2)	1.9 (1.2-3.1)	1.9 (1.2-3.0)	2.3 (1.5-3.8)
-3.5	1.7 (1.1-2.6)	1.9 (1.2-2.9)	1.9 (1.2-3.0)	2.2 (1.4-3.4)
-4.0	1.7 (1.0-2.6)	1.7 (1.1-2.6)	1.5 (1.0-2.4)	1.9 (1.2-2.9)
>4.0	1.0 (ref)	1.9 (1.1-3.1)	1.0 (0.6-1.7)	1.7 (1.1-2.8)
Hospital admissions for coronary heart disease and deaths (1235 cases)*				
-3.0	1.4 (0.8-2.4)	1.6 (0.9-2.8)	1.8 (1.0-3.2)	2.1 (1.1-3.8)
-3.5	1.3 (0.7-2.2)	1.5 (0.9-2.7)	1.5 (0.8-2.6)	1.6 (0.9-2.9)
-4.0	1.3 (0.7-2.3)	1.4 (0.8-2.4)	1.3 (0.8-.4)	1.4 (0.8-2.6)
>4.0	1.0 (ref)	1.2 (0.6-2.3)	1.1 (0.6-2.1)	1.0 (0.5-1.8)
Deaths from coronary heart disease (480 cases)*				
-3.0	1.4 (0.5-4.0)	1.8 (0.6-5.1)	2.1 (0.7-6.2)	3.0 (1.0-8.6)
-3.5	1.4 (0.5-3.9)	1.9 (0.7-5.2)	2.2 (0.8-6.1)	2.7 (1.0-7.6)
-4.0	1.9 (0.7-5.3)	1.8 (0.7-5.2)	1.7 (0.6-4.8)	1.6 (0.6-4.5)
>4.0	1.0 (ref)	1.4 (0.4-4.6)	1.6 (0.5-4.7)	1.3 (0.4-4.0)

†Odds ratios adjusted for sex and year of birth. *Hazard ratios adjusted for sex and year of birth. Reproduced from Barker DJF et al. Fetal origins of adult disease: Strength of effects and biological basis. Internat J Epidemiol 2002; 31:1235-1239, by permission of Oxford University Press.

Famine Studies

In 1944-45 part of the Netherlands suffered a period of famine when rations fell to <800 calories a day (the Dutch Hunger Winter). The population was well-nourished prior to the famine and food supplies were restored after 5 months. Pregnant mothers experienced extreme undernutrition for sharply delineated periods of gestation. Birthweights were normal among women exposed to famine in early gestation, but reduced by 350 g in those exposed in late gestation.[133] An early follow-up study showed increased obesity in young men whose mothers experienced famine in early gestation.[134] Cardiovascular disease risk factors were recently measured in 700 men and women born before, during and after the famine. Late gestation exposure was associated with glucose intolerance, increased insulin resistance, and an increase in type 2 diabetes, compared with subjects conceived after the famine.[135] Early gestation exposure was associated with higher LDL/HDL cholesterol concentrations[136] and (in women) higher BMI and waist circumference. Famine exposure was associated with increased infant mortality, but there were no effects on adult mortality (so far assessed only up to the age of 50 years).[137]

Stanner compared men and women exposed to famine during the 1941-42 Siege of Leningrad with subjects born outside the siege area.[138] The famine lasted longer and was less acute in onset and termination than the Dutch famine, and birthweights were not recorded. Those exposed to famine in utero and/or in infancy showed increased subscapular/triceps skinfold ratios, diastolic blood pressure, ischaemic changes on ECG, and PAI activity and antigen, but lower factor VII concentrations than unexposed subjects. Unlike the Dutch study, there were no effects on glucose and insulin measurements. Men and women exposed in infancy were more likely to complain of angina than the in utero exposed group.

In rural Gambia, there is an annual cycle of adequate nutrition (harvest season) alternating with severe undernutrition during the rains (hungry season). Moore et al studied CVD risk factors in young adults for whom month of birth and infant anthropometric records were available.[139] There were no differences between season of birth groups (hungry vs. harvest) in blood pressure, fasting plasma glucose, insulin, lipid, fibrinogen or cortisol concentrations, or post-load glucose and insulin concentrations in an oral glucose tolerance test. Risk factors were not related to the subjects' weight-for-age in infancy and childhood.

Follow-Up of Diet and Nutrition Surveys and Trials

Three studies suggest that the balance of maternal protein and carbohydrate intakes during pregnancy is related to blood pressure in the offspring. During 1948-54 pregnant women attending Aberdeen Maternity Hospital recorded 7-day food diaries in late gestation. The offspring were traced and studied at the age of 40 years.[140,141] Blood pressure was not directly related to maternal intakes of energy, protein, fat, carbohydrate, calcium, or a range of vitamins. At low maternal protein intakes (<50 g/day) a higher percentage calorie intake from protein was associated with lower offspring blood pressure. The reverse was true at high protein intakes.[140] In an attempt to replicate these findings, Shiell at al studied young men and women born in Motherwell, Scotland 1952-76, where mothers were advised to eat a high-meat low carbohydrate diet to prevent preeclampsia.[142] Dietary intakes of 10 foods were recorded by trained staff. Protein intakes were higher than in Aberdeen (88g v 73 g). High intakes of meat and fish were associated with higher blood pressure. The highest blood pressures were found in men and women whose mothers had high meat/fish intakes but low intakes of green vegetables (Table 3). In a study in the Philippines, maternal intakes of protein, fat and energy were measured in late gestation and the children followed up in adolescence.[143] In boys, a higher percentage of maternal energy derived from protein, and in girls a higher percentage derived from fat, were associated with lower blood pressures. Mean protein intakes were not reported but are likely to have been low in this population.

Cardiovascular risk factors were recently measured in men and women whose mothers took part in the Oxford Nutrition Survey during 1942-1944, in which blood samples were taken in the third trimester to measure maternal vitamin A, C, B1 and phosphatase status.[144] No associations were found between these indices of maternal nutrition and offspring blood pressure, glucose and insulin or lipid concentrations.

These studies have many limitations, namely crude measures of diet and nutrition and large losses to follow-up. It is difficult to make much sense of their combined results. The data suggest that the balance rather than absolute intakes of protein and carbohydrate may influence blood pressure in the offspring, and that both low and high protein intakes may have adverse effects. They also suggest there may be maternal diet effects not mediated by reductions in birthweight. So far, there is no strong evidence that normal variations in dietary intakes during pregnancy have important effects on CVD risk in the offspring, but this has not been sufficiently studied. A number of prospective studies of maternal diet in pregnancy, specifically designed to investigate long-term outcomes in the offspring, are due to report outcomes in the children in the next few years.[145-147]

Table 3. Mean systolic blood pressure of men and women, adjusted for gender, body mass index, alcohol consumption, and cuff size according to their mother's consumption of meat, fish, and green vegetables in late pregnancy[142]

Green Vegetables (portions/wk)	Meat and Fish (portions/wk) ≤11	-15	-21	>21	All	Regression Coefficient (mm Hg/portion/wk) β	95% CI for β	p
<7	120 (131)	120 (93)	123 (54)	124 (41)	121 (319)	0.26	0.03-0.50	0.03
≥7	118 (58)	120 (94)	120 (74)	121 (81)	120 (307)	0.16	-0.06-0.38	0.16
All	119 (189)	120 (187)	121 (128)	122 (122)	120* (626)	0.19	0.04-0.35	0.02

Number of subjects given in parentheses. Overall SD = 11.5 mm Hg. Adapted from Shiell et al. Hyptertension 2001; 38:1282-8.

In the only follow up study so far of a randomised controlled trial of a nutritional intervention in pregnancy, Belizan studied children born during a trial of maternal calcium supplementation in Argentina. Supplementation was associated with lower blood pressure.[148]

Maternal Diabetes and Fetal Macrosomia

Although the main focus of this chapter is fetal growth restriction, recent data show that maternal gestational diabetes, which results in fetal overgrowth (macrosomia) also has adverse long-term effects. Offspring of diabetic mothers have an increased risk of obesity and type 2 diabetes compared with offspring of nondiabetic mothers or women who became diabetic after the pregnancy (Fig. 12).[149,150] Gestational diabetes produces a U-shaped relationship between birthweight and adult type 2 diabetes.[151-153] With increasing obesity worldwide, gestational diabetes is also increasing[154] and may make an important contribution to the rising incidence of type 2 diabetes.

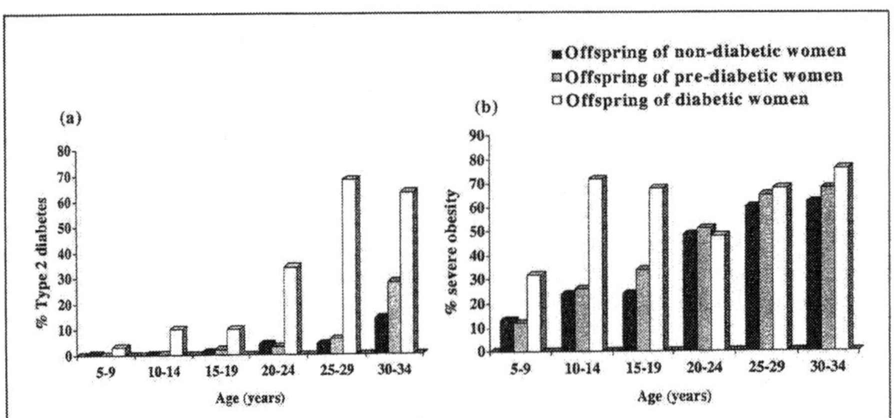

Figure 12. Prevalence of (a) type 2 diabetes and (b) obesity in offspring of nondiabetic, prediabetic and diabetic women (Pima Indians).[149] Reproduced from Dabelea et al. Effect of diabetes in pregnancy on offspring: Follow-up research in Pima Indians. J Mat Fet Med 2000; 9:83-88 with permission from Taylor & Francis Ltd (http://www.tandf.co.uk/journals).

Conclusions: Public Health Implications and Future Research

The 'developmental origins' hypothesis is attractive because it suggests that common degenerative diseases could be prevented by improving maternal health and fetal, infant and childhood development. Although it is not clear what optimal growth is and how it can be achieved, it may be more attainable than persuading middle-aged adults to return to 'primitive' lifestyles, and more positive than concluding that large numbers of people have genes incompatible with modern life. Data from experimental animals provide powerful evidence that a mother's nutrition programmes the metabolism of her offspring. However, there is currently insufficient evidence that maternal nutrition underlies CVD in humans. Future research should focus on the determinants of fetal growth, including maternal diet, and incorporate follow-up of the children for CVD-related outcomes. More research is needed into the long-term effects of weight gain in infancy. However, there is ample evidence to support efforts to prevent excessive BMI gain in childhood and this should be a public health priority. Current data suggest that the greatest benefit in terms of future risk-reduction will be in low birthweight individuals. Epidemiological research should go hand in hand with studies of the genes known to influence fetal growth, those associated with cardiovascular disease, and gene-environment interactions.

References

1. Kermack WO, McKendrick AG, McKinlay PL. Death rates in great britain and sweden; some general regularities and their significance. Lancet 1934; i:698-703.
2. Forsdahl A. Are poor living conditions in childhood and adolescence an important risk factor for arteriosclerotic disease? Br J Prev Soc Med 1977; 31:91-95.
3. Wadsworth MEJ, Cripps HA, Midwinter RE et al. Blood pressure in a national birth cohort at the age of 36 related to social and familial factors, smoking and body mass. BMJ 1985; 291:1543-1548.
4. Barker DJP, Osmond C. Infant mortality, childhood nutrition, and ischaemic heart disease in England and Wales. Lancet 1986; i:1077-1081.
5. Barker DJP, Osmond C, Winter PDW et al. Weight in infancy and death from ischaemic heart disease. Lancet 1989; ii:577-580.
6. Osmond C, Barker DJP, Winter PD et al. Early growth and death from cardiovascular disease in women. BMJ 1993; 307:1519-1524.
7. Barker DJP, Osmond C, Simmonds SJ et al. The relation of small head circumference and thinness at birth to death from cardiovascular disease in adult life. BMJ 1993; 306:422-426.
8. Martyn CN, Barker DJP, Osmond C. Mother's pelvic size, fetal growth, and death from stroke and coronary heart disease in men in the UK. Lancet 1996; 348:1264-1268.
9. Frankel S, Elwood P, Sweetnam P et al. Birthweight, body mass index and incident coronary heart disease. Lancet 1996; 348:1478-1480.
10. Rich-Edwards JW, Stampfer MJ, Mansin J et al. Birthweight and risk of cardiovascular disease in a cohort of women followed up since 1976. BMJ 1997; 315:396-400.
11. Forsen T, Eriksson JG, Tuomilehto J et al. Mother's weight in pregnancy and coronary heart disease in a cohort of Finnish men: Follow up study. BMJ 1997; 315:837-840.
12. Leon DA, Lithell HO, Vagero D et al. Reduced fetal growth rate and increased risk of death from ischaemic heart disease: Cohort study of 15,000 Swedish men and women born 1915-29. BMJ 1998; 317:241-245.
13. Forsen T, Eriksson JG, Tuomilehto J et al. Growth in utero and during childhood among women who develop coronary heart disease: Longitudinal study. BMJ 1999; 319:1403-1407.
14. Eriksson JG, Forsen T, Tuomilehto J et al. Early growth, adult income and risk of stroke. Stroke 2000; 31:869-874.
15. Hales CN, Barker DJP, Clark PMS et al. Fetal and infant growth and impaired glucose tolerance at age 64. BMJ 1991; 303:1019-1022.
16. Barker DJP, Hales CN, Fall CHD et al. Type 2 (noninsulin-dependent) diabetes mellitus, hypertension and hyperlipidaemia (syndrome X): Relation to reduced fetal growth. Diabetologia 1993; 36:62-67.
17. Valdez R, Athens MA, Thompson GH et al. Birthweight and adult outcomes in a bi-ethnic population in the USA. Diabetologia 1994; 37:624-631.
18. Lithell HO, McKeigue PM, Berglund L et al. Relation of size at birth to noninsulin dependent diabetes and insulin concentrations in men aged 50-60 years. BMJ 1996; 312:406-410.

19. Yarborough DE, Barrett-Connor E, Kritz-Silverstein D et al. Birthweight, adult weight, and girth as predictors of the Metabolic Syndrome in postmenopausal women. Diabetes Care 1998; 21:1652-1658.
20. Jie M, Law C, Zhang K-L et al. Effects of infant birthweight and maternal body mass index in pregnancy on components of the insulin resistance syndrome in China. Ann Intern Med 2000; 132:253-260.
21. Levitt NS, Lambert EV, Woods D et al. Impaired glucose tolerance and elevated blood pressure in low birthweight, nonobese, young South African adults: Early programming of cortisol axis. J Clin Endocrinol Metab 2000; 85:4611-4618.
22. Byberg L, McKeigue PM, Zethelius B et al. Birth weight and the insulin resistance syndrome: Association of low birth weight with truncal obesity and raised plasminogen activator inhibitor-1 but not with abdominal obesity or plasma lipid disturbances. Diabetologia 2000; 43:54-60.
23. Eriksson JG, Forsen T, Tuomilehto J et al. Effects of size at birth and childhood growth on the insulin resistance syndrome in elderly individuals. Diabetologia 2002; 45:342-348.
24. Huxley RR, Shiell AW, Law CM. The role of size at birth and post-natal catch-up growth in determining systolic blood pressure: A systematic review of the literature. J Hyperten 2000; 18:815-831.
25. Newsome CA, Shiell AW, Fall CHD et al. Is birthweight related to later glucose and insulin metabolism? A systematic review. Diabetic Med 2003; in press.
26. Curhan GC, Willett WC, Rimm EB et al. Birthweight and adult hypertension, diabetes mellitus, and obesity in US men. Circulation 1996; 94:3246-3250.
27. Owen CG, Whincup PH, Odoki K et al. Birthweight and blood cholesterol level; a study in adolescents and systematic review. Pediatrics 2003; 111:1081-9.
28. Fall CHD, Osmond C, Barker DJP et al. Fetal and infant growth and cardiovascular risk factors in women. BMJ 1995; 310:428-432.
29. Fall CHD, Barker DJP, Osmond C et al. Relation of infant feeding to adult serum cholesterol concentration and death from ischaemic heart disease. BMJ 1992; 304:801-805.
30. Frankel S, Elwood P, Sweetnam P et al. Birthweight, adult risk factors and incident coronary heart disease: The Caerphilly study. Public Health 1996; 110:139-143.
31. Byrne CD, Wareham NJ, Phillips DIW et al. Is an exaggerated postprandial triglyceride response associated with the component features of the insulin resistance syndrome? Diabetic Med 1997; 14:942-950.
32. Barker DJP, Meade TW, Fall CHD et al. Relation of fetal and infant growth to plasma fibrinogen and factor VII concentrations in adult life. BMJ 1992; 304:148-152.
33. Barker DJP, Martyn CN, Osmond C et al. Growth in utero and serum cholesterol concentrations in adult life. BMJ 1993; 307:1524-1527.
34. Martyn CN, Meade TW, Stirling Y et al. Plasma concentrations of fibrinogen and factor VII in adult life and their relation to intra-uterine growth. Br J Haematol 1995; 89:142-146.
35. Martyn CN, Gale CR, Jesperson S et al. Impaired fetal growth and atherosclerosis of the carotid and peripheral arteries. Lancet 1998; 352:173-178.
36. Lamont D, Parker L, White M et al. Risk of cardiovascular disease measured by carotid intima-media thickness at age 49-51: Lifecourse study. BMJ 2000; 320:273-278.
37. Martyn CN, Barker DJP, Jesperson S et al. Growth in utero, adult blood pressure, and arterial compliance. Br Heart J 1995; 73:116-121.
38. Kumaran K, Fall CHD, Martyn CN et al. Blood pressure, arterial compliance and left ventricular mass; no relation to small size at birth in South Indian adults. Heart 2000; 83:272-7.
39. Vijayakumar M, Fall CHD, Osmond C et al. Growth in infancy and adult left ventricular mass. Br Heart J 1995; 73:363-7.
40. Zureik M, Bonithon-Kopp C, Lecomte E et al. Weights at birth and in early infancy, systolic blood pressure, and left ventricular structure in subjects aged 8 to 24 years. Hypertension 1996; 27:339-45.
41. Leeson CPM, Whincup PH, Cook DG et al. Flow-mediated dilation in 9- to 11-year old children. The influence of intrauterine and childhood factors. Circulation 1997; 96:2233-2238.
42. Leeson CPM, Kattenhorn M, Morley R et al. Impact of low birth weight and cardiovascular risk factors on endothelial function in early adult life. Circulation 2001; 103:1264-1268.
43. Goodfellow J, Bellamy MF, Gorman ST et al. Endothelial function is impaired in fit young adults of low birth weight. Cardiovasc Res 1998; 40:600-606.
44. Serne EH, Stehouwer CDA, ter Maaten JC et al. Birth weight relates to blood pressure and microvascular function in normal subjects. J Hypertens 2000; 18:1421-1427.
45. Sorensen HT, Sabroe S, Rothman KJ et al. Relation between weight and length at birth and body mass index in young adulthood: Cohort study. BMJ 1997; 315:1137.

46. Phillips DIW. Relation of fetal growth to adult muscle mass and glucose tolerance. Diabetic Med 1995; 12:686-90.
47. Kahn HS, Narayan KMV, Williamson DF et al. Relation of birthweight to lean and fat thigh tissue in young men. Int J Obesity 2000; 24:667-672.
48. Weyer C, Pratley RE, Lindsay RS et al. Relationship between birthweight and body composition, energy metabolism and sympathetic nervous system activity later in life. Obesity Research 2000; 8:559-565.
49. Gale CR, Martyn CN, Kellingray S et al. Intrauterine programming of adult body composition. J Clin Endocrinol Metab 2001; 86:267-272.
50. Singhal A, Wells J, Cole TJ et al. Programming of lean body mass: A link between birth weight, obesity and cardiovascular disease? Am J Clin Nutr 2003; 77:726-30.
51. Aihie Sayer A, Syddall HE, Dennison E et al. Birth weight, weight at 1 year of age, and body composition in older men: Findings from the Hertfordshire Cohort Study. Am J Clin Nutr 2004; in press.
52. Phillips DIW, Fall CHD, Cooper C et al. Size at birth and plasma leptin concentrations in adult life. Int J Obesity 1999; 23:1025-1029.
53. Malina RM, Katzmarzyk PT, Beunen G. Birthweight and its relationship to size attained and relative fat distribution at 7-12 years of age. Obesity Research 1996; 4:385-390.
54. Okosun IS, Liao Y, Rotimi CN et al. Impact of birth weight on ethnic variations in subcutaneous and central adiposity in American children aged 5-11 years. A study from the Third National Health and Nutrition Examination Survey. Int J Obesity 2000; 24:479-484.
55. Bavdekar A, Yajnik CS, Fall CHD et al. The insulin resistance syndrome (IRS) in eight-year-old Indian children: Small at birth, big at 8 years or both? Diabetes 2000; 48:2422-2429.
56. Law CM, Barker DJP, Osmond C et al. Early growth and abdominal fatness in adult life. J Epidemiol Community Health 1992; 46:184-186.
57. Barker M, Robinson S, Osmond C et al. Birthweight and body fat distribution in adolescent girls. Arch Dis Child 1997; 77:381-383.
58. Eriksson JG, Forsen T, Tuomilehto HJ et al. Early growth and coronary heart disease in later life: Longitudinal study. BMJ 2001; 322:949-953.
59. Bhargava SK, Sachdev HPS, Fall CHD et al. Relation of serial changes in childhood body mass index to impaired glucose tolerance in young adulthood. New Eng J Med 2004; 350:865-75.
60. Law CM, Shiell AW, Newsome CA et al. Fetal, infant, and childhood growth and adult blood pressure: A longitudinal study from birth to 22 years of age. Circulation 2002; 105:1088-1092.
61. Victora CG, Barros FC, Horta BL et al. Short-term benefits of catch-up growth for small-for-gestational-age infants. Int J Epidemiol 2001; 30:1325-30.
62. Singhal A, Lucas A. Early origins of cardiovascular disease: Is there a unifying hypothesis? Lancet 2004; 363:1642-5.
63. Eriksson JG, Forsen T, Tuomilehto J et al. Catch up growth in childhood and death from coronary heart disease: Longitudinal study. Br Med J 1999; 318:427-31.
64. Eriksson JG, Forsen T, Tuomilehto HJ et al. Early growth and coronary heart disease in later life: Longitudinal study. BMJ 2001; 322:949-953.
65. Forsen T, Eriksson J, Tuomilehto J et al. The fetal and childhood growth of persons who develop type 2 diabetes. Ann Intern Med 2000; 133:176-182.
66. Eriksson JG, Forsen T, Tuomilehto J et al. Early adiposity rebound in childhood and risk of type 2 diabetes in adult life. Diabetologia 2003; 46:190-4.
67. Barker DJP, Forsen T, Uutela A et al. Size at birth and resilience to effects of poor living conditions in adult life: Longitudinal study. Br Med J 2001; 323:1-5.
68. Forrester TE, Wilks RJ, Bennett FI et al. Fetal growth and cardiovascular risk factors in Jamaican schoolchildren. British Medical Journal 1996; 312:156-60.
69. Stein C, Fall CHD, Kumaran K et al. Fetal growth and coronary heart disease in South India. Lancet 1996; 348:1269-1273.
70. Jie M, Law C, Zhang K-L et al. Effects of infant birthweight and maternal body mass index in pregnancy on components of the Insulin Resistance Syndrome in China. Ann Intern Med 2000; 132:253-260.
71. Miura K, Nakagawa H, Tabata M et al. Birthweight, childhood growth, and cardiovascular risk factors in Japanese aged 20 years. Am J Epidemiol 2001; 153:783-9.
72. Suzuki T, Minami J, Ohrui M et al. Relationship between birthweight and cardiovascular risk factors in Japanese young adults. Am J Hyperten 2000; 13:907-13.
73. Levitt NS, Lambert EV, Woods D et al. Impaired glucose tolerance and elevated blood pressure in low birthweight, nonobese, young South African adults: Early programming of cortisol axis. J Clin Endocrinol Metab 2000; 85:4611-4618.

74. Fall CHD. Developing countries and affluence in type 2 diabetes: The thrifty phenotype. Ed DJP Barker. Br Med Bulletin 2001; 60:33-50.
75. Fall CHD, Stein C, Kumaran K et al. Size at birth, maternal weight, and noninsulin-dependent diabetes (NIDDM) in South Indian adults. Diabetic Med 1998; 15:220-227.
76. Law CM, Egger P, Dada O et al. Body size at birth and blood pressure among children in developing countries. Int J Epidemiol 2000; 29:52-9.
77. Yajnik CS, Fall CHD, Coyaji KJ et al. Neonatal anthropometry: The thin-fat Indian baby; the Pune Maternal Nutrition Study. Int J Obesity 2003; 27:173-80.
78. Yajnik CS, Lubree HG, Rege SS et al. Adiposity and hyperinsulinaemia in Indians are present at birth. J Clin Endocrinol Metab 2002; 87:5575-5580.
79. Chowdhury B, Lantz L, Sjostrom L. Computed tomography - determined body composition in relation to cardiovascular risk factors in Indian and matched Swedish males. Metabolism 1996; 45:634-644.
80. Banerji MA, Faridi N, Atluri R et al. Body composition, visceral fat, leptin and insulin resistance in Asian Indian men. J Clin Endocrinol Metab 1999; 84:137-144.
81. Chandalia M, Abate N, Garg A et al. Relationship between generalized and upper body obesity to insulin resistance in Asian Indian men. J Clin Endocrinol Metab 1999; 84:2329-2335.
82. Yajnik CS, Yudkin JS. The Y-Y paradox. Lancet 2004; 363:163.
83. Barker DJP. The fetal origins of coronary heart disease. BMJ 1995; 311:171-4.
84. Lucas A. Programming by early nutrition in man. Ciba Foundation Symposium No. 156. 1991; 156:38-50.
85. Waterland RA, Garza C. Potential mechanisms of metabolic imprinting that lead to chronic disease. Am J Clin Nutr 1999; 69:179-197.
86. Phillips DIW. Insulin resistance as a programmed response to fetal undernutrition. Diabetologia 1996; 39:1119-1122.
87. Hales CN, Barker DJP. Type 2 (noninsulin-dependent) diabetes mellitus: The thrifty phenotype hypothesis. Diabetologia 1992; 35:595-601.
88. Mackenzie HS, Brenner BM. Fewer nephrons at birth: A missing link in the etiology of essential hypertension? Am J Kidney Dis 1995; 26:91-98.
89. Martyn CN, Greenwald SE. Impaired synthesis of elastin in walls of aorta and large conduit arteries during early development as an initiating event in pathogenesis of systemic hypertension. Lancet 1997; 350:953-955.
90. Phillips DIW, Barker DJP, Fall CHD et al. Elevated plasma cortisol concentrations: A link between low birthweight and the insulin resistance syndrome? J Clin Endocrinol Metab 1998; 83:757-760.
91. Phillips DIW, Barker DJP. Association between low birthweight and high resting pulse in adult life: Is the sympathetic nervous system involved in programming the insulin resistance syndrome? Diabetic Med 1997; 14:673-677.
92. Barker DJP. Mothers, babies and health in later life. London: Churchill Livingstone, 1998.
93. Leon DA, Koupilova I, Lithell HO et al. Failure to realise growth potential in utero and adult obesity in relation to blood pressure in 50 year old Swedish men. BMJ 1996; 312:410-416.
94. Lucas A, Fewtrell MS, Cole TJ. Fetal origins of adult disease - the hypothesis revisited. BMJ 1999; 319:245-249.
95. Metcalfe NB, Monaghan P. Compensation for a bad start: Grow now, pay later? Trends in Ecology and Evolution 2001; 16:254-60.
96. McCance RA. Food, growth and time. Lancet 1962; ii:621-626.
97. Lever AF, Harrap SB. Essential hypertension: A disorder of growth with origins in childhood? J Hyperten 1992; 10:101-120.
98. Fall CHD, Pandit AN, Law CM et al. Size at birth and plasma insulin-like growth factor-1 concentrations in childhood. Arch Dis Child 1995; 73:287-293.
99. Neel JV. Diabetes mellitus: A "Thrifty" Genotype Rendered Detrimental by "Progress"? Am J Hum Genet 1962; 14:353-361.
100. Morton NE. The inheritance of human birthweight. Ann Hum Genetics 1955; 20:125-134.
101. Klebanoff MA, Mednick BR, Schulsinger C et al. Father's effect on infant birth weight. Am J Obstet Gynaecol 1998; 178:1022-1026.
102. Brooks AA, Johnson MR, Steer PJ et al. Birth weight: Nature or nurture? Early Human Dev 1995; 42:29-35.
103. Frayling TM, Hattersley AT. The role of genetic susceptibility in the association of low birthweight with type 2 diabetes. In: DJP Barker, ed. Type 2 Diabetes: The Thrifty Phenotype. Br Med Bulletin, 2001:60:33-50.

104. Hattersley AT, Tooke JE. The fetal insulin hypothesis: An alternative explanation of the association of low birthweight with diabetes and vascular disease. Lancet 1999; 353:1789-1792.
105. Dunger DB, Ong KKL, Huxtable SJ et al. Association of the INS VNTR with size at birth. Nature Genetics 1998; 19:98-100.
106. Casteels K, Ong K, Phillips P. Mitochondrial 16189 variant, thinness at birth and type 2 diabetes. Lancet 1999; 353:1499-1500.
107. Vaessen N, Janssen JA, Heutink P et al. Association between genetic variation in the gene for insulin-like growth factor-I and low birthweight. Lancet 2002; 359:1036-1037.
108. Poulsen P, Vaag AA, Kyvic KO et al. Low birthweight is associated with NIDDM in discordant monozygotic and dizygotic twin pairs. Diabetologia 1997; 40:439-446.
109. Baird J, Osmond C, MacGregor A et al. Testing the fetal origins hypothesis in twins: The Birmingham twin study. Diabetologia 2001; 44:33-39.
110. Phillips DIW, Davies MJ, Robinson JS. Fetal growth and the fetal origins hypothesis in twins; problems and perspectives. Twin Research 2001; 4:327-331.
111. Davey Smith G, Hart C, Ferrell C et al. Birth weight of offspring and mortality in the renfrew and paisley study: Prospective observational study. BMJ 1997; 315:1189-1193.
112. Davey Smith G, Harding S, Rosato M. Relation between infant's birth weight and mother's mortality: Prospective observational study. BMJ 2000; 320:839-840.
113. Smith GCS, Pell JP, Walsh D. Pregnancy complications and maternal risk of ischaemic heart disease: A retrospective cohort study of 129,290 births. Lancet 2001; 357:2002-2006.
114. Lawlor DA, Davey Smith G, Ebrahim S. Birth weight of offspring and insulin resistance in late adulthood: Cross sectional survey. BMJ 2002; 325:359-362.
115. Lindsay RS, Dabelea D, Roumain J et al. Type 2 diabetes and low birth weight: The role of paternal inheritance in the association of low birth weight and diabetes. Diabetes 2000; 49:445-449.
116. Hypponen E, Davey Smith G, Power C. Parental diabetes and birth weight of offspring: Intergenerational cohort study. BMJ 2003; 326:19-20.
117. Yajnik CS, Coyaji KJ, Joglekar CV et al. Paternal insulin resistance and fetal growth: Problem for the 'fetal insulin' and the 'fetal origins' hypotheses. Diabetologia 2001; 44:1197-1201.
118. Yajnik CS, Joglekar CV, Bavdekar A et al. Parental risk of heavy birth weight child, 8 years after delivery. Ped Res 2001; 50(Suppl):4A, Conference abstract.
119. King H, Aubert RE, Herman WH. Global burden of diabetes, 1995-2025; prevalence, numerical estimates and projections. Diabetes Care 1998; 12:1414-1431.
120. Dowse GK, Zimmet PZ, Finch CF et al. Decline in incidence of epidemic glucose intolerance in Nauruans: Implications for the 'thrifty genotype'. Am J Epidemiol 1991; 133:1093-1104.
121. Morris JA. Fetal origin of maturity-onset diabetes mellitus: Genetic or environmental cause? Medical Hypotheses 1997; 51:285-288.
122. Eriksson JG, Lindi V, Uusitupa M et al. The effects of the Pro12Ala polymorphism of the peroxisome proliferator activated receptor γ2 gene on insulin sensitivity and insulin metabolism interact with size at birth. Diabetes 2002; 51:2321-2324.
123. Ozanne SE, Dorling MW, Wang CL et al. Impaired PI 3-kinase activation in adipocytes from early growth-restricted rats. Am J Physiol 2001; 280:E534-E539.
124. Wolff GL, Kodell RL, Moore SR et al. Maternal epigenetics and methyl supplements affect agouti gene expression in A^{vy}/a mice. FASEB J 1998; 12:949-957.
125. Joseph KS, Kramer MS. Should we intervene to improve fetal growth? In: Kuh D, Ben-Schlomo Y, eds. A Life Course Approach to Chronic Disease Epidemiology. Oxford: Oxford University Press, 1997.
126. Boyko EJ. Proportion of type 2 diabetes cases resulting from impaired fetal growth. Diabetes Care 2000; 23:1260-1264.
127. Huxley R, Neil A, Collins R. Unravelling the fetal origins hypothesis: Is there really an inverse association between birthweight and subsequent blood pressure? Lancet 2002; 360:659-665.
128. Harding JE. The nutritional basis of the fetal origins of adult disease. Int J Epidemiol 2001; 30:15-25.
129. Barker DJP, Eriksson JG, Forsen T et al. Fetal origins of adult disease: Strength of effects and biological basis. Int J Epidemiol 2002; 31:1235-9.
130. Godfrey KM, Forrester T, Barker DJP et al. Maternal nutritional status in pregnancy and blood pressure in childhood. Br J Obstet Gynaecol 1994; 101:398-403.
131. Adair LS, Kuzawa BBA, Borja J. Maternal energy stores and diet composition during pregnancy program adolescent blood pressure. Circulation 2001; 104:1034-1039.
132. Laor A, Stevenson DK, Shemer J et al. Size at birth, maternal nutritional status in pregnancy and blood pressure at age 17: Population based analysis. BMJ 1997; 315:449-53.

133. Susser M, Stein Z. Timing in prenatal nutrition: A reprise of the Dutch Famine Study. Nutr Rev 1994; 52:84-94.
134. Ravelli GP, Stein ZA, Susser MW. Obesity in young men after famine exposure in utero and early infancy. NEJM 1976; 7:349-54.
135. Ravelli ACJ, van der Meulen JHP, Michels RPJ et al. Glucose tolerance in adults after prenatal exposure to famine. Lancet 1998; 351:173-177.
136. Roseboom TJ, van der Meulen JHP, Osmond C et al. Plasma lipid profiles in adults after prenatal exposure to the Dutch famine. Am J Clin Nutr 2000; 72:1101-1106.
137. Roseboom TJ, van der Meulen JHP, Osmond C et al. Adult survival after prenatal exposure to the Dutch famine 1944-45. Paed Perinat Epidemiol 2001; 15:220-5.
138. Stanner SA, Bulmer K, Andres C et al. Does malnutrition in utero determine diabetes and coronary heart disease in adulthood? Results from the Leningrad Siege study, a cross-sectional study. BMJ 1997; 315:1342-9.
139. Moore SE, Halsall I, Howarth D et al. Glucose, insulin and lipid metabolism in rural Gambians exposed to early malnutrition. Diabetic Med 2001; 18:646-53.
140. Campbell DM, Hall MH, Barker DJP et al. Diet in pregnancy and the offspring's blood pressure 40 years later. Br J Obstet Gynaecol 1996; 103:273-280.
141. Shiell AW, Campbell DM, Hall MH et al. Diet in late pregnancy and glucose-insulin metabolism of the offspring 40 years later. Br J Obstet Gynaecol 2000; 107:890-5.
142. Shiell AW, Campbell-Brown M, Haselden S et al. High-meat, low-carbohydrate diet in pregnancy: Relation to adult blood pressure in the offspring. Hypertension 2001; 38:1282-1288.
143. Adair LS, Kuzawa BBA, Borja J. Maternal energy stores and diet composition during pregnancy program adolescent blood pressure. Circulation 2001; 104:1034-9.
144. Huxley RR, Neil AWN. Does maternal nutrition in pregnancy and birthweight influence levels of CHD risk factors in adult life? Br J Nutr 2004; 91:459-68.
145. Godfrey KM, Robinson S, Barker DJP et al. Maternal nutrition in early and late pregnancy in relation to placental and fetal growth. BMJ 1996; 312:410-414.
146. Inskip HI, Hammond J, Borland S et al. Determinants of fruit and vegetable consumption in 5,630 women aged 20-34 years from the Southampton Women's Survey. Ped Res 2001; 50(Suppl):58A, Conference abstract.
147. Rao S, Yajnik CS, Kanade A et al. Intake of micronutrient-rich foods in rural Indian mothers is associated with the size of their babies at birth; the Pune Maternal Nutrition Study. J Nutr 2001; 131:1217-1224.
148. Belizan JM, Villar J, Bergel E et al. Long-term effect of calcium supplementation during pregnancy on the blood pressure of offspring: Follow up of a randomised controlled trial. BMJ 1997; 315:281-5.
149. Dabelea D, Knowler WC, Pettitt DJ. Effect of diabetes in pregnancy and offspring: Follow-up research in the Pima Indians. J Matern-Fetal Med 2000; 9:83-88.
150. Dabelea D, Hanson RL, Lindsay RS et al. Intrauterine exposure to diabetes conveys risks for type 2 diabetes and obesity; a study of discordant sibships. Diabetes 2000; 49:2208-2211.
151. Wei J-N, Sung F-C, Li C-Y et al. Low birthweight and high birthweight infants are at increased risk to have type 2 diabetes among schoolchildren in Taiwan. Diabetes Care 2003; 26:343-8.
152. McCance DR, Pettitt DJ, Hanson RL et al. Birth weight and noninsulin dependent diabetes: Thrifty genotype, thrifty phenotype, or surviving small baby genotype? BMJ 1994; 308:942-945.
153. Rich-Edwards JW, Colditz GA, Stampfer MJ et al. Birthweight and the risk of type 2 diabetes in adult women. Ann Int Med 1999; 130:278-284.
154. King H. Epidemiology of glucose intolerance and gestational diabetes in women of childbearing age. Diabetes Care 1998; 21(Suppl 2):B9-B13.

Chapter 3

Studies of Twins:
What Can They Tell Us about the Developmental Origins of Adult Health and Disease?

Ruth Morley,* Terence Dwyer and John B. Carlin

Abstract

There is still limited understanding of the causal pathways underlying the observed association between exposures during fetal life and later health and disease in humans. Without better understanding we cannot estimate public health implications and assess the potential for intervention.

Study of twins should help us understand more about the role of factors shared by both twins versus factors affecting the individual fetus, the role of genetic factors, the role of placental factors, and which aspects or consequences of postnatal growth are associated with increased risk of later cardiovascular disease.

Generalisability of data from twin studies is open to question, but there is evidence that birth size—cardiovascular disease risk associations are similar in twins to those generally observed in singletons, suggesting that similar causal pathways are involved and study of twins will be informative.

There is an extensive body of literature relating to the association between size at birth and risk of later disease "the fetal origins of adult disease hypothesis",[1] and increasing understanding that growth throughout the developmental phase of life may be important. However, understanding of the underlying causal pathways is still limited and we need to understand these before we can estimate public health implications and assess the potential for intervention.

We consider:

1. Possible causal pathways underlying these observations.
2. How study of twins might shed light on these causal pathways.
3. Whether findings in twins are generalisable.

Possible Causal Pathways

Figure 1 shows some of the possible pathways and factors that may be involved in links between early development and risk of later disease.

Maternal Factors

Some maternal factors (A) will not affect later health. Others (B) may influence fetal growth and birth size, leading to risk of later disease. In other cases (C) gestational exposures may

*Corresponding Author: Ruth Morley—Department of Paediatrics, University of Melbourne, and Murdoch Childrens Research Institute, Royal Children's Hospital, Flemington Road, Parkville, Victoria 3052, Australia. Email: morleyr@unimelb.edu.au

Early Life Origins of Health and Disease, edited by E. Marelyn Wintour and Julie A. Owens.

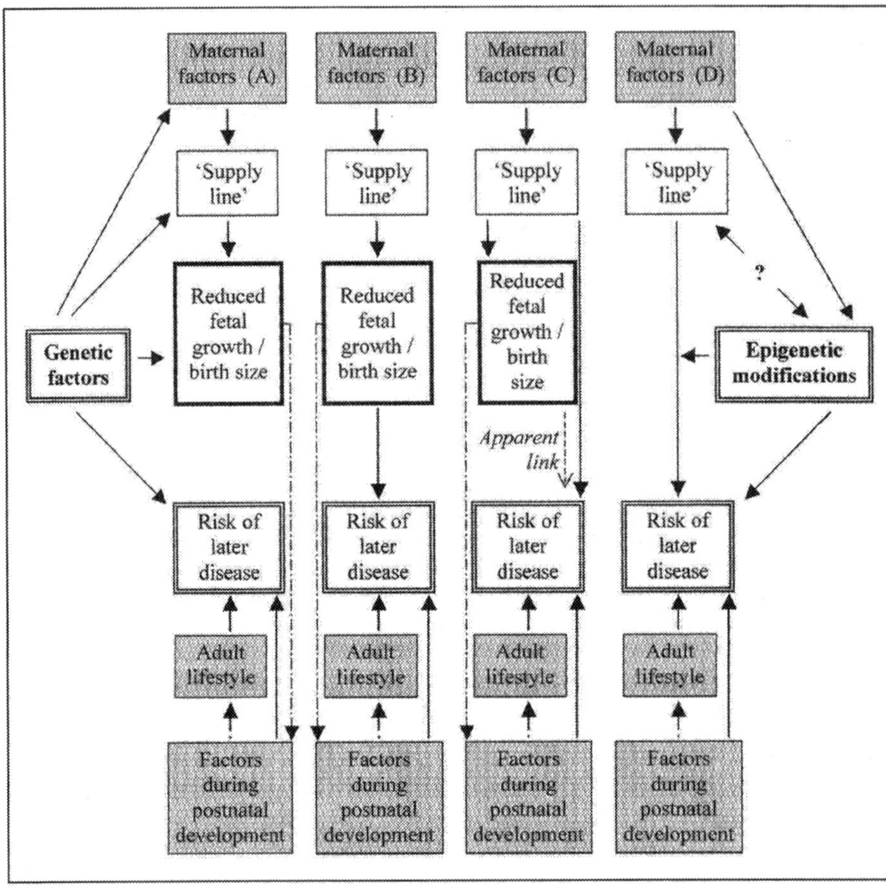

Figure 1. Pathways that may be involved in links between early development and later disease risk. Shading denotes factors that are currently potentially modifiable.

influence both fetal growth and disease risk, so there is an apparent link between size at birth and risk of later disease. Finally, some factors may influence disease risk with no influence on birth size.[2,3] Maternal factors may be of any of these types, and it is possible that the same exposure could act in more than one of these four different ways, depending on the timing, severity and duration of the exposure, and the risk factor or disease outcome examined. Evidence from the Dutch famine provides some support for this concept (summarised in ref. 2).

The "Supply Line" from Mother to Fetus

Blood supply to the uterus and placental development, health and function are important factors in the fetal supply line and in the human it is likely that intrauterine growth restriction relates to placental abnormality or insufficiency rather than maternal nutritional factors.[4] The placenta transports oxygen and nutrients from the maternal to the fetal circulation, metabolises key nutrients that are then supplied to the fetus (e.g., amino acids, especially glycine, and the active form of vitamin D) and produces hormones like placental lactogen and growth hormone that may influence fetal and maternal nutritional supply.[5] It also acts as an important barrier between maternal and fetal circulations, with enzymes that reduce fetal exposure to, for example, maternal glucocorticoids[6] and androgens,[7] as well as some xenobiotics.[8]

Post-natal factors, including nutrition and lifestyle factors, are clearly important determinants of health.[9] There has been recent interest in the role of postnatal growth rate as a determinant of later overweight/obesity, other cardiovascular risk factors, or adult cardiovascular disease. However, there is some inconsistency in the now extensive literature on this subject. For example, men with CHD had similar birthweight but lower weight at a year than others in a British cohort[10] and in a Swedish cohort low weight gain in the first year was associated with increased risk of CHD.[11] Conversely there are publications demonstrating that accelerated growth in infancy is disadvantageous,[12-14] and increased growth or size during childhood or adolescence are disadvantageous.[15-18]

Genetic Factors

Genetic endowment may be linked to both birth weight and later disease risk, so that there is an apparent link between birth weight and disease risk.[19-21] Since the placenta is part of the conceptus, genes affecting fetal growth could potentially act via an influence on placental development. There is also some evidence of interaction between genes and birth size. For example, interaction has been observed between birthweight and (i) angiotensin converting enzyme (ACE) polymorphism with respect to insulin response to glucose load,[22] (ii) apolipoprotein (apo) E genotypes and plasma lipid levels,[23] (iii) polymorphism of the peroxisome proliferator-activated receptor (PPAR)-gamma2 gene and insulin sensitivity and metabolism,[24] (iv) K121Q polymorphism of the plasma cell glycoprotein-1 gene and type 2 diabetes and hypertension[25] and (v) vitamin D receptor genotype and adult bone size and mineral density.[26] Mechanisms underlying such interactions are not understood.

Epigenetic Factors

Epigenetic modifications (discussed elsewhere in this book, Chs. 6, 7) represent a potential but largely unproven mechanism for programming the fetus or developing child in response to environmental factors.

DNA is associated with proteins called histones to form a complex substance known as chromatin. DNA methylation or histone acetylation/deacetylation alter the structure of chromatin to silence (switch off) genes without changing DNA sequence. Such modifications are described as epigenetic.[27] Imprinted genes are switched on or off according to parent of origin so are functionally haploid (only one copy is functional) so they are vulnerable to genetic or epigenetic mutations, or to epigenetic modification in response to environmental factors.[28] However, imprinting defects in humans generally cause clinically recognisable abnormal phenotypes, such as Prader-Willi and Beckwith-Wiedemann syndromes.

Evidence has emerged recently that heritable epigenetic changes in expression of nonimprinted genes may result from early nutritional or other exposures and could potentially affect disease risk.[29-31] In rats there is also evidence relating to early postnatal exposures to maternal behaviour.[32] Whether imprinting can similarly be modified by environmental factors has been little investigated, though circumstantial and animal evidence suggest this is a possibility.[33,34]

How Study of Twins Might Shed Light on the Underlying Causal Pathways

A natural starting point is the question of whether an association between birth size and later disease risk is seen in populations of twins. If there is no such association, then study of twins is unlikely to be helpful. A negative relationship between birth weight and blood pressure is seen in most (but not all) studies of twins,[35-46] as in studies of singletons, though there are a number of problems with comparing results of the various studies, including differences in statistical methodology that can affect both regression coefficients and 95% confidence intervals. In a recent large study of twins[47] the association between birth weight and risk of type 2 diabetes was similar in magnitude to that seen in populations of singletons.[48]

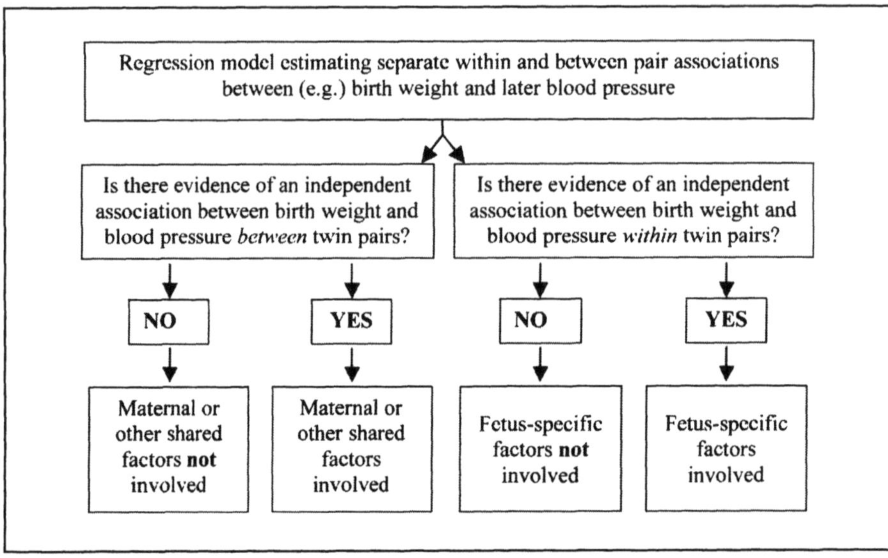

Figure 2. Between versus within pair analyses.

The Role of Shared Factors

We suggested comparing the association between birth weight (or birth weight standard deviation score for gestation and sex) and cardiovascular risk factors in twins treated as individuals, with that estimated within pairs.[35] Within pairs analysis assesses whether pair difference in birth weight is related to pair difference in cardiovascular risk, and so controls perfectly for all shared factors (identified and unidentified). If an association is seen in the cross-sectional analyses (which essentially ignore the cotwin's data) but is not seen or is substantially reduced in within-pair analyses then one could conclude that shared factors were involved in the underlying causal pathways. Conversely if it remains, we can conclude that factors affecting the individual fetus are involved.

However, the statistical issues involved in making these arguments precise are more involved than we have space to cover adequately here. Twins are effectively "clusters" within a dataset and there are a number of ways in which recognising the paired structure of the data can be incorporated into the analysis. When treating twins as individuals, one at least needs to recognise the lack of independence between cotwins, since this is necessary to obtain valid estimates of precision (standard errors) and statistical significance. Gains in efficiency of estimation can also be obtained by using generalised estimating equations (GEE) or other varieties of multilevel modelling.[41,43] However, a broader issue is that the simple "treated-as-individuals" regression model can be generalised to a form that estimates two distinct coefficients which more explicitly represent "within-pair" and "between-pair" regression effects.[44,47,49,50] The coefficient for the "between pairs" association may be interpreted to assess the role of shared factors whereas the "within pair" coefficient assesses the role of factors affecting the individual fetus (Fig. 2). This is the most appropriate statistical approach, permitting direct assessment of the relative strength of these associations.

Shared Factors will include:

- Maternal factors, including maternal nutrition, lifestyle factors, hormonal and metabolic status
- Paternal factors, almost always shared[51]
- Placental imprinting (apart from stochastic differences, the frequency of which are not known, or any resulting from environmental exposures)

- DNA sequence in monozygotic (MZ or "identical") twins
- Many factors in the postnatal environment

Factors influencing the individual fetus will include:

- The "supply line" to the individual fetus
- Genetic factors in dizygotic (DZ or "nonidentical") twins

Comparison of within and between Pair Differences in Monozygotic and Dizygotic Twins

It is clear that genetic factors are shared in MZ twins but differ in DZ twins. Thus if an association seen between individual twins is substantially weaker when estimated in a paired analysis within MZ twin pairs than it is when estimated within DZ twin pairs, the most obvious conclusion is that genes are involved in the underlying causal pathway. However, the possibility of greater sharing by MZ twins of other factors in the feto-placental unit should be considered.

Interpretation of such differences has proved challenging in practice. Findings to date have been somewhat contradictory for blood pressure (e.g., IJzerman's[38] versus Christensen's findings[41]) and there is only one reported study of angina[52] and of acute myocardial infarction,[53] suggesting genetic factors are involved in the relationship between coronary heart disease and birth weight.

In the large study of birth weight and type 2 diabetes, the between pair association in MZ and DZ twins was of similar magnitude and statistically significant in both cases, suggesting a role for shared factors.[47] However, the within pair association appeared to be reduced more (relative to the between pair association) in DZ than in MZ twins. The mechanism for this, if it reflected a real effect, is not obvious. It should be remembered here that around two thirds of MZ twins will share a placenta, whereas DZ twins have separate placentas (see Fig. 3, originally published in ref. 35). This study did not have information on chorionicity (placentation) but these findings may point to a possible key role for the placenta.

The Role of Postnatal Growth

Within pair analyses in large cohorts present an opportunity to estimate the role of aspects of postnatal growth as determinants of later health, without confounding by gestation length and largely unconfounded by external factors influencing child growth. These include parental eating habits, activity level and lifestyle (likely to influence similar factors in the child), as well as attitudes to child feeding and rearing.[54,55]

Comparison of Twins with Singletons

In general, twins have lower birth weight than singletons. This is partly because of higher risk of preterm delivery, and partly because twins are on average smaller for gestational age than singletons.[56] It might therefore be expected that twins, as a group, would have higher risk of cardiovascular disease than singletons. Studies to date have demonstrated no difference,[57-62] other than in one small study.[63,64] If the larger studies are generally correct, this suggests that birth weight per se is not involved in the observed associations, and that whatever causes the general constraint on intrauterine growth of twins is not involved in causal pathways underlying the association between size at birth and risk of later cardiovascular disease. However, it is clear from Iliadou's study[47] that among twins there is a similar association between birth weight and type 2 diabetes to that seen in the general population.[48]

A note of caution is that 'ecological' gross twin-singleton comparisons have not taken into account all relevant confounding factors so that inferences about birth weight-related pathways (or different postnatal growth) can only be made with limited confidence.

Another implication of these findings is that accelerated post-natal growth per se may not be related to risk of later cardiovascular disease. Twins in general have accelerated growth relative to singletons. Wilson followed twin children to age 9 and showed that they "caught up" to singletons over that period.[65,66] Study of postnatal growth in twins versus singletons may provide important clues as to what aspect or consequence of accelerated post-natal growth is involved in programming of later cardiovascular risk.

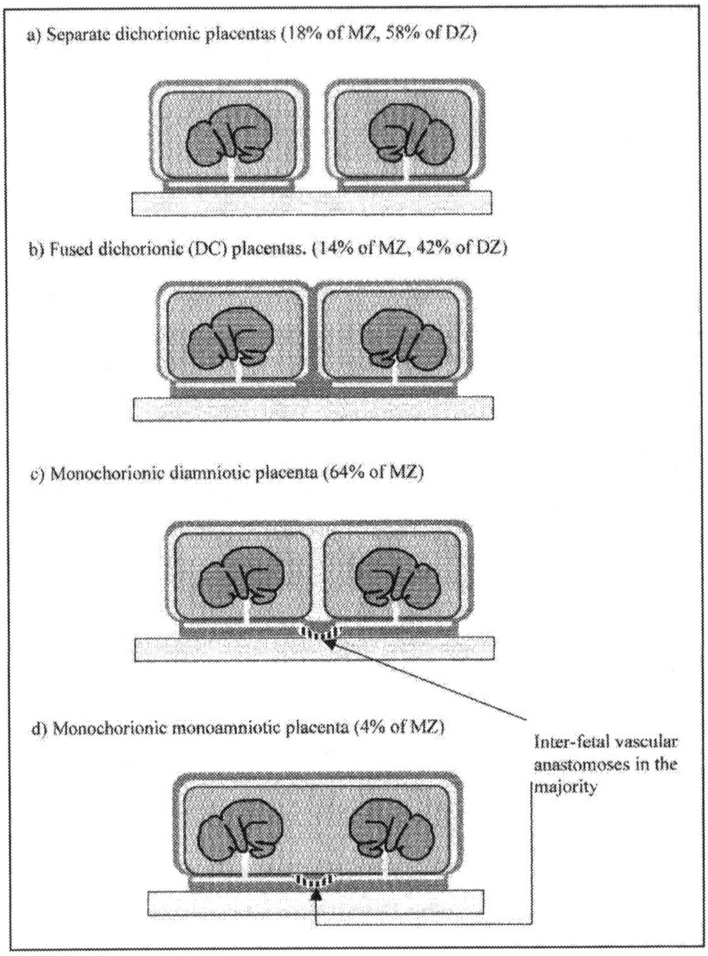

Figure 3. Different types of twin chorionicity (placentation). MZ = monozygotic. DZ = dizygotic.

The Role of the Placenta

Birth weight discordance is common in twin pairs, suggesting that within pairs the intrauterine environment can differ, and confirming that aspects of placental structure or function are likely to be important determinants of fetal growth. There is an important opportunity, not yet exploited, to use twins to study the role of the placenta. Maternal factors are perfectly controlled and postnatal environment is much more similar than that of unrelated singletons. Furthermore, both maternal and postnatal factors are better controlled than between siblings. Twins therefore provide a particular opportunity to investigate the role of placental differences (in structure, function, disease and possibly epigenetic modifications) in terms of both birth size and later health. In the case of monozygotic twins genetic factors will also be controlled.

Around two thirds of MZ twins are monochorionic (share one placenta), and it seems intuitively unlikely that epigenetic modifications will differ between two parts of one placenta. However, this has not been investigated and several studies have reported discordance for Beckwith-Wiedemann syndome in MZ twin pairs,[67] some monochorionic.[68,69]

Large prospective twin studies with good biological sample collection will be needed to elucidate the role of placental factors.

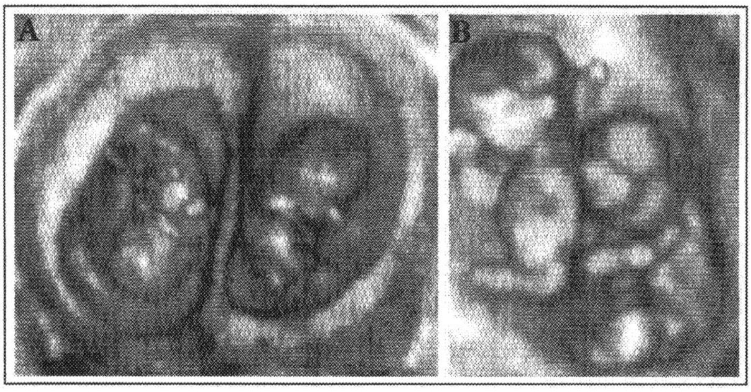

Figure 4. 3D ultrasound scans of twins at 12 weeks of gestation. A) Dichorionic twins (courtesy of Dr. Martin Metzenbauer, Donauspital, Vienna, Austria). B) Monochorionic diamnionic twins (courtesy of Dr. Philippa Ramsay, Ultrasound for Women, Sydney Adventist Hospital, Australia).

Are Findings in Twins Generalisable?

There are important differences between twin and singleton pregnancies, and some researchers believe that we cannot extrapolate from twins to the largely singleton population[48,70,71] because of "the substantially different biology of fetal growth in twins".[70] If this were true it might shed doubt on aspects of evidence from studies of multifetal species like rodents. Furthermore, and as we have indicated elsewhere, fetal growth per se may not be in the causal pathways linking gestational factors to later health, and in any case if it were, the public health implications of the association would be limited.[2]

Issues regarding generalisability of twin data are:

1. **Shorter median gestation.**
   Evidence on the relationship between gestation length and cardiovascular disease or its risk factors is inconclusive and most recent large studies (with reliable gestation length estimates) failed to demonstrate an association independent of birth size.[72-77] There is insufficient information to determine whether gestation length modifies the relationship between birth size and later health.
2. **Bias in volunteer twin cohorts.**
   Representative sampling is not a requirement for aetiologic studies.[78] A cohort with a wide spread of exposures (risk factors) to ensure adequate control of any known confounding factors that might obscure associations of interest is the main requirement, and this requirement is generally met in twin cohorts.
3. **Recalled birth weights in some twin cohorts.**
   Recalled birth weight has also been used in some singleton cohort studies, for example the Caerphilly Study.[79] We do not know whether there is a systematic difference in recall accuracy between twin and singleton cohorts.
4. **Zygosity and chorionicity (placentation).**
   In one study MZ twins had a more adverse lipid profile and higher fasting plasma glucose and insulin concentrations than DZ. twins,[80,81] though these findings await confirmation in other cohorts. Conversely, in another study mortality among female dizygotic twins was 1.77 times higher than among monozygotic twins at age 30 - 59 years.[82] However, the issue of chorionicity (shared placenta versus separate placentae in MZ twins, see Figs. 3, 4), was not taken into account by any of these studies because of lack of information.

   Two thirds of MZ twins share a placenta and there are vascular communications between their circulations. If circulating factors mediate the relationship between intrauterine compromise and later health, then such factors could pass from a compromised twin to its uncompromised cotwin, thus blunting the association between birth weight and later health

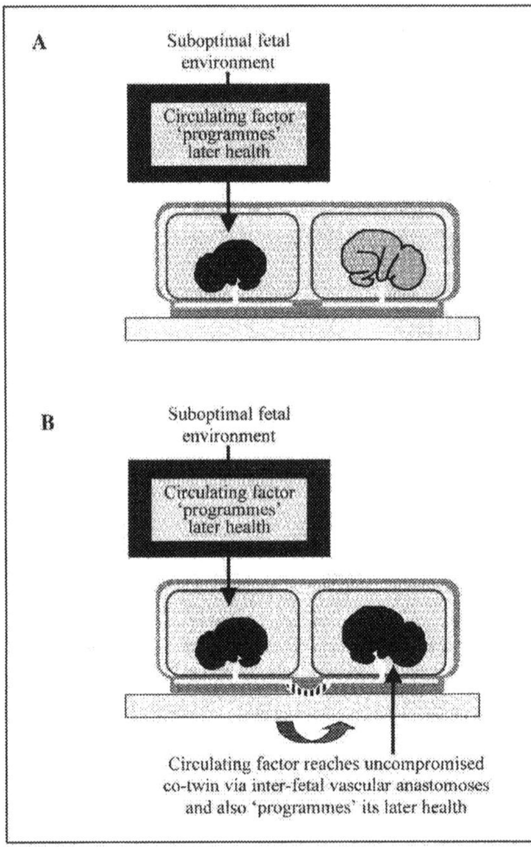

Figure 5. Monochorionic twins: potential for circulating factors from compromised twin to 'programme' otherwise uncompromised co-twin.

in the "uncompromised" twin (Fig. 5). This remains speculative and Iliadou's findings (of at least as strong within pair association between birth weight and type II diabetes in MZ versus DZ twins) suggest such 'blunting' is unlikely, at least in the programming of later risk of type II diabetes.

5. **Gender mix.**
   It has been suggested that exposure of a female twin to testosterone from a male cotwin might affect her later health.[70] There is little good evidence to support this,[83-94] though it is noteworthy that there are no data on cardiovascular disease or its risk factors. Many investigators have avoided this issue by studying only same-sex pairs.[37,38,40,42-45]
6. **Maternal subfertility and assisted reproduction technology.**
   Assisted reproduction technology (ART) is associated with an increased twinning rate[95] and there is evidence of a small influence on human fetal growth.[96] No study to date has investigated cardiovascular outcomes in relation to ART or maternal subfertility but there are reports of imprinting defects in children conceived by ART, in particular by ICSI (intracytoplasmic sperm injection).[97] Whether these defects relate to the causes of infertility or to the ICSI procedure (or both) is not known. However, there is evidence that abnormal spermatogenesis (leading to low sperm counts) is associated with an increase in defective methylation of the *H19* locus,[98] potentially leading to the presence of two inactive *IGF2* genes in the placenta, with implications for embryo development.

Summary

Study of twins should help us understand more about:

- The role of shared factors versus factors affecting the individual fetus.
- The role of genetic factors and possibly epigenetic ones.
- The role of placental factors.
- Which aspects or consequences of postnatal growth are associated with increased risk of later cardiovascular disease.

References

1. Barker DJP. Programming the baby. Mothers, babies and health in later life. Edinburgh: Churchill Livingstone, 1998:13-41.
2. Morley R, Owens J, Blair E et al. Is birthweight a good marker for gestational exposures that increase risk of adult disease? Paediatr Perinat Epidemiol 2002; 16:194-199.
3. Morley R, Carlin JB, Dwyer T. Maternal calcium supplementation and cardiovascular risk factors in twin offspring. Int J Epidemiol (doi:10.1093/ije/dyh284).
4. Henriksen T, Clausen T. The fetal origins hypothesis: Placental insufficiency and inheritance versus maternal malnutrition in well-nourished populations. Acta Obstet Gynecol Scand 2002; 81:112-114.
5. Harding JE. The nutritional basis of the fetal origins of adult disease. Int J Epidemiol 2001; 30:15-23.
6. Speirs HJ, Seckl JR, Brown RW. Ontogeny of glucocorticoid receptor and 11beta-hydroxysteroid dehydrogenase type-1 gene expression identifies potential critical periods of glucocorticoid susceptibility during development. J Endocrinol 2004; 181:105-116.
7. Meinhardt U, Mullis PE. The essential role of the aromatase/p450arom. Semin Reprod Med 2002; 20:277-284.
8. Syme MR, Paxton JW, Keelan JA. Drug transfer and metabolism by the human placenta. Clin Pharmacokinet 2004; 43:487-514.
9. Lawlor DA, Davey Smith G, Ebrahim S. Life course influences on insulin resistance: Findings from the British Women's Heart and Health Study. Diabetes Care 2003; 26:97-103.
10. Fall CH, Vijayakumar M, Barker DJ et al. Weight in infancy and prevalence of coronary heart disease in adult life. BMJ 1995; 310:17-19.
11. Eriksson JG, Forsen T, Tuomilehto J et al. Early growth and coronary heart disease in later life: Longitudinal study. BMJ 2001; 322:949-953.
12. Stettler N, Zemel BS, Kumanyika S et al. Infant weight gain and childhood overweight status in a multicenter, cohort study. Pediatrics 2002; 109:194-199.
13. Singhal A, Lucas A. Early origins of cardiovascular disease: Is there a unifying hypothesis? Lancet 2004; 363:1642-1645.
14. Ong KK, Ahmed ML, Emmett PM et al. Association between postnatal catch-up growth and obesity in childhood: Prospective cohort study. BMJ 2000; 320:967–971.
15. Forsen T, Eriksson J, Tuomilehto J et al. The fetal and childhood growth of persons who develop type 2 diabetes. Ann Intern Med 2000; 133:176-182.
16. Dwyer T, Blizzard L, Venn A et al. Syndrome X in 8 year-old Australian children: Stronger associations with current body fatness than with infant size or growth. Int J Obes Relat Metab Disord 2002; 26:1301-1309.
17. Cole TJ. Modelling postnatal exposures and their interactions with birth size. J Nutr 2004; 134:201-204.
18. Monteiro PO, Victora CG, Barros FC et al. Birth size, early childhood growth, and adolescent obesity in a Brazilian birth cohort. Int J Obes Relat Metab Disord 2003; 27:1274-1282.
19. Wannamethee SG, Lawlor DA, Whincup PH et al. Birthweight of offspring and paternal insulin resistance and paternal diabetes in late adulthood: Cross sectional survey. Diabetologia 2004; 47:12-18.
20. Lawlor DA, Davey Smith G, Ebrahim S. Birth weight of offspring and insulin resistance in late adulthood: Cross sectional survey. BMJ 2002; 325:359.
21. Li HL, Liu DP, Liang CC. Paraoxonase gene polymorphisms, oxidative stress, and diseases. J Mol Med 2003; 81:766-779.
22. Cambien F, Leger J, Mallet C et al. Angiotensin I-converting enzyme gene polymorphism modulates the consequences of in utero growth retardation on plasma insulin in young adults. Diabetes 1998; 47:470-475.
23. Garces C, Benavente M, Ortega H et al. Influence of birth weight on the apo E genetic determinants of plasma lipid levels in children. Pediatr Res 2002; 52:873-878.

24. Eriksson JG, Lindi V, Uusitupa M et al. The effects of the Pro12Ala polymorphism of the peroxisome proliferator-activated receptor-gamma2 gene on insulin sensitivity and insulin metabolism interact with size at birth. Diabetes 2002; 51:2321-2324.
25. Kubaszek A, Markkanen A, Eriksson JG et al. The association of the K121Q polymorphism of the plasma cell glycoprotein-1 gene with type 2 diabetes and hypertension depends on size at birth. J Clin Endocrinol Metab 2004; 89:2044-2047.
26. Dennison EM, Arden NK, Keen RW et al. Birthweight, vitamin D receptor genotype and the programming of osteoporosis. Paediatr Perinat Epidemiol 2001; 15:211-219.
27. Turner BM. Chromatin and gene regulation: Molecular mechanisms of epigenetics. Oxford UK, Blackwell Science, 2001.
28. Young LE. Imprinting of genes and the Barker hypothesis. Twin Res 2001; 4:307-317.
29. Kaati G, Bygren LO, Edvinsson S. Cardiovascular and diabetes mortality determined by nutrition during parents' and grandparents' slow growth period. Eur J Hum Genet 2002; 10:682-688.
30. Pembrey ME. Time to take epigenetic inheritance seriously. Eur J Hum Genet 2002; 10:669-671.
31. Waterland RA, Jirtle RA. Transposable elements: Targets for early nutritional effects on epigenetic gene regulation. Mol Cell Biol 2003; 23:5293-5300.
32. Weaver IC, Cervoni N, Champagne FA et al. Epigenetic programming by maternal behavior. Nat Neurosci 2004, [Epub ahead of print].
33. Abu-Amero SN, Ali Z, Bennett P et al. Expression of the insulin-like growth factors and their receptors in term placentas: A comparison between normal and IUGR births. Mol Reprod Dev 1998; 49:229-235.
34. Young LE, Fernandes K, McEvoy TG et al. Epigenetic change in IGF2R is associated with fetal overgrowth after sheep embryo culture. Nature Genet 2001; 27:153-154.
35. Dwyer T, Morley R, Blizzard L. Twins and fetal origins hypothesis: Within-pair analyses. Lancet 2002; 359:2205-6.
36. Dwyer T, Blizzard L, Morley R et al. Within pair association between birth weight and blood pressure at age 8 in twins from a cohort study. British Medical Journal 1999; 319:1325-1329.
37. Poulter NR, Chang CL, MacGregor AJ et al. Association between birth weight and adult blood pressure in twins: Historical cohort study. British Medical Journal 1999; 319:1330-1333.
38. IJzerman RG, Stehouwer CD, Boomsma DI. Evidence for genetic factors explaining the birth weight-blood pressure relation. Analysis in twins. Hypertension 2000; 36:1008-1012.
39. Baird J, Osmond C, MacGreggor A et al. Testing the fetal origins hypothesis in twins: The Birmingham study. Diabetalogia 2001; 44:33-39.
40. Zhang J, Brenner RA, Klebanoff MA. Differences in birth weight and blood pressure at age 7 years among twins. American Journal of Epidemiology 2001; 153:779-782.
41. Christensen K, Stovring H, McGue M. Do genetic factors contribute to the association between birth weight and blood pressure? Journal of Epidemiology and Community Health 2001; 55:583-587.
42. Loos RJ, Fagard R, Beunen G et al. Birth weight and blood pressure in young adults: A prospective twin study. Circulation 2001; 104:1633-1638.
43. Nowson CA, MacInnis RJ, Hopper JL et al. Association of birth weight and current body size to blood pressure in female twins. Twin Research 2001; 4:378-384.
44. Johansson-Kark M, Rasmussen F, De Stavola B et al. Fetal growth and systolic blood pressure in young adulthood: The Swedish Young Male Twins Study. Paediatric and Perinatal Epidemiology 2002; 16:200-209.
45. McNeill G, Tuya C, Campbell DM et al. Blood pressure in relation to birth weight in twins and singleton controls matched for gestational age. Am J Epidemiol 2003; 158:150-155.
46. McNeill G, Tuya C, Smith W. The role of genetic and environmental factors in the association between birthweight and blood pressure: Evidence from meta-analysis of twin studies. Int J Epidemiol 2004; 33:995-1001.
47. Iliadou A, Cnattingius S, Lichtenstein P. Low birthweight and Type 2 diabetes: A study on 11,162 Swedish twins. Int J Epidemiol 2004; 33:948-53.
48. Phillips DI. Commentary: Twins, low birthweight and type 2 diabetes. Int J Epidemiol 2004; 33:953-4.
49. Hopper JL, Seeman E. The bone density of female twins discordant for tobacco use. N Engl J Med 1994; 330:387-392.
50. Mann V, De Stavola BL, Leon DA. Separating within and between effects in family studies: An application to the study of blood pressure in children. Stat Med 2004; 23:2745-56.
51. Wenk RE, Houtz T, Brooks M et al. How frequent is heteropaternal superfecundation? Acta Genet Med Gemellol (Roma) 1992; 41:43-47.
52. Hubinette A, Cnattingius S, Johansson AL et al. Birth weight and risk of angina pectoris: Analysis in Swedish twins. Eur J Epidemiol 2003; 18:539-544.

53. Hubinette A, Cnattingius S, Ekbom A et al. Birthweight, early environment, and genetics: A study of twins discordant for acute myocardial infarction. Lancet 2001; 357:1997-2001.
54. Brown R, Ogden J. Children's eating attitudes and behaviour: A study of the modelling and control theories of parental influence. Health Educ Res 2004; 19:261-271.
55. Trost SG, Sallis JF, Pate RR et al. Evaluating a model of parental influence on youth physical activity. Am J Prev Med 2003; 25:277-282.
56. Taylor GM, Owen P, Mires GJ. Foetal growth velocities in twin pregnancies. Twin Research 1998; 1:9-14.
57. Vagero D, Leon D. Ischaemic heart disease and low birth weight: A test of the fetal-origins hypothesis from the Swedish Twin Registry. Lancet 1994; 343:260-263.
58. Christensen K, Vaupel JW, Holm NV et al. Mortality among twins after age 6: Fetal origins hypothesis versus twin method. British Medical Journal 1995; 310:432-436.
59. Christensen K, Wienke A, Skytthe A et al. Cardiovascular mortality in twins and the fetal origins hypothesis. Twin Research 2001; 4:344-349.
60. Poulter NR, Chang CL, MacGregor AJ et al. Association between birth weight and adult blood pressure in twins: Historical cohort study. British Medical Journal 1999; 319:1330-1333.
61. de Geus EJ, Posthuma D, Ijzerman RG et al. Comparing blood pressure of twins and their singleton siblings: Being a twin does not affect adult blood pressure. Twin Research 2001; 4:385-391.
62. Tuya C, Mutch WJ, Broom I et al. Size at birth, fasting glucose and insulin levels and insulin resistance in adult twins. Twin Research 2003; 6:302-306.
63. Jefferies CA, Hofman PL, Wong W et al. Increased nocturnal blood pressure in healthy prepubertal twins. Journal of Hypertension 2003; 21:1319-324.
64. Jefferies CA, Cutfield WS, Robinson EM et al. Twin children are insulin resistant and have ambulatory BP abnormalities irrespective of birth weight or gestational age. Pediatric Research 2003; 53(suppl):9A.
65. Wilson RS. Twin growth: Initial deficit, recovery, and trends in concordance from birth to nine years. Ann Hum Biol 1979; 6:205-220.
66. Wilson RS. Growth and development of human twins. In: Falkner F, Tanner JM, eds. Human Growth: A Comprehensive Treatise. 2nd ed. Methodology, Ecological, Genetic, and Nutritional Effects on Growth. New York: Plenum Press, 1986:197-211.
67. Weksberg R, Shuman C, Caluseriu O et al. Discordant KCNQ1OT1 imprinting in sets of monozygotic twins discordant for Beckwith-Wiedemann syndrome. Hum Mol Genet 2002; 11:1317-1325.
68. Orstavik RE, Tommerup N, Eiklid K et al. Nonrandom X chromosome inactivation in an affected twin in a monozygotic twin pair discordant for Wiedemann-Beckwith syndrome. Am J Med Genet 1995; 56:210-214.
69. Chien CH, Lee JS, Tsai WY et al. Wiedemann-Beckwith syndrome with congenital central hypothyroidism in one of monozygotic twins. J Formos Med Assoc 1990; 89:132-136.
70. Phillips DI, Davies MJ, Robinson JS. Fetal growth and the fetal origins hypothesis in twins problems and perspectives. Twin Research 2001; 4:327-331.
71. Jefferies CA, Hofman PL, Knoblauch H et al. Insulin resistance in healthy prepubertal twins. J Pediatr 2004; 144:608-613.
72. Rona RJ, Qureshi S, Chinn S. Factors related to total cholesterol and blood pressure in British 9 year olds. Journal of Epidemiology and Community Health 1996; 50:512-518.
73. Whincup PH, Bredow M, Payne F et al. Size at birth and blood pressure at 3 years of age. The Avon Longitudinal Study of Pregnancy and Childhood (ALSPAC). American Journal of Epidemiology 1999; 149:730-739.
74. Whincup PH, Cook DG, Papacosta O. Do maternal and intrauterine factors influence blood pressure in childhood? Archives of Disease in Childhood 1992; 67:1423-1429.
75. Leon DA, Johansson M, Rasmussen F. Gestational age and growth rate of fetal mass are inversely associated with systolic blood pressure in young adults: An epidemiologic study of 165,136 Swedish men aged 18 years. American Journal of Epidemiology 2000; 152:597-604.
76. Yiu V, Buka S, Zurakowski D et al. Relationship between birthweight and blood pressure in childhood. American Journal of Kidney Disease 1999; 33:253-260.
77. Leon DA, Lithell HO, Vagero D et al. Reduced fetal growth rate and increased risk of death from ischaemic heart disease: Cohort study of 15,000 Swedish men and women born 1915-29. British Medical Journal 1998; 317:241-245.
78. Rothman KJ, Greenland S. Modern Epidemiology. Philadelphia: Lippincott-Raven, 1998.
79. Frankel S, Elwood P, Sweetnam P et al. Birthweight, adult risk factors and incident coronary heart disease: The Caerphilly Study. Public Health 1996; 110:139-143.
80. Poulsen P, Vaag A, Beck-Nielsen H. The influence of zygosity status on blood pressure and on lipid profiles in male and female twins. Journal of Hypertension 2002; 20:645-649.

81. Poulsen P, Vaag A, Beck-Nielsen H. Does zygosity influence the metabolic profile of twins? A population based cross sectional study. British Medical Journal 1999; 319:151-154.
82. Christensen K, Vaupel JW, Holm NV et al. Mortality among twins after age 6: Fetal origins hypothesis versus twin method. British Medical Journal 1995; 310:432-436.
83. Dempsey PJ, Townsend GC, Richards LC. Increased tooth crown size in females with twin brothers: Evidence for hormonal diffusion between human twins in utero. American Journal of Human Biology 1999; 11:577-586.
84. Elkadi S, Nicholls ME, Clode D. Handedness in opposite and same-sex dizygotic twins: Testing the testosterone hypothesis. Neuroreport 1999; 10:333-336.
85. Gaist D, Bathum L, Skytthe A et al. Strength and anthropometric measures in identical and fraternal twins: No evidence of masculinization of females with male cotwins. Epidemiology 2000; 11:340-343.
86. Henderson BA, Berenbaum SA. Sex-typed play in opposite-sex twins. Developmental Psychobiology 1997; 31:115-123.
87. Loehlin JC, Martin NG. A comparison of adult female twins from opposite-sex and same-sex pairs on variables related to reproduction. Behavior Genetics 1998; 28:21-27.
88. Loehlin JC, Martin NG. Dimensions of psychological masculinity-femininity in adult twins from opposite-sex and same-sex pairs. Behavior Genetics 2000; 30:19-28.
89. McFadden D. A masculinizing effect on the auditory systems of human females having male cotwins. Proceedings of the National Academy of Science USA 1993; 90:11900-11904.
90. MacFadden D, Lochlin JC, Pasanen EG. Additional findings on heritability and prenatal masculinization of cochlear mechanisms: Click-evoked otoacoustic emissions. Hearing Research 1996; 97:102-119.
91. Miller EM. Reported myopia in opposite sex twins: A hormonal hypothesis. Optometry and Vision Science 1995; 72:34-36.
92. Miller EM, Martin N. Analysis of the effect of hormones on opposite-sex twin attitudes. Acta Geneticae Medicae Gemellogiae (Roma) 1995; 44:41-52.
93. Resnick SM, Gottesman II, McGue M. Sensation seeking in opposite-sex twins: An effect of prenatal hormones? Behavior Genetics 1993; 23:323-329.
94. Rose RJ, Kaprio J, Winter T et al. Feminity and fertility is sisters with twin brothers: Prenatal androgenization? Cross-sex socialization? Psychological Science 2002; 13:263-267.
95. Schieve LA, Ferre C, Peterson HB et al. Perinatal outcome among singleton infants conceived through assisted reproductive technology in the United States. Obstet Gynecol 2004; 103:1144-1153.
96. Cox GF, Burger J, Lip V et al. Intracytoplasmic sperm injection may increase the risk of imprinting defects. Am J Hum Genet 2002; 71:162-164.
97. De Rycke M, Liebaers I, Van Steirteghem A. Epigenetic risks related to assisted reproductive technologies: Risk analysis and epigenetic inheritance. Hum Reprod 2002; 17:2487-2494.
98. Marques CJ, Carvalho F, Sousa M et al. Genomic imprinting in disruptive spermatogenesis. Lancet 2004; 363:1700-1702.

Chapter 4

Prenatal Programming of Human Motor Function

Julia B. Pitcher,* David J. Henderson-Smart and Jeffrey S. Robinson

Abstract

In a world in which athletic skill is often valued more highly than intellectual prowess, we know surprisingly little about the development of the human motor system. Even less is known about how an adverse intrauterine event or environment might program motor learning, memory and function throughout the lifespan. We are only beginning to investigate how events during development of the brain and central nervous system might predispose some individuals to older age onset of some common neurological disorders such as Parkinson's and Alzheimer's diseases. Anecdotal and empirical evidence suggests that one or more adverse events occurring in utero may result in long-term changes in neuromotor development. These changes may be evident from infancy, or not become apparent until later in life. This chapter reviews this evidence. We suggest that the research focus must shift towards neurophysiological rather than neurodevelopmental paradigms, if these programmed changes in neuromotor function and the mechanisms responsible are to be fully understood. We also introduce the idea that the impact on the individual of maladaptive neuromotor programming might be reduced by the development of early intervention therapies, which utilise the developing nervous system's capacity for plastic change.

Introduction

There is emerging evidence that one of the developing systems that may be programmed by an adverse intrauterine environment is the human motor system. Programming has been described as a process whereby a disturbance of the environment, at critical stages of development of regulatory systems and their target tissues, alters development in such a way as to permanently change functional capacity and predispose the individual to disease in later life. Adverse conditions in utero have been implicated in long-term alterations in brain structure and function, and possibly the later development of neurological diseases. There is increasing evidence that motor and vision disorders, schizophrenia (discussed elsewhere in this book, Ch. 17), epilepsy and autism may have their origins, at least in part, in altered prenatal neurodevelopment.[1-2] Infants whose growth before birth has been restricted have increased rates of perinatal mortality and morbidity and increasingly, evidence of long-term neurological and neuromotor problems into childhood.[3] The neurodevelopmental studies performed to date examining the relationship between low birthweight and neuromotor function have provided some descriptive information. However, it is not known if the motor deficits they have

*Corresponding Author: Julia B. Pitcher—Department of Obstetrics and Gynaecology, University of Adelaide, Women's and Children's Hospital, 72 King William Road, North Adelaide SA 5006, Australia. Email: julia.pitcher@adelaide.edu.au

Early Life Origins of Health and Disease, edited by E. Marelyn Wintour and Julie A. Owens.

identified persist into adulthood, what their extent and impact on neuromotor function is for the individual, and what underlying physiological determinants of that function have been impaired. Determining how aspects of the prenatal and postnatal environments alter normal development of motor function is complex, particularly as the human brain is highly plastic during development of the central nervous system and indeed throughout life.

This review will focus on and critically evaluate the evidence for altered motor function. We address some of the possible sources of altered programming of motor function in utero, the evidence for specific alterations in motor physiology from animal studies, and suggest future directions for research into the physiology of programming of altered neuromotor development. The techniques for both induction and experimental measurement of experience or activity dependent motor plasticity are introduced, and we outline how these techniques might be developed for therapeutic intervention.

Intrauterine Growth Retardation (IUGR) versus Being Born Small for Gestational Age (SGA)

Because of the relative inaccessibility of the human fetus, surrogate markers of the prenatal environment, primarily fetal growth, as indicated by size at birth for gestational age, have been used in most studies of programming in humans. Suboptimal fetal growth is referred to as either intrauterine growth restriction (IUGR), or being born small for gestational age (SGA). IUGR and SGA are often used interchangeably, but by definition, they are different conditions. SGA is defined by the World Health Organisation as a birthweight below the tenth percentile for gestational age, whereas IUGR is a pathological condition where the fetus is unable to grow to its genetically-determined potential size, due either to abnormal genetic or environmental conditions that impair growth. So while all IUGR infants are SGA, not all SGA infants are IUGR. Both SGA and IUGR are recognised as a key risk factors for neurological deficits[4] and have been associated with learning difficulties, delayed language skills, behavioural problems[5] and motor co-ordination and psychomotor development problems in children.[6] The likelihood of these deficits is amplified in those IUGR babies whose in utero conditions are so compromised that they are delivered before term, adding the consequences of prematurity to IUGR.[7]

Critical Periods in Fetal Brain and Nervous System Development

Human brain and nervous system development begins about 2 weeks after conception and may not be completed until the third decade of life. Prenatally, it consists of a series of precisely timed events that occur during discrete time windows.[8,9] These prenatal events include neural induction, neurulation, proliferation, migration, axon and dendrite formation and outgrowth, synaptogenesis, differentiation and apoptosis (see ref. 9 for review). At each of these stages, the developing brain is vulnerable to a range of insults that can lead to long-term abnormalities and dysfunction. Therefore, the nature of any neurological abnormality is likely to depend upon the type and severity of the insult and the neurodevelopmental stage at which it occurs.[10] During the first and second trimesters while neurogenesis of the cerebral cortex is underway, early insults or epigenetically induced malfunctions during specific stages can result in cortical hypoplasia (reduced cell number), cortical ectopias (abnormal migration) or cortical dysplasias (abnormal dendritic shape or number).[8] Abnormalities of cortical development are being increasingly implicated in developmental delay, cognitive deficits and epilepsy in children and young adults.[11] During the third trimester, the developing human brain is particularly vulnerable to inflammatory, hypoxic, ischemic and infectious insults that predominantly result in lesions. In the early part of the third trimester the white matter appears most at risk, while later in the third trimester, the cortical or deep grey matter (particularly the basal ganglia and thalamus) become most vulnerable.[12,13] The cerebellar cortex also develops in the third trimester and is particularly vulnerable to undernutrition associated with placental insufficiency during this time.[14]

Intrauterine Growth Restriction and Corticospinal Development

IUGR infants have been shown to have abnormal brain structure, including reduced whole brain weight,[15] cerebellar weight[16,17] and hemispheric volume.[18] Rat and guinea pig studies have shown a significant association between IUGR birthweight, brain weight and reduced numbers of CA 1 pyramidal neurons in the hippocampus (thought to have a key role in learning and memory),[19,20] reduced number and disturbed migration of cerebellar cells,[19,21,22] reduced cerebellar apoptosis threshold,[23] and the amount and distribution of nerve growth factor throughout the brain.[24] Growth retarded fetal sheep have been shown to have thinner motor and visual cortices and reduced cortical synaptogenesis than their normally-grown counterparts.[25,26] Normal function of the cerebral cortex and the cerebellum is essential for normal movements. The alterations in brain structure in IUGR infants and experimental animal paradigms, the frequent anecdotal observations of clumsiness and poor motor skill development in IUGR children, and the subtle neurological signs reported in IUGR infants[5,6] suggest that human subjects who were growth-restricted in utero may have measurable deficits in corticomotor function.

Further support for this notion comes from the consequences of prenatal and post-natal malnutrition in the rat, where neuronal loss, reduced brain weight, and a 15% reduction of the velocity at which action potentials are conducted along the corticospinal tract have been reported.[27] These changes have also been associated with decreased diameter of corticospinal axons and reduced myelination of corticospinal fibres.[28,29] These authors also reported a pronounced loss of large compared with small corticospinal fibres in rats nutritionally restricted from conception to adulthood.[28] While one must be cautious in extrapolating from animal studies, these results clearly suggest the possibility that corticospinal and intracortical structure and function may be adversely affected in IUGR children and adults. This is likely to be manifest as poor motor development and poor motor control, particularly of fine and dextrous movements that rely heavily on the motor cortex and cerebellum, and on continuing afferent feedback to these brain structures from the periphery. However, these effects may be very subtle and difficult to detect without specific, physiological assessment of motor pathways.

Assessing the Impact of Growth Restriction on Motor Development in Infants, Children and Adults

Neurodevelopmental versus Neurophysiological Approaches

No study has examined the neurophysiological nature of motor deficits following IUGR or SGA in humans, although a number of studies have assessed aspects of neurodevelopment. There are key differences between neurodevelopmental versus neurophysiological assessments. In simple terms, neurodevelopmental tests assess the age-appropriate appearance of broad cognitive and/or physical milestones with reference to the "normal" population, whereas neurophysiological tests provide direct and/or indirect measurements of the physiological function of one or more specific neural pathways. Hence neurodevelopmental assessments aim to determine whether an individual is developing at the expected rate for their age, but unlike neurophysiological assessments, they provide little or no quantitative data about which neural pathway(s) might be expressing any abnormal development. With neurophysiological assessments, it is possible to narrow down the possibilities by examining a particular motor pathway and a motor behaviour or physiological outcome controlled primarily by that pathway.[30]

The results from the neurodevelopmental studies of IUGR/SGA have been equivocal. This is partly because a proportion of IUGR infants are also born prematurely, either due to premature onset of labour, or by obstetric intervention in the best interests of the infant or mother's survival. However, there is also considerable inherent variability in outcomes depending on the age of the cohort studied, the types of assessments used, and general study design. Most studies have used various scales of infant or child motor development to determine deviations from age-appropriate development. Developmental behavioural testing essentially examines sensorimotor function.

The development of the sensorimotor system is dependent upon the rate and degree of maturation of both the motor and somatosensory systems; not only do the central and peripheral branches of these tracts develop at different rates, alterations in the development of one or both tracts will affect sensorimotor development.[31] Normal inter-individual variability in these rates of development add a not insignificant degree of heterogeneity to outcomes and reduce the sensitivity of these tests, particularly in identifying more subtle deficits that might exist in one or both systems. So while these studies have been very useful in helping to identify the nature of the impact of IUGR on neuromotor development, we know virtually nothing of the physiological mechanisms responsible for the alterations.

The Fetus—Motor Behaviour in Utero

Early postnatal motor experiences have been shown to be important for the development of motor skills in humans and animals. However, little is known about the role of early postnatal motor experience in motor development[30] and even less of the influence, if any, of in utero motor experience. Longitudinal studies using real-time ultrasound have revealed that the fetus begins moving as early as the seventh week of gestation with slow neck extensions, then startle-type movements.[32,33] A few days later, more general movements begin appearing including breathing and jaw movements. By about 13 weeks, swallowing and rhythmical sucking appears that is of the same rate as those observed in breast-feeding term infants.[34] Rather than being voluntary movements under cortical control, there is evidence from animal studies that these movement patterns may originate endogenously from central pattern generators.[34] In the adult, for example, walking is thought to be under the general control of central pattern generators for locomotion located in the spinal cord and possibly utilising afferent feedback.[35]

In the only study published of motor behaviour in IUGR fetuses in utero, Bekedam et al[36] made hour-long real-time ultrasound assessments of ten IUGR and ten AGA fetuses. Compared to AGA fetuses, IUGR fetuses tended to display a limited repertoire of slow movements, with little variability in movement velocity or amplitude, although there were some AGA fetuses with a similar movement profile. In particular, IUGR fetuses displayed distinctly reduced startle, twitching and isolated limb movements. However, there is no direct evidence that this reduced activity is not simply an energy saving strategy rather than reflecting a pathological alteration in neuromotor function.

The Neonate and Older Infant—General Movements

There are a handful of studies of the neurodevelopmental outcomes of IUGR in neonates. Most of these have come from Heinz Prechtl's group in the Netherlands and the principal assessment of choice has been Prechtl's Method, a qualitative assessment of "general movements" that is believed to reflect early postnatal brain development and predict later neurodevelopmental outcome in high-risk infants[32,33,37] (see ref. 38 for review). General movements are endogenously generated, frequently occurring movement patterns of the head, trunk and limbs that are observable from birth until approximately 20 weeks postnatal age.[39] They are divided into two postnatal periods: the "writhing" period (birth to 8 weeks) and the "fidgety movements" period (approximately 8 to 20 weeks). Writhing refers to movements that are often elliptical in form, are small to moderate in amplitude, of slow to medium speed and generally performed close to the body.[39,40] Fidgety movements are characterised by small, circular, low amplitude movements of moderate speed and variable acceleration in a range of directions.[39,40] They are continually present except when the infant is focussing their attention, crying or is distracted. Their frequency of occurrence during the 12-week period has a parabolic trajectory; they begin appearing as isolated movements, gradually become more frequent and then slowly disappear.[39] The absence of fidgety movements has been shown to predict the later development of cerebral palsy with a sensitivity of 95% and a specificity of 96%.[32] The presence of abnormal fidgety movements is strongly predictive of neurodevelopmental outcomes at 2 years of age.[33,40,41]

Compared with AGA infants matched for age, gender and socio-economic status, IUGR infants exhibit more abnormal general movements, regardless of their gestational age at birth.[82-84] Abnormal movements, particularly those that occur in the early and late fidgety periods and that do not normalise soon after birth, are strongly predictive of a poor or abnormal neurodevelopmental outcome at 2 years of age.[82-84] General movement quality in both AGA and IUGR infants appears not to be affected by prematurity, which suggests that normal motor outcomes in early childhood are critically dependent upon normal fetal growth.[82] Cramped synchronised general movements, which are abnormal and predictive of cerebral palsy, tend to appear later postnatally in IUGR than in AGA infants.[85] Abnormalities or the absence of fidgety movements has been associated with poor neurodevelopmental outcomes in term and preterm babies with perinatal brain lesions of a range of severity.[74] However, most IUGR infants who demonstrate abnormal general movements have no evidence of lesions or other structural abnormalities on ultrasound scan.[83,84] This suggests that a suboptimal intrauterine substrate supply to the fetus has a longer term influence on neurodevelopment that is not necessarily due to lesion-type injuries from hypoxic-ischaemic or haemorrhagic insults in utero[6] One hypothesis might be that IUGR perturbs normal development of central pattern generators. However, abnormal movement patterns in IUGR infants will often normalise after they reach their term age,[83,84] with normal neurodevelopmental outcomes at 2 years of age, suggesting that the development of normal early postnatal movements can "catch up", at least in some infants.

While the results from the general movements studies have been relatively homogeneous, the findings from studies using different infant development instruments have been equivocal. Lacey and colleagues (in ref. 42) compared infants born small for gestational age with AGA infants on development of state stability, posture and movement. SGA infants scored significantly lower than AGA infants on every item, including stage of independent feeding, leg antigravity posturing, limb resistance strength and traction, arm movements and head turning. However, Newman and colleagues[5] found no differences in motor abilities when comparing 65 4-month-old SGA with 71 AGA babies. Motor function was assessed on two instruments: the Neuromotor and Motor Development Assessment and the Griffiths Scale of Infant Development. Similarly, in a much larger study of 265 SGA and 329 AGA babies, SGA babies scored equally well on the motor scale of the Bayley Scales of Infant Development at 13 months of age as their AGA controls, but scored poorly on the mental development scale.[43] Martikainen's[44] study further muddies the waters; she classified SGA infants as either symmetric or asymmetric (based on a proportionate or disproportionate biparietal diameter to crown-rump length at birth) and compared them with preterm and term AGA babies at 18 months of age using the Denver Development Screening test. Unfortunately, this version of the test has very limited value as a research instrument as it has been shown to have poor sensitivity and specificity.[45] The symmetrically growth restricted babies performed more poorly than any other group on fine and gross motor development, speech and social abilities. There is considerable debate as to the significance of differentiating morphometric discrepancies in fetal growth restriction, and this will not be discussed in detail here.

The development of clinical imaging techniques has provided the opportunity to examine alterations in brain structure and activation in vivo, although to date their use in assessing the short and long-term effects of IUGR has been very limited. Roelants-van Rijn and colleagues[46] recently investigated 14 preterm SGA neonates to ascertain whether IUGR associated with placental insufficiency alters cerebral metabolism, and whether this precedes the adverse neurodevelopmental outcome often seen at 2 years of age (using the Griffiths Scales of Infant Development). Compared with preterm AGA infants, the SGA infants had similar cerebral metabolism and neurodevelopment at 2 years. However, the authors admit that the study was significantly underpowered; they would have required a minimum of 110 SGA babies to detect a difference in the N-acetylaspartate:choline ratio on proton magnetic resonance spectroscopy. A reduced N-acetylaspartate signal can indicate diminished neuronal density and possibly altered neuronal mitochondrial function.[46]

A proportion of growth-restricted infants will be so compromised in utero that they are delivered preterm. In assessing the neurodevelopmental outcomes in these children it is often hard to differentiate the short and/or longer-term effects due to IUGR from those due to preterm delivery.[47,48] Being able to differentiate these effects can assist in identifying the nature of longer-term risks associated with IUGR earlier in gestation, compared with those experienced by the full- or near-term IUGR individual. These former babies also present a dilemma for the obstetrician, since delaying the delivery could increase the risk of hypoxia, haemorrhage and further neurological compromise, while preterm delivery is associated with the risks of intraventricular haemorrhage, respiratory distress syndrome and cerebral palsy. The outcome also differs between males and female neonates, with males being more likely to die if born preterm and IUGR.[47] The evidence to date suggests that while compromised cognitive and IQ development at 2 years of age is associated with low birthweight rather than gestational age, motor development is compromised by low birthweight, reduced birthweight ratio (the actual birthweight compared with the expected birthweight for gestational age) and a shorter gestational age.[47,49]

Children and Adolescents—Neurodevelopmental Milestones

The literature regarding neuromotor outcomes in school-age children born IUGR is sparse and somewhat variable, both methodologically and in outcomes. This is partly because no two studies appear to have used the same assessment instrument, and the findings have been inconsistent. They range from little or no increase in the likelihood of suboptimal motor performance to significant decrements in a wide range of fine and gross motor skills. Again, none have included neurophysiological assessments of motor function.

In one of the earliest studies, a group of 60 full term SGA children, whose intrauterine growth had been serially assessed by measuring head circumference growth on ultrasound, were assessed at 4 years using the Griffiths Extended Scales for Children.[50] If head growth had slowed prior to 34 weeks, children tended to be shorter and lighter. However, onset of IUGR (i.e., slowed head growth) prior to 26 weeks was additionally associated with much poorer neurodevelopmental outcomes. Harvey and colleagues[51] followed up a similar group of children at age 5 years using the McCarthy Scales of Children's Abilities and found that in the pre-26 week IUGR children, motor function, particularly coordination and balance, was significantly poorer when compared to IUGR children whose head growth had not begun to slow until after 26 weeks. Interestingly, there was no difference between the groups on verbal or memory tasks. The children's teachers expressed the view that boys fared worse than girls and were more "clumsy".[51]

Neligan et al[52] reported gross motor function but not manual dexterity to be significantly reduced in SGA compared to AGA children assessed using the Ozeretsky-Stott test, while others have reported difficulties with coordination, visuo-spatial and writing skills at age 6-7 years.[53] However, this contrasts with two other studies that have found either no motor deficits at 6 years of age,[54] or only a slightly increased probability of reduced manual dexterity.[55] Importantly, SGA children who demonstrated impaired motor performance were also more likely to exhibit cognitive deficits than those children with normal motor function. Strauss and Deitz[6] concluded that IUGR does not alter motor development in 7-year-old children, but based these findings on the Bender-Gestalt test, which is a psychometric test of visuo-motor maturity, emotional disturbance and visual-perception skills; it does not assess gross or fine motor function.

There have only been two studies of IUGR/SGA neuromotor outcomes in adolescents and both have reported poorer motor function at 12-14 years of age.[56,57] Both studies assessed motor function using the Movement Assessment Battery for Children (or Movement ABC). Unfortunately, in the Indian study, all the SGA children had been born preterm and all the controls were full-term normal birthweight babies, so it is difficult to dissect out the relative impact of IUGR, although preterm AGA children were also assessed.[56] Both the SGA and preterm-AGA children had poorer writing skills and visuo-motor skills compared to controls, but no difference in reading skills. Both groups also scored poorly on the Movement ABC compared to controls of the same sex. While girls performed more poorly than boys on balance and ball skills, their manual dexterity skills were superior to the boys.[56]

Similarly, a Norwegian study[57] found that 14 year olds born SGA at term had five-times the likelihood of exhibiting motor deficits than term AGA control children, particularly poor manual dexterity and poor balance and only among the boys. This was compared with very low birthweight (VLBW) preterm children, in whom motor function was globally reduced across all tests of the Movement ABC and seen equally in boys and girls. The reduced motor function could not be explained by poor postnatal growth in either group. Poor postnatal growth has previously been associated with poor motor function (Movement ABC) and cognitive impairment in 7-year-old children who were born preterm.[58]

Adults

To date there has been no published study of the neuromotor outcomes in adults who were growth restricted in utero. There is one study currently underway examining a cohort of adults aged 28 years, for which only preliminary results have yet been reported.[59]

The Case for Neurophysiological Assessment

Determining on the basis of a single assessment whether a child's sub-optimal motor performance is due to developmental delay, long-term neurological impairment or variance within the normal range, is fraught with the high likelihood of a false-positive assessment when using motor assessments that rely on age-appropriate performance. The "clumsy child" is not a rare phenomenon, but the origins of this are poorly understood. Epidemiological studies estimate the international prevalence of "developmental co-ordination disorder" among children aged between 5 and 11 years to be 6%.[60] It is likely that at least a proportion of these children will have been growth restricted in utero, although this has not been investigated. While there is some evidence that motor impairments may persist into adulthood, the prevalence of this is unknown. Even in supposedly normally developing children, the trajectory of motor development is not necessarily linear. For example, Darrah et al[61] followed the motor development of 47 infants from the age of 2 weeks until the age of 18 months, after all had begun walking. Motor development was assessed repeatedly on the Alberta Infant Motor Scale (AIMS) and 31% of the infants studied scored below the suggested cut-off (i.e., the 10th percentile) for normal development on at least one occasion. This inter- and intra-individual variability in motor development trajectory is not an uncommon finding in children, particularly under 18 months of age.[61,62]

Developmental course, differences between males and females, and laterality differences vary significantly between different types of motor tasks.[63] Subtle motor impairments or "soft neurological signs" are not uncommon in young children, but are difficult to identify from developmental assessments.[63] Many are broad-based assessments where a composite score for a given "skill" is derived from several individual tests. These may be good overall screening tools, but they provide no information about any underlying pathophysiology.[64] Hence, identifying particularly very young children in need of intervention can be difficult. It is arguably more valuable to assess the physiological function of the individual motor and sensory pathways as well. While these too can sometimes show a relatively wide variability in the normal range, the probability of identifying a specific pathology is probably higher than a composite skills-for-age assessment alone. Two of the main parameters that can be measured in neonates, infants and children are motor conduction time and corticomotor threshold. Both measurements utilize a technique known as transcranial magnetic stimulation (TMS).

Transcranial Magnetic Brain Stimulation

Transcranial magnetic brain stimulation (TMS) was developed in the 1980s and for the past decade has been widely used in both clinical assessment of neurological patients and in

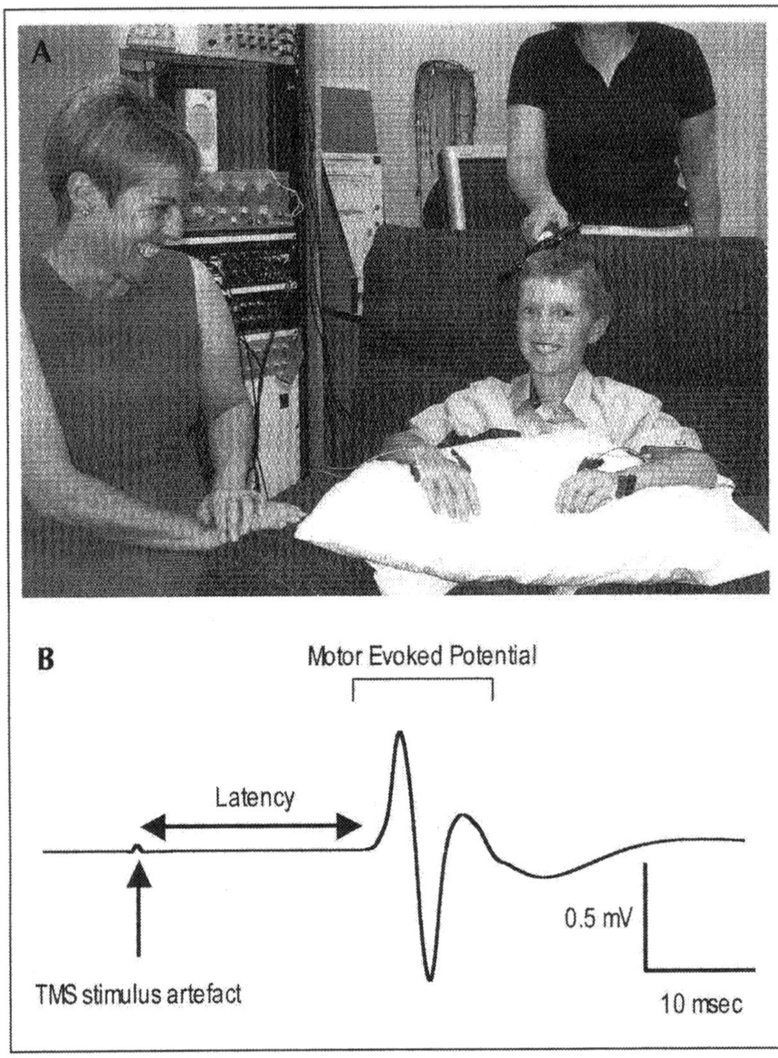

Figure 1. A) Tom (aged 8 years) undergoes transcranial magnetic brain stimulation while his mother looks on. The coil is held against his head over the area of the cortical representation of his left hand. (Photograph used with child and parental consent.) B) An example of a motor evoked potential recorded from the first dorsal interosseous (index finger) muscle. The latency corresponds to the total motor conduction time.

research using normal and clinical populations. The technique is non-invasive, nonpainful, very well-tolerated, safe and allows assessment of the human corticospinal system in vivo in awake infants, children and adults. Briefly, the magnetic stimulator consists of several large capacitors connected to an insulated "coil" that is held over the subject's scalp. When the capacitors are discharged, a very brief but strong magnetic pulse flows through the coil. This pulse passes painlessly through the scalp and activates some of the underlying brain cells. The cells of the motor cortex lie very close to the surface of the brain and if the coil is held over the area of the motor cortex containing motoneuron cells that activate, for example, the muscles of the hand, the effects of the stimulation can be examined by recording the responses in the hand muscles (a technique known as surface electromyography [EMG]). In simple terms, TMS allows us to examine the integrity and function of the corticospinal pathway from the motor cortex to the muscle by analysing various characteristics of the electrical potentials evoked in the muscles (called motor evoked potentials, or MEPs). Different TMS techniques also allow examination of the function and integrity of connections between cells in the motor cortex. Interhemispheric or transcallosal inhibition can also be readily investigated.

TMS and EMG can be used to create a "motor map" of a muscle by stimulating with a figure-of-8 coil at multiple scalp sites (usually delineated by a grid) and ascertaining the magnitude (peak-to-peak amplitude or area) of the MEPs. The cortical surface map that is generated is related to the representation of the target muscle in the contralateral motor cortex.[65-67] Cortical mapping has been used extensively to determine changes in the area and excitability of cortical representations of muscles under a normal conditions[65,68-69] and conditions including amputation,[70-71] local anaesthesia,[70] in musicians,[72] in Braille readers,[73-75] practising motor tasks[76] with peripheral nerve stimulation[77,78] and pharyngeal stimulation.[79]

Motor Conduction Time

TMS has been shown to be a valuable tool in not only evaluating corticospinal development, but also in assessing children with disordered motor control or developmental motor delay. It has been used successfully to evoke responses in term and preterm neonates.[80] As there is little or no data regarding abnormalities in these parameters following IUGR, we have discussed them with reference to normal chronological development. The most frequently researched parameter is the efferent central motor conduction time (CMCT), which is calculated as the difference between the latency of the EMG response to motor cortex stimulation (the total motor conduction time [TMCT]) and the latency of the EMG response to cervical spinal root stimulation (the peripheral conduction time [PMCT]).[81] Both latencies can be evoked with TMS. Since TMS, particularly at low to moderate stimulus intensities, tends to activate the large, fast-conducting corticospinal fibres of the corticomotoneuronal tract, the evoked latencies tend to reflect conduction times in these fibres, rather than the smaller, slower fibres.

There is a different development trajectory in normal children, for both upper and lower extremities, when CMCT and PMCT are compared.[151] CMCT does not reach adult values until approximately 10 years of age[151,152] while PMCT matures much more rapidly, reaching adult values by approximately 3 years of age.[151] Conduction in the afferent pathways shows a similar maturation profile to the efferent pathways in the periphery, but central afferent conduction shows a faster maturation than the efferent CMCT, reaching adult values by the age of 5-7 years.[82] In the neonate, slow corticospinal axonal conduction in the large fibres in both humans[80,83] and macaque monkeys[84] tends to indicate that these axons are poorly myelinated at birth, although there is considerable individual variability in this. Morphological and magnetic resonance imaging studies have shown that the human pyramidal tracts are well-myelinated by about 2-3 years of age[85] although myelination of the corticospinal tracts themselves continues into late adolescence.[86] It has been suggested that the relatively protracted development of CMCT for the next 7-8 years may be due to synaptogenesis and the concomitant process of developing and refining cortical synaptic efficacy, rather than ongoing myelination.[87]

There is limited evidence that motor nerve conduction is altered in term IUGR babies. But the results of the one study to date are confounded by a design that compared these babies with preterm AGA infants.[88] It is also not clear if these authors corrected the preterm infants ages for post-conceptional age when tested. The more severe the growth restriction, the slower the conduction velocity in the IUGR babies. But preterm AGA infants had slower conduction velocities than IUGR babies when the two groups were compared. A cautious interpretation here might be that shorter gestation and low birthweight are both potential risk factors for reduced postnatal conduction velocity.

Motor Threshold

The lowest TMS intensity required for evoking motor potentials in the relaxed muscle of interest is known as the resting motor threshold. In adults, the motor threshold reflects the membrane characteristics principally of the cortical, but also the spinal motoneurons. In children, motor threshold reflects the developmental stage of myelination of the corticospinal tracts (the less myelinated the tracts, the higher the threshold) and the membrane characteristics and synaptic efficacy of the cortical and spinal motoneurons.[89] Motor thresholds are high in neonates compared to adults, and fall exponentially with increasing age until reaching adult values at approximately 13 years of age.[80,87,89,90] Motor threshold tends to be lower in the dominant hemisphere of the motor cortex (i.e., the left motor cortex in right-handers) and the magnitude of the difference between the left and right sides is thought to be related, at least in part, to hand dominance in adults.[91] Garvey and colleagues[89] showed that this asymmetry exists in children (although they did not test children younger than 6 years of age) and decreases with increasing age (6.4% in 6-8 year olds versus 4.6% in adults). Pitcher et al have recently reported that a larger asymmetry in motor threshold is associated with a lower birthweight when assessed in the cortical representations of hand muscles of young adult humans of the same chronological age.[59] This is related to a more left-handed laterality quotient with decreasing birthweight in males, but not in females (unpublished findings). While these latter results were not associated with premature birth, increased left hand preference has been associated with very premature birth, very low birthweight and reduced motor abilities at age 4-6 years.[92,93]

Induced Cortical Plasticity as Therapy

One possible therapeutic avenue for ameliorating the effects of an adverse fetal environment on subsequent motor development may be by inducing beneficial plastic changes in the motor areas of the brain and/or spine. While it may be difficult to prevent some types of fetal growth restriction, it may be possible to exploit the immature brain's extraordinary capacity for reorganisation postnatally, to effectively reprogram the maladaptive motor programming induced in utero. Plasticity, in its broadest sense, can be defined as the process by which synapses, cells or tissues alter their structure and/or function in response to altered central or peripheral input.

Adaptive plasticity refers to the processes by which neural circuits in the brain and spinal cord reorganise the strength and nature of their functional synaptic connectivity to improve function in response to an alteration in sensory input or injury.[94] Repetitive performance of voluntary movements, such as practising a musical instrument or a motor skill, is associated with plasticity in the motor cortex.[76] Some examples of this adaptive plasticity include increases in the size of the representation of the left hand fingers in the right motor cortex of violinists[72] and increased excitability of the "reading" finger's representation in the motor cortex of Braille readers.[74,75,95] Recent studies in normal humans and stroke patients using single pulse or rapid rate, repetitive TMS (rTMS) to stimulate the brain and/or electrical pulses to stimulate the muscles, have shown that a similar pattern of plastic changes can be induced in the brain that outlasts the period of stimulation.[77,78,96-98] This has lead to an increasing number of studies exploring the use of imposed stimulation protocols to maximise functional recovery of motor control following stroke and other lesional brain injuries in adults. For example, Pitcher et al[97] recently showed that low frequency peripheral stimulation of the motor nerve reduces motor cortex excitability while high frequency stimulation increases it. Others have shown it is possible to induce similar changes using rTMS.[99,100] Hence it is theoretically feasible that different frequencies of afferent stimulation or TMS could be used therapeutically to selectively drive abnormally increased or decreased motor cortex excitability towards more "normal" levels. The brain of the infant and young child has approximately twice as many synapses as the adult brain, which may account for its superior ability to reorganise its synaptic connections. This also probably contributes to the ability of young children to recover from early brain injuries.[101] However, the danger for the scientist or clinician endeavouring to exploit the developing brain's large capacity for plastic reorganisation for therapeutic purposes lies in the fact that such interventions may well induce maladaptive outcomes. This is certainly possible in the human motor system, since our knowledge of its development is surprisingly sparse.

Maternal Stress and Motor Development in the Offspring

Apart from sub-optimal nutrition in utero, another source for altered prenatal programming of motor development is maternal stress. Maternal stress during pregnancy has also been shown to impair neuromotor development in rats and non-human primates, although the number of studies is relatively small. The gestational timing at which maternal stress occurs appears to determine the type of adverse affect on neuromotor development and outcomes.[102-104] In rats, acquisition of a range of precocious reflexes in the pups was delayed if the dams were exposed to the stressor at gestational day 10, but not when exposed at gestational day 14.[105] These differences resolved by the end of the second postnatal week, suggesting any deleterious effect of the stressor on development had recovered. Infant squirrel monkeys whose mothers' had been repeatedly stressed during pregnancy had poorer motor abilities and impaired balance. If stressed during early gestation, the effects on infant neuromotor outcomes were more severe and persistent than in infants whose mothers were stressed during mid-to-late gestation.[111] Only those infants stressed in early gestation had lower birthweight, reduced muscle tone, poor co-ordination, impaired righting reflexes and decreased post-rotary nystagmus.

Schneider's[104] findings also raise another possibility. Unlike the control infants who showed rapid and linear motor development postnatally, both early and mid-gestation prenatally stressed infant monkeys showed either a flat or much more variable trajectory of development, suggesting that the discrepancy in motor abilities between the stressed and control infants increased postnatally. Hence one explanation for the abnormal postnatal motor development trajectories is that the stressed infants had significant motor learning difficulties that limited their ability to overcome their deficits with postnatal activity-dependent experience.

Programming Adult Neurological Disease: Parkinson's Disease

A long-term effect of fetal neuromotor programming may be an increased predisposition to adult or later onset neurological disorders, for example, Parkinson's disease. Parkinson's disease is a progressive motor pathology characterised by the loss of capacity to initiate and control appropriate voluntary movements. Patients do not become symptomatic until they have lost more than 60-80% of their dopamine-producing neurons in the substantia nigra pars compacta of the basal ganglia, a process that can take over a decade. Nigral dopamine cell numbers decline during normal ageing, but not everyone develops the disease before they die, presumably because the cell loss "threshold" has not been reached; in Parkinson's disease, this rate of cell loss is accelerated significantly by an unknown mechanism. It is possible that individuals born with fewer dopamine-producing nigral cells are more likely to become symptomatic than someone of the same age and sex who starts life with a normal or optimal nigral cell number.

There is emerging evidence that alterations to prenatal dopaminergic neurogenesis might be linked to the onset of Parkinson's disease, at least in some cases.[106] This has come largely from the identification and role of the transcription factors, including Nurr1, that regulate dopaminergic neurogenesis during development of the brain. Alterations in the timing of onset and/or the level of expression of these transcription factors during fetal life may predispose mature dopaminergic nigral neurons to earlier and possibly accelerated cell death in at least some cases of idiopathic Parkinson's disease.[106] The current understanding of the cellular regulation of trophic factors important in the development and maintenance of healthy synapses is relatively sparse. However, immature neurons are thought to be apoptosed by trophic factor deprivation if they do not form functional synapses with their targets. While this is a normal and important part of proper development of the nervous system, it may also adversely reduce cell numbers when interference alters synaptic development. Therefore, the number of dopaminergic nigral neurons produced that survive to maturity, and the lifespan of those that do mature, may be compromised by an adverse in utero environment during dopaminergic neurogenesis that alters the time of onset, effectiveness or availability of trophic factors and/or transcription factors.[106] For example, it is known that an excess of retinoic acid (part of the Vitamin A molecule) inhibits the transcriptional activity of Nurr1 and is teratogenic. Nurr1 is thought to play a key role in differentiation, maturation and maintenance of brain dopaminergic neurons and neural circuits.[107] There is also evidence that it has an instructive role in dopamine synthesis and storage in the adult brain.[108] In rats exposed prenatally to excess retinoic acid between gestational days 14-16, cerebellar weight was reduced and the cluster of behavioural and motor effects was consistent with disturbance in the mesolimbic dopamine system.[109] Disturbances of the development of the dopaminergic neural system have also been associated with attention deficit disorder and schizophrenia, although direct evidence for the latter has yet to be presented.[110]

Where to Now?

There is a reasonable body of literature on the neurodevelopmental outcomes of IUGR/SGA, but few studies have addressed the likely sources and none the specific neurophysiological changes associated with the prenatal programming of human neuromotor function. We also do not know if degenerative neurological disorders associated with ageing and with motor dysfunction as a symptom, such as Parkinson's disease, have at least some of their origins in altered development due to an adverse environment in utero, that may or may not be evident in motor function during life. There is either contradictory or no evidence for overt motor dysfunction in childhood and no published studies in adults. However, as with schizophrenia reviewed in the chapter by Bennet and Gunn, the evidence suggests there are long-term adverse outcomes of an adverse environment in utero on human neuromotor function. Interestingly, schizophrenia also has motor dysfunction as a relatively prominent symptom, often well before overt psychiatric manifestations of the disease (see refs. 111, 112).

What this review has not addressed is the role played by socio-economic circumstances in either prenatal or postnatal motor development, why males and females appear to respond differently, and the role of the presence or absence of catch-up growth in postnatal motor development. In addition, there is considerable debate within the fetal programming literature regarding possible inter-generational effects of fetal growth restriction, as well as the differential effects of symmetric and asymmetric growth restriction that have not been discussed here. Lastly, the effects of undernutrition on specific neurotransmitters at different stages of motor development are likely to have a significant influence on the nature and severity of the outcomes, and require a more comprehensive review than is possible here. Prenatal undernutrition, malnutrition of specific substrates (particularly amino acids) and maternal stress that alters hypothalamic pituitary function are all likely to disrupt the timetable of expression of neurotransmitters, neuromodulators and their receptors.[113] Evidence from animal and pharmacological studies strongly suggests that apart from GABA, perturbation particularly of glycine, glutamate, dopamine, serotonin and acetylcholine are also likely to have long-term adverse outcomes on neuromotor development and function.[113-115]

Future studies are needed to address firstly, the normal pre and postnatal development of the human motor systems, of which we know surprisingly little, and secondly, the timing and nature of a range of perturbations in utero that may alter the normal motor developmental trajectory. Studies of human cohorts at all ages, that combine epidemiological and neurophysiological approaches, are required to establish the physiological consequences of programming on motor function, and how this affects the health of the individual throughout the lifespan. Because the human brain is so highly plastic, particularly in the first few years postnatally but also throughout life, there is a huge potential for the development of early intervention therapies that could ameliorate at least some of the acquired effects of an adverse intrauterine environment during gestation.

Acknowledgments

The financial support of the South Australian Channel 7 Children's Research Foundation is gratefully acknowledged. JBP is an Australian NH&MRC Peter Doherty Research Fellow.

References

1. Bui BV, Rees SM, Loeliger M et al. Altered retinal function and structure after chronic placental insufficiency. Invest Ophthalmol Vis Sci 2002; 43(3):805-812.
2. Zubrick SR, Kurinczuk JJ, McDermott BM et al. Fetal growth and subsequent mental health problems in children aged 4 to 13 years. Dev Med Child Neurol 2000; 42(1):14-20.
3. Spinillo A, Stronati M, Ometto A et al. Infant neurodevelopmental outcome in pregnancies complicated by gestational hypertension and intra-uterine growth retardation. J Perinat Med 1993; 21(3):195-203.
4. Dunn HG, Crichton JU, Grunau RV et al. Neurological, psychological and educational sequelae of low birth weight. Brain Dev 1980; 2(1):57-67.
5. Newman DG, O'Callaghan MJ, Harvey JM et al. Characteristics at four months follow-up of infants born small for gestational age: a controlled study. Early Human Dev 1997; 49(3):169-181.
6. Strauss R, Dietz W. Growth and development of term children born with low birth weight: Effects of genetic and environmental factors. J Pediatr 1998; 133(1):67-72.
7. Bos AF, Einspieler C, Prechtl HF. Intrauterine growth retardation, general movements, and neurodevelopmental outcome: a review. Dev Med Child Neurol 2001; 43(1):61-68.
8. Berger-Sweeney J, Hohmann CF. Behavioral consequences of abnormal cortical development: insights into developmental disabilities. Behav Brain Res 1997; 86(2):121-142.
9. Monk CS, Webb SJ, Nelson CA. Prenatal neurobiological development: molecular mechanisms and anatomical change. Dev Neuropsychol 2001; 19(2):211-236.
10. Rees S, Harding R. Brain development during fetal life: influences of the intra-uterine environment. Neurosci Lett 2004; 361(1-3):111-114.
11. Kuzniecky RI, Barkovich AJ. Malformations of cortical development and epilepsy. Brain Dev 2001; 23(1):2-11.
12. Barkovich AJ, Kuzniecky RI, Jackson GD et al. Classification system for malformations of cortical development: Update 2001. Neurology 2001; 57(12):2168-2178.
13. Krageloh-Mann I. Imaging of early brain injury and cortical plasticity. Exp Neurol 2004; 190(Suppl 1):84-90.
14. Gramsbergen A. Clumsiness and disturbed cerebellar development: insights from animal experiments. Neural Plast 2003; 10(1-2):129-140.
15. Chase HP, Welch NN, Dabiere CS et al. Alterations in human brain biochemistry following intrauterine growth retardation. Pediatrics 50(3):403-411.
16. Chase HP, Lindsley WF Jr, O'Brien D. Undernutrition and cerebellar development. Nature 1969; 221(180):554-555.
17. Neville HE, Chase HP. Undernutrition and cerebellar development. Exp Neurol 1971; 33(3):485-497.
18. Toft PB, Leth H, Ring PB et al. Volumetric analysis of the normal infant brain and in intrauterine growth retardation. Early Human Dev 1995; 43(1):15-29.
19. Mallard C, Loeliger M, Copolov D et al. Reduced number of neurons in the hippocampus and the cerebellum in the postnatal guinea-pig following intrauterine growth-restriction. Neuroscience 2000; 100(2):327-333.
20. Rees S, Mallard C, Breen S et al. Fetal Brain Injury Following Prolonged Hypoxemia and Placental Insufficiency: A Review. Comp Biochem Physiol - A: Molec Integr Physiol 1998; 119(3):653-660.
21. Sasaki J, Fukami E, Mimura S et al. Abnormal cerebral neuronal migration in a rat model of intrauterine growth retardation induced by synthetic thromboxane A2. Early Human Dev 2000; 58(2):91-99.
22. Tashima L, Nakata M, Anno K et al. Prenatal influence of ischemia-hypoxia-induced intrauterine growth retardation on brain development and behavioral activity in rats. Biol Neonate 2001; 80(1):81-87.
23. Lane RH, Ramirez RJ, Tsirka AE et al. Uteroplacental insufficiency lowers the threshold towards hypoxia-induced cerebral apoptosis in growth-retarded fetal rats. Brain Res 2001; 895(1-2):186-193.
24. Sakamoto H, Kuzuya H, Tamaru M et al. Developmental changes in the NGF content in the brain of young, growing, low-birth-weight rats. Neurochem Res 1998; 23(1):115-120.
25. Bisignano M, Rees S. The effects of intrauterine growth retardation on synaptogenesis and mitochondrial formation in the cerebral and cerebellar cortices of fetal sheep. Int J Dev Neurosci 1988; 6(5):453-460.

26. Rees S, Bocking AD, Harding R. Structure of the fetal sheep brain in experimental growth retardation. J Dev Physiol 1988; 10(3):211-225.
27. Quirk GJ, Mejia WR, Hesse H et al. Early malnutrition followed by nutritional restoration lowers the conduction velocity and excitability of the corticospinal tract. Brain Res 1995; 670(2):277-282.
28. Sima A, Sourander P. The effect of pre and postnatal undernutrition on the calibre growth of central motor fibres: a morphometric ultrastructural study on rat cortico-spinal tract. Acta Neuropathol (Berl). 1976; 34(2):105-114.
29. Sima A, Sourander P. The effect of pre and postnatal undernutrition on axonal growth and myelination of central motor fibers. A morphometric study on rat cortico-spinal tract. Acta Neuropathol (Berl). 1978; 42(1):15-18.
30. Martin JH, Choy M, Pullman S et al. Corticospinal system development depends on motor experience. J Neurosci 2004; 24(9):2122-2132.
31. Wood SL, Beyer BK, Cappon GD. Species comparison of postnatal CNS development: functional measures. Birth Defects Res Part B Dev Reprod Toxicol 2003; 68(5):391-407.
32. Prechtl HF, Einspieler C, Cioni G et al. An early marker for neurological deficits after perinatal brain lesions. Lancet 1997; 349(9062):1361-1363.
33. Prechtl HFR. The importance of fetal movements. In: Connolly KJ, Forssberg H, eds. Neurophysiology and Neuropsychology of Motor Development. Series No. 143 ed. London: Mac Keith Press; 1997:42-53.
34. Forssberg H. Neural control of human motor development. Curr Opin Neurobiol 1999; 9(6):676-682.
35. Pearson KG. Generating the walking gait: role of sensory feedback. Prog Brain Res 2004; 143:123-129.
36. Bekedam DJ, Visser GH, de Vries JJ et al. Motor behaviour in the growth retarded fetus. Early Hum Dev 1985; 12(2):155-165.
37. Prechtl HF, Fargel JW, Weinmann HM et al. Postures, motility and respiration of low-risk preterm infants. Dev Med Child Neurol 1979; 21(1):3-27.
38. Hadders-Algra M. General movements: a window for early identification of children at high risk for developmental disorders. J Pediatr 2004; 145(Suppl 1):S12-S18.
39. Prechtl HFR. General movement assessment as a method of developmental neurology: new paradigms and their consequences The 1999 Ronnie MacKeith Lecture. Dev Med Child Neurol 2001; 43(12):836-842.
40. Zuk L, Harel S, Leitner Y et al. Neonatal general movements: an early predictor for neurodevelopmental outcome in infants with intrauterine growth retardation. J Child Neurol 2004; 19(1):14-18.
41. Bos AF, van Loon AJ, Martijn A et al. Spontaneous motility in preterm, small-for-gestational age infants I. Quantitative aspects. Early Human Dev 1997; 50(1):115-129.
42. Henderson-Smart DJ. Postnatal consequences of chronic intrauterine compromise. Reprod Fertil Dev 1995; 7(3):559-565.
43. Markestad T, Vik T, Ahlsten G et al. Small-for-gestational-age (SGA) infants born at term: growth and development during the first year of life. Acta Obstet Gynecol Scand Suppl 1997; 165:93-101.
44. Martikainen MA. Effects of intrauterine growth retardation and its subtypes on the development of the preterm infant. Early Hum Dev 1992; 28(1):7-17.
45. Glascoe FP, Byrne KE, Ashford LG et al. Accuracy of the Denver-II in developmental screening. Pediatrics 1992; 89(6 Pt 2):1221-1225.
46. Roelants-van Rijn AM, van der Grond J, Stigter RH et al. Cerebral structure and metabolism and long-term outcome in small-for-gestational-age preterm neonates. Pediatr Res 2004; 56(2):285-290.
47. Hediger ML, Overpeck MD, Ruan WJ et al. Birthweight and gestational age effects on motor and social development. Paediatr Perinat Epidemiol 2002; 16(1):33-46.
48. Thornton JG, Hornbuckle J, Vail A et al.Gs. Infant wellbeing at 2 years of age in the Growth Restriction Intervention Trial (GRIT): multicentred randomised controlled trial. Lancet 2004; 364(9433):513-520.
49. Hutton JL, Pharoah POD, Cooke RWI et al. Differential effects of preterm birth and small gestational age on cognitive and motor development. Arch Dis Child Fetal Neonatal Ed 1997; 76(2):F75-81.
50. Fancourt R, Campbell S, Harvey D et al. Follow-up study of small-for-dates babies. Br Med J 1976; 1(6023):1435-1437.
51. Harvey D, Prince J, Bunton J et al. Abilities of children who were small-for-gestational-age babies. Pediatrics 1982; 69(3):296-300.
52. Neligan GA, Kolvin I, Scott DM et al. Born too soon or born too small: A follow-up study to seven years of age. Vol 61. London: William Heinemann Medical Books, Spastics International Medical Publications, 1976.
53. Leitner Y, Fattal-Valevski A, Geva R et al. Six-year follow-up of children with intrauterine growth retardation: long-term, prospective study. J Child Neurol 2000; 15(12):781-786.

54. Low JA, Galbraith RS, Muir D et al. Intrauterine growth retardation: a study of long-term morbidity. Am J Obstet Gynecol 1982; 142(6 Pt 1):670-677.
55. Sommerfelt K, Sonnander K, Skranes J et al. Neuropsychologic and motor function in small-for-gestation preschoolers. Pediatr Neurol 2002; 26(3):186-191.
56. Chaudhari S, Otiv M, Chitale A et al. Pune low birth weight study—cognitive abilities and educational performance at twelve years. Indian Pediatr 2004; 41(2):121-128.
57. Evensen KAI, Vik T, Helbostad J et al. Motor skills in adolescents with low birth weight. Arch Dis Child Fetal Neonatal Ed 2004; 89(5):F451-455.
58. Cooke RWI, Foulder-Hughes L. Growth impairment in the very preterm and cognitive and motor performance at 7 years. Arch Dis Child 2003; 88(6):482-487.
59. Pitcher JB, Robertson AL, Miles TS et al. The influence of birthweight on neuromotor outcomes in adult humans. Paper presented at: 31st Fetal & Neonatal Physiological Society Annual Meeting, 2004; Castelvecchio Pascoli, Italy.
60. Mandich A, Polatajko HJ. Developmental coordination disorder: Mechanisms, measurement and management. Human Mov Sci 2003; 22(4-5):407-411.
61. Darrah J, Redfern L, Maguire TO et al. Intra-individual stability of rate of gross motor development in full-term infants. Early Human Dev 1998; 52(2):169-179.
62. Swanson MW, Bennett FC, Shy KK et al. Identification of neurodevelopmental abnormality at four and eight months by the movement assessment of infants. Dev Med Child Neurol 1992; 34(4):321-337.
63. Largo RH, Fischer JE, Rousson V. Neuromotor development from kindergarten age to adolescence: developmental course and variability. Swiss Med Wkly 2003; 133(13-14):193-199.
64. Georgieff M. Intrauterine growth retardation and subsequent somatic growth and neurodevelopment. J Pediatr 1998; 133(1):3-5.
65. Wassermann EM, McShane LM, Hallett M et al. Noninvasive mapping of muscle representations in human motor cortex. Electroencephalogr Clin Neurophysiol 1992; 85(1):1-8.
66. Wilson SA, Thickbroom GW, Mastaglia FL. Transcranial magnetic stimulation mapping of the motor cortex in normal subjects—the representation of 2 intrinsic hand muscles. J Neurol Sci 1993; 118(2):134-144.
67. Thickbroom GW, Sammut R, Mastaglia FL. Magnetic stimulation mapping of motor cortex - factors contributing to map area. Electromyography and Motor Control-Electroencephalography and Clinical Neurophysiology 1998; 109(2):79-84.
68. Mortifee P, Stewart H, Schulzer M et al. Reliability of transcranial magnetic stimulation for mapping the human motor cortex. Electroencephalogr Clin Neurophysiol 1994; 93(2):131-137.
69. Taylor JL, Fogel W, Day BL et al. Ipsilateral cortical stimulation inhibited the long-latency response to stretch in the long finger flexors in humans. J Physiol 1995; 488(Pt 3):821-831.
70. Ridding MC, Rothwell JC. Reorganisation in Human Motor Cortex. Can J Physiol Pharmacol 1995; 73(2):218-222.
71. Chen R, Corwell B, Yaseen Z et al. Mechanisms of cortical reorganization in lower-limb amputees. J Neurosci 1998; 18(9):3443-3450.
72. Elbert T, Pantev C, Wienbruch C et al. Increased cortical representation of the fingers of the left hand in string players. Science 1995; 270(5234):305-307.
73. Cohen LG, Celnik P, Pascual Leone A et al. Functional relevance of cross-modal plasticity in blind humans. Nature 1997; 389(6647):180-183.
74. Pascual Leone A, Torres F. Plasticity of the sensorimotor cortex representation of the reading finger in Braille readers. Brain 1993; 116(Pt 1):39-52.
75. Pascual Leone A, Wassermann EM, Sadato N et al. The role of reading activity on the modulation of motor cortical outputs to the reading hand in Braille readers. Ann Neurol 1995; 38(6):910-915.
76. Classen J, Cohen LG. Practice-induced plasticity in the human motor cortex. In: Boniface S, Ziemann U, eds. Plasticity in the Human Nervous System: Investigation with Transcranial Magnetic Stimulation. Cambridge: Cambridge University Press; 2003:90-106.
77. Ridding MC, Brouwer B, Miles TS et al. Changes in muscle responses to stimulation of the motor cortex induced by peripheral nerve stimulation in human subjects. Exp Brain Res 2000; 131(1):135-143.
78. Ridding MC, McKay DR, Thompson PD et al. Changes in corticomotor representations induced by prolonged peripheral nerve stimulation in humans. Clin Neurophysiol 2001; 112(8):1461-1469.
79. Hamdy S, Rothwell JC, Aziz Q et al. Long-term reorganization of human motor cortex driven by short-term sensory stimulation. Nature Neurosci 1998; 1(1):64-68.
80. Eyre JA, Miller S, Clowry GJ et al. Functional corticospinal projections are established prenatally in the human foetus permitting involvement in the development of spinal motor centres. Brain 2000; 123(Pt 1):51-64.

81. Nezu A, Kimura S, Takeshita S. Topographical differences in the developmental profile of central motor conduction time. Clin Neurophysiol 1999; 110(9):1646-1649.
82. Muller K, Ebner B, Homberg V. Maturation of fastest afferent and efferent central and peripheral pathways: no evidence for a constancy of central conduction delays. Neurosci Lett 1994; 166(1):9-12.
83. Eyre J, Taylor J, Villagra F et al. Exuberant ipsilateral corticospinal projections are present in the human newborn and withdrawn during development probably involving an activity-dependent process. Dev Med Child Neurol 2000; 82:12.
84. Olivier E, Edgley SA, Armand J et al. An electrophysiological study of the postnatal development of the corticospinal system in the macaque monkey. J Neurosci 1997; 17(1):267-276.
85. Holland BA, Haas DK, Norman DV et al. MRI of normal brain maturation. AJNR Am J Neuroradiol 1986; 7(2):201-208.
86. Paus T, Zijdenbos A, Worsley K et al. Structural maturation of neural pathways in children and adolescents: in vivo study. Science 1999; 283(5409):1908-1911.
87. Nezu A, Kimura S, Uehara S et al. Magnetic stimulation of motor cortex in children: maturity of corticospinal pathway and problem of clinical application. Brain Dev 1997; 19(3):176-180.
88. Bhatia BD, Prakash U. Electrophysiological studies in preterm and growth retarded low birth weight babies. Electromyogr Clin Neurophysiol 1993; 33(8):507-509.
89. Garvey MA, Ziemann U, Bartko JJ et al. Cortical correlates of neuromotor development in healthy children. Clin Neurophysiol 2003; 114(9):1662-1670.
90. Masur H, Althoff S, Kurlemann G et al. Inhibitory period and late muscular responses after transcranial magnetic stimulation in healthy children. Brain Dev 1995; 17(2):149-152.
91. Triggs WJ, Calvanio R, Macdonell RA et al. Physiological motor asymmetry in human handedness: evidence from transcranial magnetic stimulation. Brain Res 1994; 636(2):270-276.
92. O'Callaghan MJ, Burn YR, Mohay HA et al. The prevalence and origins of left hand preference in high risk infants, and its implications for intellectual, motor and behavioural performance at four and six years. Cortex 1993; 29(4):617-627.
93. O'Callaghan MJ, Burn YR, Mohay HA et al. Handedness in extremely low birth weight infants: aetiology and relationship to intellectual abilities, motor performance and behaviour at four and six years. Cortex 1993; 29(4):629-637.
94. Johnston MV. Brain plasticity in paediatric neurology. Eur J Paediatr Neurol 2003; 7(3):105-113.
95. Pascual Leone A, Cammarota A, Wassermann EM et al. Modulation of motor cortical outputs to the reading hand of braille readers. Ann Neurol 1993; 34(1):33-37.
96. Nudo RJ, Plautz EJ, Frost SB. Role of adaptive plasticity in recovery of function after damage to motor cortex. Musc Nerve 2001; 24(8):1000-1019.
97. Pitcher JB, Ridding MC, Miles TS. Frequency-dependent, bi-directional plasticity in motor cortex of human adults. Clin Neurophysiol 2003; 114(7):1265-1271.
98. Siebner HR, Rothwell J. Transcranial magnetic stimulation: new insights into representational cortical plasticity. Exp Brain Res 2003; 148(1):1-16.
99. Gangitano M, Valero-Cabre A, Tormos JM et al. Modulation of input-output curves by low and high frequency repetitive transcranial magnetic stimulation of the motor cortex. Clin Neurophysiol 2002; 113(8):1249-1257.
100. Muellbacher W, Ziemann U, Boroojerdi B et al. Effects of low-frequency transcranial magnetic stimulation on motor excitability and basic motor behavior. Clin Neurophysiol 2000; 111(6):1002-1007.
101. Johnston MV. Clinical disorders of brain plasticity. Brain Dev 2004; 26(2):73-80.
102. Schneider ML, Coe CL, Lubach GR. Endocrine activation mimics the adverse effects of prenatal stress on the neuromotor development of the infant primate. Dev Psychobiol 1992; 25(6):427-439.
103. Schneider ML, Coe CL. Repeated social stress during pregnancy impairs neuromotor development of the primate infant. J Dev Behav Pediatr 1993; 14(2):81-87.
104. Schneider ML, Roughton EC, Koehler AJ et al. Growth and Development Following Prenatal Stress Exposure in Primates: An Examination of Ontogenetic Vulnerability. Child Dev 1999; 70(2):263-274.
105. Patin V, Vincent A, Lordi B et al. Does prenatal stress affect the motoric development of rat pups? Dev Brain Res 2004; 149(2):85-92.
106. Barzilai A, Melamed E. Molecular mechanisms of selective dopaminergic neuronal death in Parkinson's disease. Trends Molec Med 2003; 9(3):126-132.
107. Eichele G. Retinoids: from hindbrain patterning to Parkinson disease. Trends Genet 1997; 13(9):343-345.
108. Hermanson E, Joseph B, Castro D et al. Nurr1 regulates dopamine synthesis and storage in MN9D dopamine cells. Exp Cell Res 2003; 288(2):324-334.
109. Holson RR, Adams J, Ferguson SA. Gestational stage-specific effects of retinoic acid exposure in the rat. Neurotoxicol Teratol 1999; 21(4):393-402.

110. Perrone-Capano C, Da Pozzo P, di Porzio U. Epigenetic cues in midbrain dopaminergic neuron development. Neurosci Biobehav Rev 2000; 24(1):119-124.
111. Neumann CS, Walker EF. Motor dysfunction in schizotypal personality disorder. Schizophr Res 1999; 38(2-3):159-168.
112. Neumann CS, Walker EF. Neuromotor functioning in adolescents with schizotypal personality disorder: associations with symptoms and neurocognition. Schizophr Bull 2003; 29(2):285-298.
113. Herlenius E, Lagercrantz H. Neurotransmitters and neuromodulators during early human development. Early Human Dev. 2001; 65(1):21-37.
114. Herlenius E, Lagercrantz H. Development of neurotransmitter systems during critical periods. Exp Neurol 2004; 190(Suppl 1):8-21.
115. Singer JH, Berger AJ. Development of inhibitory synaptic transmission to motoneurons. Brain Res Bull 2000; 53(5):553-560.

Chapter 5

Adaptive Responses of Early Embryos to Their Microenvironment and Consequences for Post-Implantation Development

Jeremy Thompson,* Michelle Lane and Sarah Robertson

Abstract

Early embryos are adaptive to the environment they encounter during development and this facilitates embryo resilience to environmental insults. However, it is clear from findings in nonhuman species that adaptive plasticity during early development can have adverse consequences manifesting over the long-term in formation of the post-natal and adult phenotype, and that this occurs via partially characterized programming phenomena. Environmental effectors resulting in adaptive changes to embryos discovered to date include amino acid nutrition, oxygen concentration (both hypoxic and hyperoxic) and growth factor exposure. Other environmental factors known to influence programming of early development have yet to be fully evaluated for their long-term consequences, an area which still requires much work. The mechanisms involved in translation of environmental adaptations to long-term programming are only now being elucidated. Epigenetic mechanisms, especially DNA methylation of imprinted genes, have been associated with adaptive responses to altered environments. Other proposed mechanisms involve temporal gene expression patterns perturbed at critical events during development, such as implantation and early placental morphogenesis. The contribution of programming in early embryos to phenotypic variation and, more importantly, potential health status of resulting offspring has particular relevance to health and diet at the time of conception and to children born following assisted reproductive technologies, especially where embryonic manipulations in artificial environments are involved.

Introduction

Early mammalian embryo development spans the fertilized zygote to the blastocyst stage, just prior to embryo implantation. While the duration of development varies, the morphological changes that occur are relatively constant across most mammalian species. There are considerable technical challenges to studying this process in vivo, and much of our knowledge about early development stems from in vitro studies.

Application of rapid advances in this knowledge has led to in vitro embryo production and development. Infertility treatment using in vitro fertilization (IVF) is now considered a

*Corresponding Author: Jeremy Thompson—Research Centre for Reproductive Health, Department of Obstetrics and Gynaecology, The University of Adelaide, Adelaide SA 5005, Australia. Email: jeremy.thompson@adelaide.edu.au

Early Life Origins of Health and Disease, edited by E. Marelyn Wintour and Julie A. Owens.

routine clinical procedure, with many countries reporting that between 1-3% of children born in developed countries are conceived from in vitro fertilization (IVF) procedures.[1] Furthermore in vitro embryo production is increasingly utilized in agriculture for improving genetic selection for desirable production traits, as well as in conservation of exotic species and propagation of companion animals. Against this technological advance are the reports of perturbed fetal growth following such early embryonic manipulations. Rare syndromes, such as the Beckwith-Wiedemann Syndrome has been linked to IVF-derived conception[2,3] and the Large Offspring Syndrome is well characterized in ruminant embryos following in vitro embryo culture and transfer.[4,5]

How are potentially abnormal embryos produced and why cannot they be recognized prior to implantation? The environment encountered by early embryos in the laboratory is far removed from that of the reproductive tract. Bowman and McLaren in 1970[6] reported that although in vitro embryo development to the blastocyst stage can be readily achieved, this was accompanied by a reduction in fetal viability when compared to in vivo derived counterparts recovered from the reproductive tract and immediately transferred to surrogate recipients. We now recognize such perturbed fetal development as manifestation of the phenomenon of 'embryonic programming', where the environment encountered during in vitro development, despite supporting apparently normal morphological development, alters the phenotype of the embryo and of the ensuing progeny in fetal and postnatal life. The concept of embryonic programming is recognised to extend to in vivo environments, whereby nutritional perturbations and other stressors impact early embryo development in such a manner as to permanently skew fetal and post natal growth trajectory and metabolic phenotype.

The purpose of this chapter is to describe recent advances in our knowledge of the key environmental determinants of preimplantation development, specifically including physiochemical, nutritional and growth factor parameters. The mechanism(s) by which environment acts to impart programming in the embryo in such a way as to constrain subsequent growth and development is also explored.

Regulation of Embryo Development

Early embryo development is distinguished by two distinct morphological developmental periods (Fig. 1). A reductive cleavage process, whereby the fertilized zygote cleaves into smaller cells, marks the precompaction period. There is no net growth during this period, indeed a reduction in protein[7] and mRNA content[8,9] occur over this period of development, both of which were laid down in the maturing oocyte and therefore represent primarily maternal transcription and translation products. During precompaction, the embryonic genome is significantly activated[9] after a period of transcriptional inactivity.[10] This is a significant nuclear reprogramming event, involving a number of structural changes (e.g., histone acetylation) and recruitment of maternal mRNA to establish functional reprogramming,[9] so that new protein synthesis occurs which signals the morphological transition of the embryo from a collection of discrete pluripotent cells to that of a differentiated entity, containing more than one cell type. Indeed, recent genomic analysis of gene expression during mouse embryo development reveals that there are 2 separate phases in differential gene expression, one associated with oocyte activation and embryonic genome activation and the second involved in the initiation of cellular differentiation between the 4-cell stage and 8-cell stage.[8,11] Another molecular event known to occur during early development is global demethylation and remethylation of the parental genomes,[12] although the degree of de- and remethylation does appear to vary between species.[13] Nevertheless, early embryonic development is associated with significant genetic regulation, with several genes already known to be essential for normal development as their inhibition (by RNA*i*), or absence causes early embryonic mortality prior to or around the time of implantation (e.g., Nek2,[14] B-myb,[15] for an early review see ref. 16).

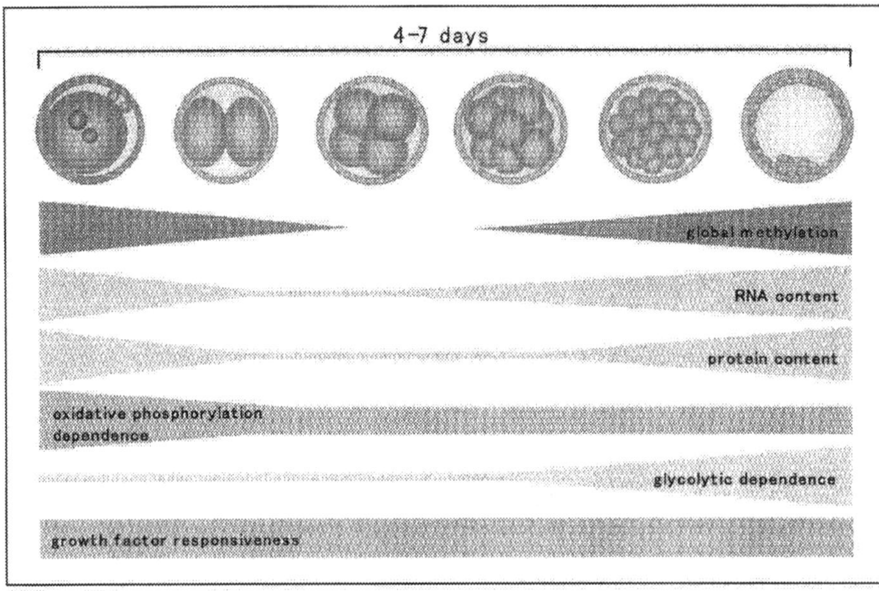

Figure 1. Early development of the mammalian embryo (4-7 days of development, dependent on species), incorporating the major molecular and biochemical features during development. Fertilization of an oocyte results in the pronuclear stage zygote. Reductive cleavage then occurs, which coincides with global demethylation, decreasing RNA and protein content and a dependency on oxidative phosphorylation. After major activation of the embryonic genome, increasing levels of RNA synthesis occurs followed by increasing levels of protein synthesis that includes signalling the morphological changes within the embryo (compaction of cells leading to the morula stage and then differentiation into the inner cells mass and trophectoderm, as well as the formation of the blastocoel cavity, at the blastocyst stage). This is also associated with remethylation of the genome and an increasing dependence on glycolytic activity. The embryo expresses growth factor receptors and is responsive to growth factors throughout early development, however, there is temporal sensitivity to different growth factors during this period.

Plasticity of Embryo Development

Zygotes and early embryos are not bound to an absolute set of physiochemical or nutritional conditions for successful development. For example, there are several commercial manufacturers of culture media systems for human embryo development in clinical infertility treatment, all of which differ in concentrations of key constituents. The ability of embryos to develop adequately under varying conditions has been referred to the 'plasticity' of embryo development. Nevertheless, we now recognise that there is an associated cost of the inherent requirements for adaptation to less than optimal environmental parameters.[17,18] While adaptation is often not reflected in easily measurable immediate changes in embryo morphology or viability, there is increasing evidence it leads to perturbation of the programming process which underpins phenotype in fetal and postnatal life.[19]

The capacity for an early embryo to continue 'normal' development is currently assessed using morphological techniques, involving the kinetics of cell replication (the number of cell divisions visible by a certain time period following fertilization) and qualitative measures such as size, shape and colour of blastomeres, degree of degeneration (usually associated with cellular fragmentation) and organization of blastomeres. The precision and predictive value such surrogate measures of viability is, unsurprisingly, weak.[20] Biochemical measures have been mooted (e.g., refs. 21, 22), but none have been integrated routinely at this time. Furthermore, as the composition of the extracellular environment clearly interacts with biochemical parameters, application of such measures would require individual validation in each media system.

Determining developmental capacity and degree of adaptation to early environments will ultimately reside with measuring the orchestrated temporal gene transcription profile that occurs. Despite such global analyses emerging,[23] this remains unlikely to be achievable in the immediate term as a noninvasive technology. It remains that effects of adaptation on developmental competence are best measured retrospectively following transfer of embryos and analysis of post-implantation outcomes.

Physiochemical Parameters Affecting Long-Term Development

Ionic Composition and pH

Mammalian embryos are sensitive to both osmolarity[24] and intracellular pH flux, especially to acidification.[25,27] This is associated with relatively undeveloped homeostatic mechanisms for regulating intracellular osmolarity and buffering pH. For example, the generally ubiquitous Na^+/H^+ exchanger can be nonfunctional in early cleaving mouse embryos.[26,27] A mild acidification of the culture medium can also result in changes in the ultrastructure of the early embryo.[28] However, there is little data concerning the long-term consequences of altering either pH or osmolarity. Of significance is the study by Schultz and colleagues,[29] demonstrating that embryo culture media with a relatively high ionic concentration of Na^+ caused perturbation of gene expression.

Oxygen

Oxygen concentration in the surrounding atmosphere is a well-characterised factor regulating early embryo development. It has been shown in a wide range of species that oxygen concentration around 5-7% is optimal for embryo development in vitro and this corresponds well with the reported values found in the oviduct of mammals. Normal atmosphere (21%) has been found to inhibit early embryo development, although different species have differing sensitivities. For example, the embryos of F1 generation hybrid strains of mice[30] and even early human embryos[31] appear mostly unaffected by atmospheric O_2 concentrations, yet other species, such as sheep embryos[32] and some outbred strains of mice[30] are exquisitely sensitive and fail to develop beyond the stage that is associated with embryonic genome activation in atmospheric O_2 concentrations. The mechanism for such perturbed early development has been associated with increased production of reactive oxygen species (ROS), although there is little direct evidence for increased intracellular ROS production under atmospheric O_2 (reviewed by ref. 33). Nevertheless, atmospheric O_2 can increase aneuploidy rates in mouse embryos[34] and has recently been shown that incubation under atmosphere causes perturbed fetal development following transfer in the mouse.[35]

Although relevant to in vitro culture, elevated O_2 concentrations are unlikely to affect in vivo development. However lifestyle factors such as smoking or extreme exercise might well provide the opportunity for embryos to be exposed to hypoxic conditions. The little data collected suggest there is a decreasing O_2 gradient from the oviductal environment to the uterus[36] and that the uterine lumen environment is one of low oxygen (around 2-5%,[36]). Furthermore, human conceptuses are known to spend much of the first 7 weeks of life under low oxygen conditions, until maternal blood flow through the spiral arteries occurs at around week 8 of gestation.[37] We have recently examined the role of hypoxia (2%) in mouse embryos during post-compaction development and have found that O_2-sensitive gene expression is significantly up-regulated when compared to embryos obtained in vivo or embryos cultured under 7% O_2.[38] Furthermore, preliminary data suggests that the Hypoxia Inducible Factor (HIF) transcription factor family are involved in this regulation of gene expression and that hypoxic post-compaction stage embryos are subsequently programmed for perturbed fetal development (Kind, Feil, Lane and Thompson, unpublished observations). Interestingly, the same O_2 concentration in cattle embryos is not considered hypoxic and when bovine embryos are exposed to 2% oxygen only small increases in gene expression are observed[39] and are associated with improved embryo quality.[39,40]

Nutritional Parameters Affecting Long-Term Development

Amino Acids and Ammonia

Amino acids are known to play a variety of roles during early embryo development. These include actions as intracellular osmolytes, energy substrates and protein synthesis substrates.[41] Differences in the type and quantity of amino acid transporters are known to occur during early development, suggesting that the relative importance of different amino acids also varies with developmental stage,[41] a postulate supported by the observation that different groups of amino acids are required for pre and post-compaction development.[42] Surprisingly, the importance of amino acids as media additives for in vitro culture has only been recognized recently. This is partly a reflection of plasticity of embryo development, where blastocyst formation occurs independently of amino acid provision but deprivation during culture in vitro leads to reduced viability and developmental competence at implantation.[43]

Gardner and Lane demonstrated that amino acid inclusion, especially that of glutamine, significantly increases the level of ammonia within embryo culture media systems.[44] It was shown that the benefits of amino acid addition could be annulled by the effect of ammonia build-up, partly from degradation of glutamine over the course of embryo culture, and partly as a result of deamination of amino acids during metabolism.[44] Early embryos appear to be sensitive to levels of ammonia as low as 100 μM and levels above 300 μM yield significant detrimental effects.[44] This can be overcome by either enzymatically removing the ammonia from the media during culture or by sequential removal of embryos to "fresh" medium over the course of development (usually every 48 h).[44] The latter has been coupled with a change in the energy substrate and ionic composition of medium that reflects the transition from oviduct to uterine environments, producing what is now referred to as "Sequential Culture Systems". More recently, the same authors[45] have demonstrated that ammonia causes significant shifts in a number of morphological and biochemical measures in mouse embryos, including an intracellular acidification of approximately 0.2 pH units with 300 μM NH_4^+. Furthermore, these changes were associated with a decreased fetal survival following transfer to recipient mothers and reduced fetal weight of surviving fetuses. Significantly, expression of the paternally imprinted gene *H19* was also increased in blastocysts following embryo culture in ammonia, suggesting perturbed methylation of imprinted control regions.

The "Large Offspring Syndrome" in ruminants has also been linked with high ammonia levels during the early development period, within both in vivo treatments[46] and also in vitro culture,[47] demonstrating that the responses of embryos to environmental stressors are species specific.

Hexoses

Glucose concentration in the extracellular environment surrounding early embryos is known to significantly influence developmental capacity, especially during the precompaction stage.[17] Embryos of several species are sensitive to glucose levels above approximately 3mM, but this too varies with different species. Early embryos of several species cannot utilize glucose as their sole energy substrate during in vitro culture, indeed levels of glucose uptake and utilization are low.[48] This situation rapidly changes with compaction, when a shift in metabolic preferences, causes post-compaction stage embryos to become increasingly glycolytic in their activity.[48,49] Conversely, hyperglycaemia can also detrimentally affect post-compaction embryos,[50] manifesting in delayed development[51] and increased levels of cellular apoptosis.[52,53] It is therefore surprising that in the mouse, hyperglycaemia-exposed embryos when transferred into normoglycaemic surrogates appear unaffected by their high glucose exposure.[51] However, embryos from different mouse strains have differential sensitivities to glucose.[54]

Lipids

There is little known of the lipid requirements of early embryos during development. Nevertheless, saturated lipid can be easily incorporated into the embryo from the surrounding environment.[55] Thompson and colleagues[56] reported that incubation of sheep embryos in the presence of serum, particularly human serum, promoted significant incorporation of lipid into the embryo forming large vesicles. The lipid was subsequently identified as triglyceride,[55] which significantly alters the biochemical composition,[55] biochemical activity[57] and freezing ability[58] of embryos. It remains to be determined if such incorporation alters homeostatic mechanisms that lead to altered embryonic programming. Nevertheless, serum addition in both ruminants and mouse embryos has been associated with perturbed imprinting and subsequent fetal development.[56,59,60,61]

Growth Factor and Cytokine Environment and Long-Term Development

In vivo, the growth and development of the preimplantation embryo as it traverses the female reproductive tract is influenced by cytokines and growth factors secreted from epithelial cells lining the oviduct and uterus. The identity and biological effects of growth factors and cytokines targeting the preimplantation embryo have been reviewed.[62-66] An array of different factors are secreted in precise spatial and temporal patterns such that the profile of factors experienced by the embryo would flux over the course of development. Expression is regulated principally by ovarian steroid hormones and factors present in seminal plasma.[61,67] Furthermore there is extraordinary sensitivity of cytokine synthesis to systemic and environmental regulation with nutrition and endocrine status, stress, inflammation and infection all strongly impacting the balance of cytokines synthesised. This notion illustrates the means by which cytokines likely comprise a sensing system that compliments the actions of nutrient availability in providing the embryo with information on the external environment. Through both positive and negative effects on timing and extent of proliferation and differentiation, these factors presumably contribute to synchronising embryo growth with the maternal changes that lead to uterine receptivity.

Experiments largely in mouse embryos but more recently in human and other species show embryos express cytokine and growth factor receptors from conception until implantation.[66-68] Some growth factors are synthesized by the embryo itself and have autocrine actions during early development. These factors were originally identified in experiments showing that embryos develop better in small volumes of culture medium.[69] In vitro experiments show that supplementation of culture media with cytokines and growth factors can promote embryo growth and development, affecting proliferation, viability and differentiation of blastomeres into trophectoderm and inner cell mass lineages.

The relative effects of different factors, and the immediate consequences of their deprivation for blastocyst development, have been studied using experimental strategies including addition of exogenous recombinant growth factors to the embryo culture, neutralization of ligand or receptors with specific antibodies or antisense oligonucleotides. Knockout mice with null mutations in ligand or receptor genes have also provided important insights.

Several studies have linked their impact on blastocyst development with implantation success. However the significance of depriving the embryo of growth factors for long-term health and viability of the fetus is more difficult to evaluate and has just begun to be explored. It seems imperative to address this since in human IVF and assisted reproductive technologies in animals, preimplantation embryos are generally cultured in media that do not contain growth factors. Their absence is clearly a major deficit when the in vitro and physiological environments are compared. Experimental strategies employing embryo transfer after manipulation of the cytokine environment in vitro, or in vivo through use of genetic strategies, provide the best approach. This overcomes the confounding issue of endogenous manipulation during the preimplantation period having indirect influences on other determinants of fetal growth, for example in endometrial receptivity and placental development.

Well-designed embryo transfer experiments have lead to compelling evidence for long-term effects of early growth factor environment on fetal and post-natal development, and metabolic programming in progeny. Insulin has been found to stimulate cell proliferation in the inner cell mass in vitro in the rat[70] and to increase the post-transfer rates of implantation, fetal survival, and weight at birth both in rats[71] and in mice.[72] Similarly, platelet activating factor (PAF) exerts embryotrophic effects in vitro and increases the proportion of embryos that develop normally after transfer in the mouse.[73] In contrast, exposure to TNFα in the preimplantation period is implicated as a negative determinant of early development with embryotoxic actions and detrimental consequences in viable embryos after implantation, including decreased fetal weight.[74]

GM-CSF is another cytokine identified as promoting normal blastocyst development and subsequent viability.[75] The physiological significance of exposure to GM-CSF during early preimplantation embryo development for subsequent fetal and post-natal growth and body composition has been investigated in a comprehensive embryo transfer study.[76] Addition of GM-CSF to culture medium improves implantation rate, corrects culture induced deficiencies in placental structure and fetal growth trajectory, and partly alleviates the long-term adverse consequences for postnatal growth in adult mice.

Mechanisms for Altered Phenotype

Although there is now compelling evidence to demonstrate that exposure to an adverse environment during early development can lead to altered phenotypes during fetal development, the mechanisms linking adaptive responses with long-term consequences remain elusive.

Disrupted Cell Allocation

Alterations in the abundance and proportion of inner cell mass and trophectoderm cell lineages in the blastocyst at implantation consistently correlate with changes in subsequent embryonic development and it has been speculated that preimplantation stress might result in inappropriate stem-cell allocation for normal growth.[77] Fewer inner cell mass cells resulting from manipulation of maternal nutrition,[78] culture environment,[42] in vivo or in vitro depletion of growth factors[75,76] or chemical reduction in the number of blastomeres[79] have all been shown to cause retardation of fetal growth. A similar linkage is seen in diabetic rats, where blastocysts with a decreased proportion of cells allocated to the inner cell mass are associated with subsequent inhibition of fetal growth.[80] However this explanation falls short in reconciling reduced ICM size with later disturbances in placental structure and it might be argued that changes in blastocyst cell allocation are symptomatic of a common underlying molecular etiology as opposed to a causal connection.

The means by which growth factors program long-term effects in embryos remains unknown, but must be the consequence of immediate effects of growth factors on cell viability and function. It is possible that growth factors alter metabolic activity, acting to influence expression of genes that influence uptake and processing of glucose or other metabolic substrates. Indeed it seems reasonable that disruptions in the metabolic status of the preimplantation embryo could explain the converging effects of growth factors and nutrition on blastocyst development. An alternative interpretation more consistent with evidence that low metabolic activity favors optimal embryo development[18] is that growth factors such as GM-CSF function simply as cell survival signals to prevent cellular stress and activation of the apoptotic cascade in blastomeres.

Epigenetic Hypothesis

Emerging studies favor mechanisms involving aberrations in epigenetic modification, especially altered methylation state of paternally imprinted genes.[81] This is particularly the case where growth trajectory is altered, as it appears the imprinted genes that are involved with growth regulation, such as *H19* and *Igf2R* are the most affected by adaptive stress responses, rather than other classes of imprinted genes.[82] When mouse embryos are cultured in the presence of fetal calf serum, expression of the growth related imprinted genes *Igf2, H19, Grb10* and *Grb7* is affected and fetal weight is reduced.[82] Aberrant fetal growth and development in cattle and sheep after embryo culture is attributed to imprinting errors in *Igf2R* after metabolic stress.[83,84] Disrupted imprinting concurs with the changes in placental development seen in several model systems, since a number of imprinted genes have important roles in placental development. One example is *Mash-2,* which maps to chromosome 7 in the mouse and is one of an imprinted cluster of 6 genes including *Igf2* and *H19.*[85] The post implantation effects of *Mash-2* include promoting development of the spongiotrophoblast cell layer within the junctional zone of the placenta and imprinting aberrations in *Mash-2* is a candidate mechanism for the placental abnormalities related to Large Offspring Syndrome.[86,88]

As with developmental progression, a common attribute with stress mediated responses is a perturbation in metabolic state, such as alterations in intracellular pH or reduction-oxidation state, and this may be responsible for altered methylation patterning. Furthermore, it is yet to be determined if there are differences in sensitivity during different stages of early development. Early embryos appear increasingly more adaptive and more resilient as they progress through development. It seems reasonable that heightened sensitivity during very early cleavage would be reflected in increased perturbations in methylation. Little is known of chromatin stability during this period, although it is clear that aneuploidy levels are significantly increased by in vitro manipulation of embryos.[88,89] Whether and how cytokine regulation of viability and metabolic status in blastomeres is linked with the process of epigenetic modification remains to be determined. Since expression of methyltransferases and other methylation machinery are cell cycle-regulated, disruption in their function is proposed to occur after the timing of embryo development is slowed in culture,[90] whereas embryrotrophic growth factors that accelerate blastomere cell division would oppose this effect.

Causal Pathway Hypothesis

The causal pathway hypothesis predicts that fetal programming hinges on the state of the transcriptome of the embryo at critical stages of development. In particular, early implantation events are likely to be significantly affected by adaptive gene expression responses that may inhibit or perturb the quality of implantation and early placental morphogenesis, limiting access to maternal nutrient supply. Although an attractive explanation for growth retardation in children derived from assisted reproductive technologies[91,92] and animal models yielding small or large fetuses, there is little direct evidence that supports this hypothesis. The strongest evidence is via indirect association, especially where post-compaction development is altered or restricted via hypoxia or growth factor restriction, which may subsequently alter the embryo's potential to optimally implant. Within the sheep model, both asynchronous transfer of embryos[93] (where advanced embryos were placed into nonadvanced reproductive tracts), or early pregnancy progesterone therapy[94] also perturb subsequent fetal development, strongly hinting at a casual pathway hypothesis, rather than an epigenetic mechanism, although this possibility cannot be ruled out. We therefore predict that factors affecting post-compaction development are more likely to influence fetal outcomes via a causal model, whereas stress-induced responses during early cleavage are more likely to operate through epigenetic mechanisms.

Conclusion

It is clear that early embryos exhibit plasticity and are adaptive to the environment they encounter during development. Questions remain about the extent of environmental perturbation required to cause irreversible change in subsequent fetal phenotype, temporal sensitivity to stress-induced changes and mechanisms linking early events with subsequent alterations in placental morphogenesis and fetal development.

Considerable research effort has been applied to understanding the impact of altered physicochemical parameters and cellular nutrient supply during early development in an attempt to improve embryo quality and viability. The relevance of this to in vivo environments has yet to be determined, although there are indications that alterations in these parameters during in vivo development may occur in the event of nutritional or other forms of maternal stress, and have similar consequences to those described during in vitro development.

A better understanding of the roles of growth factors in protecting embryos from environmental stress holds great promise for practical application in assisted reproductive technologies and may also have implications for healthy natural pregnancy. However an important consideration in addition of growth factors to embryo culture media is the context-dependence of their function. Cytokines do not work in isolation, but rather interact within a network to amplify, modulate, or antagonise each other's activities. The response of embryos to a given cytokine, as with any cell lineage, will be dependent on the local microenvironment, most notably the concentration of other cytokines and growth factors, availability of nutrients and other signalling substances, including extracellular matrix moieties. The ultimate reaction of the embryo to its cytokine environment no doubt depends on the sum total of signals converging at the cell surface. This urges caution in the extent to which we extrapolate from in vitro experiments, since it is difficult to replicate the full range of environmental influences to which embryos would be exposed in vivo.

Follow up studies of children conceived through assisted reproductive technologies as they grow older will answer another critical question: Are human embryos also programmable during early development? However, this will prove logistically challenging and raises ethical issues over the rights of children understanding their genetic heritage and the method by which they were conceived. Thus long-term, large cohort studies on adults conceived through assisted reproduction are likely to remain unrealistic. Studies will increasingly focus on the mechanisms of embryonic programming as more descriptive work exploring outcomes of altered early environments in animal models come to light. This research emphasis will furthermore inform which aspects of human physiology might most fruitfully be interrogated after either in vitro manipulation or alteration of in vivo environment.

References

1. Land JA, Evers JLH. Risks and complications in assisted reproductiontechniques: Report of an ESHRE consensus meeting. Hum Reprod 2003; 18:455-457.
2. Halliday J, Oke K, Breheny S et al. Beckwith-wiedemann syndrome and IVF: A case-control study. Am J Hum Genet 2004; 75:526-528.
3. Maher ER, Afnan M, Barratt CL et al. Beckwith-Wiedemann syndrome and assisted reproduction technology (ART). J Med Genet 2003; 40:55-61.
4. Walker SK, Hartwich KM, Seamark RF. The production of unusually large offspring following embryo manipulation: Concepts and challanges. Theriogenology 1996; 45:111-120.
5. Kruip TAM, den Daas JHG. In vitro produced and cloned embryos: Effects on pregnancy, parturition and offspring. Theriogenology 1997; 47:43-52.
6. Bowman P, McLaren A. Viability and growth of mouse embryos after in vitro culture and fusion. J Embryol Exp Morphol 1970; 23:693-704.
7. Thompson JG, Sherman AN, Allen NW et al. Total protein content and protein synthesis within preelongation stage bovine embryos. Mol Reprod Dev 1998; 50:139-45.
8. Hamatani T, Carter MG, Sharov AA et al. Dynamics of global gene expression changes during mouse preimplantation development. Dev Cell 2004; 6:117-131.
9 Schultz RM. The molecular foundations of the maternal to zygotic transition in the preimplantation embryo. Hum Reprod Update 2002; 8:323-331.
10. Forlani S, Bonnerot C, Capgras S et al. Relief of a repressed gene expression state in the mouse 1-cell embryo requires DNA replication. Development 1998; 125:3153-3166.
11. Wang QT, Piotrowska K, Ciemerych MA et al. A genome-wide study of gene activity reveals developmental signaling pathways in the preimplantation mouse embryo. Dev Cell 2004; 6:133-144.
12. Howlett SK, Reik W. Methylation levels of maternal and paternal genomes during preimplantation development. Development 1991; 113:119-127.

13. Young LE, Beaujean N. DNA methylation in the preimplantation embryo: The differing stories of the mouse and sheep. Anim Reprod Sci 2004; 82-83:61-78.
14. Sonn S, Khang I, Kim K et al. Suppression of Nek2A in mouse early embryos confirms its requirement for chromosome segregation. J Cell Sci 2004; 117:5557-5566.
15. Tanaka Y, Patestos NP, Maekawa T et al. B-myb is required for inner cell mass formation at an early stage of development. J Biol Chem 1999; 274:28067-70.
16. Rinkenberger JL, Cross JC, Werb Z. Molecular genetics of implantation in the mouse. Dev Genet 1997; 21:6-20.
17. Bavister BD. Culture of preimplantation embryos: Facts and artifacts. Hum Reprod Update 1995; 1(2):91-148.
18. Leese HJ. Quiet please, do not disturb: A hypothesis of embryo metabolism and viability. Bio Essays 2002; 24:845-849.
19. Thompson JG, Kind KL, Roberts CT et al. Epigenetic risks related to assisted reproductive technologies: Short- and long-term consequences for the health of children conceived through assisted reproduction technology: More reason for caution? Hum Reprod 2002; 17:2783-2786.
20. Gardner DK, Sakkas D. Assessment of embryo viability: The ability to select a single embryo for transfer - a review. Placenta 2003; 24:S5-S12.
21. Lane M, Gardner DK. Selection of viable mouse blastocysts prior to transfer using a metabolic criterion. Hum Reprod 1996; 11(9):1975-1978.
22. Houghton FD, Hawkhead JA, Humpherson PG et al. Noninvasive amino acid turnover predicts human embryo developmental capacity. Hum Reprod 2002; 17(4):999-1005.
23. Zeng F, Baldwin DA, Schultz RM. Transcript profiling during preimplantation mouse development. Dev Biol 2004; 272:483-496.
24. Dawson KM, Collins JL, Baltz JM. Osmolarity-dependent glycine accumulation indicates a role for glycine as an organic osmolyte in early preimplantation mouse embryos. Biol Reprod 1998; 59:225-232.
25. Baltz JM. Intracellular pH regulation in the early embryo. Bioessays 1993; 15(8):523-539.
26. Lane M, Baltz JM, Bavister BD. Na^+/H^+ antiporter activity in hamster embryos activated during fertilization. Dev Biol 1998; 208:244-252.
27. Steeves CL, Lane M, Bavister BD et al. Differences in intracelluar pH regulation by Na^+/H^+ antiporter among two-cell mouse embryos derived from females of different strains. Biol Reprod 2001; 65:14-22.
28. Squirrell JM, Lane M, Bavister BD. Altering intracellular pH disrupts development and cellular organization in preimplantion hamster embryos. Biol Reprod 2001; 64:1845-1854.
29. Ho Y, Dohert AS, Schultz RM. Mouse preimplantation embryo development in vitro: Effect of sodium concentration in culture media on RNA synthesis and accumulation and gene expression. Mol Reprod Dev 1994; 38:131-141.
30. Payne SR, Munday R, Thompson JG. Addition of superoxide dismutase and catalase does not necessarily overcome developmental retardation of one-cell mouse embryos during in- vitro culture. Reprod Fertil Dev 1992; 4:167-74.
31. Dumoulin JC, Meijers CJ, Bras M et al. Effect of oxygen concentration on human in vitro fertilization and embryo culture. Hum Reprod 1999; 14:465-469.
32. Thompson JG, Simpson AC, Pugh PA et al. Effect of oxygen concentration on in-vitro development of preimplantation sheep and cattle embryos. J Reprod Fertil 1990; 89:573-8.
33. Harvey AJ, Kind KL, Thompson JG. REDOX regulation of early embryo development. Reproduction 2002; 123:479-486.
34. Bean CJ, Hassold TJ, Judis L et al. Fertilization in vitro increases nondisjunction during early cleavage divisions in a mouse model system. Hum Reprod 2002; 17:2362-2367.
35. Karagenc L, Sertkaya Z Ciray N et al. Impact of oxygen concentration on embryonic development of mouse zygotes. Reprod BioMed Online 2004; 9:409–417.
36. Fischer B, Bavister BD. Oxygen tension in the oviduct and uterus of rhesus monkeys, hamsters and rabbits. J Reprod Fertil 1993; 99:673-679.
37. Jauniaux E, Watson AJ, Burton G. Evaluation of respiratory gases and acid-base gradients in human fetal fluids and uteroplacental tissue between 7 and 16 weeks' gestation. Am J Obstet Gynecol 2001; 184:998-1003.
38. Kind KL, Collett RA, Harvey AJ et al. Oxygen-regulated expression of GLUT-1, GLUT-3, and VEGF in the mouse blastocyst. Mol Reprod Dev 2005; 70:37-44.
39. Harvey AJ, Kind KL, Pantaleon M et al. Oxygen-regulated gene expression in bovine blastocysts. Biol Reprod 2004; 71:1108-1119.

40. Thompson JG, McNaughton C, Gasparrini B et al. Effect of inhibitors and uncouplers of oxidative phosphorylation during compaction and blastulation of bovine embryos cultured in vitro. J Reprod Fertil 2000; 118:47-55.
41. Van Winkle IJ. Amino acid transport regulation and early embryo development. Biol Reprod 2001; 64:1-12.
42. Lane M, Gardner DK. Differential regulation of mouse embryo development and viability by amino acids. J Reprod Fertil 1997; 109:153-164.
43. Lane M, Gardner DK. Amino acids and vitamins prevent cultureinduced metabolic perturbations and associated loss of viability of mouse blastocysts. Hum Reprod 1998; 13:991-997.
44. Lane M, Gardner DK. Increase in postimplantation development of cultured mouse embryos by amino acids and induction of fetal retardation and exencephaly by ammonium ions. J Reprod Fertil 1994; 102:305-312.
45. Lane M, Gardner DK. Ammonium induces aberrant blastocyst differentiation, metabolism, pH regulation, gene expression and subsequently alters fetal developmentin the mouse. Biol Reprod 2003; 69:1109-1117.
46. McEvoy TG, Robinson JJ, Aitken RP et al. Dietary excess of urea influence the viability and metabolism of preimplantation sheep embryos and may affect fetal growth among survivors. Anim Reprod Sci 1997; 47:71-90.
47. Sinclair KD, McEvoy TG, Carolan C et al. Conceptus growth and development following in vitro culture of ovine embryos in media supplemented with bovine sera. Theriogenology 1998; 49:218.
48. Leese HJ. Metabolism of the preimplantation mammalian embryo. Oxford Rev Reprod Biology 1991; 13(CHAP 2):35-72.
49. Thompson JG. Comparison between in vivo-derived and in vitro-produced preelongation embryos from domestic ruminants. Reprod Fertil Dev 1997; 9:341-54.
50. Moley KH. Hyperglycemia and apoptosis: Mechanisms for congenital malformations and pregnancy loss in diabetic women. TRENDS in Endocrinology and Metabolism 2001; 12:78-82.
51. Beebe LFS. The effect of maternal diabetes on the preimplantation mouse embryo. Ph.D. Thesis University of Queensland 1993.
52. Hinck L, Thissen JP, De Hertogh R. Identification of caspase-6 in rat blastocysts and its implication in the induction of apoptosis by high glucose. Biol Reprod 2003; 68:1808-1812.
53. Keim AL, Chi MM-Y, Moley KH. Hyperglycemia-induced apoptotic cell death in the mouse blastocyst is dependent on experession of p53. Mol Reprod Dev 2001; 60:214-224.
54. Scott L, Whittingham DG. Influence of genetic background and media components on the development of mouse embryos in vitro. Mol Reprod Dev 1996; 43:336-346.
55. Ferguson EM, Leese HJ. Triglyceride content of bovine oocytes and early embryos. J Reprod Fertil 1999; 116:373-378.
56. Thompson JG, Gardner DK, Pugh PA et al. Lamb birth weight is affected by culture system utilized during in vitro pre-elongation development of ovine embryos. Biol Reprod 1995; 53:1385-91.
57. Krisher RL, Lane M, Bavister BD. Developmental competence and metabolism of bovine embryos cultured in semi-defiend and defined culture media. Biol Reprod 1999; 60:1345-1352.
58. Pollard JW, Leibo SP. Chilling sensitivity of mammalian embryos. Theriogenology 1994; 41:101-106.
59. Khosla S, Dean W, Reik W et al. Epigenetic and experimental modifications in early mammalian development: Part II Culture of preimplantation embryos and its long-term effects on gene expression and phenotype. Hum Reprod Update 2001; 7:419-427.
60. Sinclair KD, McEvoy TG, Maxfield EK et al. Aberrant fetal growth and development after in vitro culture of sheep zygotes. J Reprod Fertil 1999; 116:177-86.
61. Wrenzycki C, Herrmann D, Carnwath JW et al. Alterations in the relative abundance of gene transcripts in pre-implantation bovine embryos cultured in medium supplemented with either serum or PVA. Mol Reprod Dev 1999; 53:8-18.
62. Pampfer S, Arceci RJ, Pollard JW. Role of colony stimulating factor-1 (CSF-1) and other lympho-hematopoietic growth factors in mouse preimplantation development. Bioessays 1991; 13:535-540.
63. Kaye PL, Harvey MB. The role of growth factors in preimplantation development. Prog Growth Factor Res 1995; 6:1-24.
64. Kane MT, Morgan PM, Coonan C. Peptide growth factors and preimplantation development. Hum Reprod Update 1997; 3:137-57.
65. Diaz-Cueto L, Gerton GL. The influence of growth factors on the development of preimplantation mammalian embryos. Arch Med Res 2001; 32:619-26.
66. Hardy K, Spanos S. Growth factor expression and function in the human and mouse preimplantation embryo. J Endocrinol 2002; 172:221-36.

67. Robertson SA, Seamark RF, Guilbert LJ et al. The role of cytokines in gestation. Crit Rev Immunol 1994; 14:239-92.
68. Sharkey AM, Dellow K, Blayney M et al. Stage-specific expression of cytokine and receptor messenger ribonucleic acids in human preimplantation embryos. Biol Reprod 1995; 53:974-981.
69. Paria BC, Dey SK. Preimplantation embryo development in vitro: Cooperative interactions among embryos and role of growth factors. Proc Natl Acad Sci USA 1990; 87:4756-4760.
70. De Hertogh R, Vanderheyden I, Pampfer S et al. Stimulatory and inhibitory effects of glucose and insulin on rat blastocyst development in vitro. Diabetes 1991; 40:641-7.
71. Zhang X, Armstrong DT. Presence of amino acids and insulin in a chemically defined medium improves development of 8-cell rat embryos in vitro and subsequent implantation in vivo. Biol Reprod 1990; 42:662-8.
72. Kaye PL, Gardner HG. Preimplantation access to maternal insulin and albumin increases fetal growth rate in mice. Hum Reprod 1999; 14:3052-9.
73. Ryan JP, O'Neill C, Ammit AJ et al. Metabolic and developmental responses of preimplantation embryos to platelet activating factor (PAF). Reprod Fertil Dev 1992; 4:387-98.
74. Wuu YD, Pampfer S, Becquet P et al. Tumor necrosis factor alpha decreases the viability of mouse blastocysts in vitro and in vivo. Biol Reprod 1999; 60:479-83.
75. Robertson SA, Sjoblom C, Jasper MJ et al Granulocyte-macrophage colony-stimulating factor promotes glucose transport and blastomere viability in murine preimplantation embryos. Biol Reprod 2001; 64:1206-15.
76. Sjoblom C, Roberts CT, Wikland M et al. GM-CSF alleviates adverse consequences of embryo culture on fetal growth trajectory and placental morphogenesis. Endocrinology 2005; 146:2142-53.
77. Fleming TP, Ghassemifar MR, Sheth B. Junctional complexes in the early mammalian embryo. Semin Reprod Med 2000; 18:185-93.
78. Kwong WY, Wild AE, Roberts P et al Maternal undernutrition during the preimplantation period of rat development causes blastocyst abnormalities and programming of postnatal hypertension. Development 2000; 127:4195-202.
79. Tam PP. Postimplantation development of mitomycin C-treated mouse blastocysts. Teratology 1988; 37:205-12.
80. Pampfer S. Peri-implantation embryopathy induced by maternal diabetes. J Reprod Fertil Suppl 2000; 55:129-39.
81. Maher ER, Afnan M, Barratt CL. Epigenetic risks related to assisted reproductive technologies: Epigenetics, imprinting, ART and icebergs? Hum Reprod 2003; 18(12):2508-2511.
82. Khosla S, Dean W, Brown D et al. Culture of preimplantation mouse embryos affects fetal development and the expression of imprinted genes. Biol Reprod 2001; 64:918-26.
83. Young LE, Fairburn HR. Improving the safety of embryo technologies: Possible role of genomic imprinting. Theriogenology 2000; 53:627-48.
84. Young LE, Fernandes K, McEvoy TG et al. Epigenetic change in IGF2R is associated with fetal overgrowth after sheep embryo culture. Nat Genet 2001; 27:153-4.
85. Caspary T, Cleary MA, Baker CC et al. Multiple mechanisms regulate imprinting of the mouse distal chromosome 7 gene cluster. Mol Cell Biol 1998; 18:3466-74.
86. McLaughlin KJ, Szabo P, Haegel H et al. Mouse embryos with paternal duplication of an imprinted chromosome 7 region die at midgestation and lack placental spongiotrophoblast. Development 1996; 122:265-70.
87. Tanaka M, Gertsenstein M, Rossant J et al. Mash2 acts cell autonomously in mouse spongiotrophoblast development. Dev Biol 1997; 190:55-65.
88. Viuff D, Richards L, Offenberg H et al. A high proportion of bovine blastocysts produced in vitro are mixoploid. Biol Reprod 1999; 60:1273-1278.
89. Yadav BR, King WA, Betteridge KJ. Relationship between the completion of first cleavage and the chromosomal complement, sex, and developmental rates of bovine embryos generated in vitro. Mol Reprod Dev 1993; 36:434-439.
90. De Rycke M, Liebaers I, Van Steirteghem A. Epigenetic risks related to assisted reproductive technologies: Risk analysis and epigenetic inheritance. Hum Reprod 2000; 17:2487-94.
91. Koudstaal J, Braat DD, Bruinse HW et al. Obstetric outcome of singleton pregnancies after IVF: A matched control study in four Dutch university hospitals. Hum Reprod 2000; 15(8):1819-1825.
92. Wang JX, Clark AM, Kirby CA et al. The obstetric outcome of singleton pregnancies following in-vitro fertilization/gamete intra-fallopian transfer. Hum Reprod 1994; 9:141-146.
93. Wilmut I, Sales DI. Effect of an asynchronous environment on embryonic development in sheep. J Reprod Fertil 1981; (61):179-184.
94. Kleemann DO, Walker SK, Seamark RF. Enhanced fetal growth in sheep administered progesterone during the first three days of pregnancy. J Reprod Fertil 1994; 102:411-417.

Chapter 6

Modification of Epigenetic State through Dietary Manipulation in the Developing Mammalian Embryo

Nicola Vickaryous and Emma Whitelaw*

Abstract

The unraveling of the human genome is revealing the intertwined nature of epigenetic phenomena and the genetic code. Epigenetic modifications can alter patterns of gene expression, independent of DNA sequence mutation. Instead, epigenetic marks modify the existing DNA sequence. These modifications include cytosine methylation in the promoter region of genes, a phenomena associated with transcriptional silencing, and recruitment of chromatin remodeling complexes. Epigenetic marks are established early in development, are mitotically, and in some cases meiotically heritable, and can have a profound impact on an organism's phenotype. Thus, epigenetic modifications provide a mechanism by which the permanent changes associated with fetal programming can occur.

Introduction

Increasingly, there are reports that transient exposure of the fetus to an aberrant maternal environment can have long-term health effects on the offspring. Human epidemiological studies have shown a correlation between maternal malnutrition, often associated with low birth weight due to intrauterine growth restriction (IUGR), and an increased risk of adult onset diseases.[1] Collectively known as metabolic syndrome, these include enhanced susceptibility to hypertension, cardiovascular disease, and type 2 diabetes. The fetal origins hypothesis predicts that insults suffered by the fetus in utero such as limited nutrient supply, can cause permanent alterations to the individual, a phenomena known as fetal programming.[1] The thrifty gene hypothesis has been used to explain the benefit of fetal programming; the maternal environment is a reflection of the post-partum environment, and any cellular and metabolic adaptations of the fetus will allow it the best survival under these conditions. The hypothesis proposes that problems arise when the postnatal environment differs from that in utero, hence the fetal adaptations are no longer beneficial and in fact deleterious, resulting in metabolic syndrome.[2]

Epidemiological studies are useful in identifying health tendencies in human populations, but they do not reveal the underlying molecular events. Researchers interested in the mechanistic basis behind fetal programming have turned to animal models, thereby reducing the problems which plague human epidemiological studies including social, cultural and genetic heterogeneity. Studies, primarily carried out in rats, have shown that IUGR can result when the

*Corresponding Author: Emma Whitelaw—School of Molecular and Microbial Biosciences, University of Sydney, NSW, 2006, Sydney, Australia. Email: e.whitelaw@mmb.usyd.edu.au

Early Life Origins of Health and Disease, edited by E. Marelyn Wintour and Julie A. Owens.

developing fetus has impaired nutrient uptake. This can result from maternal malnutrition following overall protein/caloric or micronutrient deficiency. Secondary nutritional deprivation in the offspring can also result from placental insufficiency.[3] The effects of IUGR in rats appears to mimic that of humans, with increased blood pressure, cardiovascular disease and glucose tolerance occurring in individuals nutritionally deprived in utero.[1] Indeed, these effects extend across a variety of species including rat, mice, and sheep.[3,4]

The molecular mechanisms behind fetal programming are not clear, however there is an emerging interest in the role that epigenetic modifications may play in this process.[5] Epigenetic mechanisms make excellent candidates for fetal programming as these marks can permanently modify gene expression, without altering DNA sequence. They are established in early development and, once set, they are mitotically heritable. Thus epigenetic modifications provide a mechanism by which permanent changes to the fetus can occur. This chapter will address the potential contribution of epigenetic mechanisms to maternal diet-induced fetal programming. Some of the issues covered in this chapter overlap with those discussed in the chapter by Robert Waterland.

Epigenetic Modifications

Epigenetic modifications include methylation of cytosine residues in DNA and modifications (including methylation, acetylation, phosphorylation) of the proteins packaging the DNA, the histones, and they affect the transcriptional activity of genes. Covalent modification of DNA, through cytosine methylation, is perhaps the best studied type of epigenetic modification. DNA methylation occurs at CpG dinucleotides through the transfer of methyl groups by DNA methyltransferases (DNMTs). Hypermethylation of promoter regions of genes usually, though not always, correlates with a transcriptionally silenced state. This is thought to occur primarily as a result of the recruitment of methyl binding proteins. These proteins recognize methylated CpG's and bind to them through a methyl binding domain. In mammals, methyl binding proteins, including methyl-CpG-binding protein 2 (MECP2), other methyl-CpG-binding domain proteins (MBD), such as, MBD1, MBD2, MBD3, MBD4 and Kaiso, can recruit proteins involved in chromatin remodelling and transcriptional regulation to the locus. Histone modifications including histone acetylation, deacetylation, methylation and phosphorylation are additional epigenetic modifications involved in chromatin packaging and are believed to act through recruitment of chromatin remodeling complexes.[6] These changes to the chromatin often occur in concert with changes in cytosine methylation and in most situations, it remains unclear which is the primary event.

Examples of phenomena which involve epigenetic silencing include X-chromosome inactivation and parental imprinting. In mammals, dosage compensation of genes on the X-chromosome is accomplished by epigenetic silencing of one of the two X-chromosomes in females.[6] The decision about which (ie, the paternal or the maternal) X-chromosome is to be silenced is random and occurs in the early post-implantation embryo. Once made, the decision is inherited through subsequent rounds of cell division. The inactive X-chromosome is heavily methylated at cytosine residues and packaged into heterochromatin. Parental imprinting refers to the process by which a small number of genes (~100) are monoallelicly expressed as a result of gene silencing of one of the parental alleles. This silencing is determined by parental origin, with some genes undergoing paternal imprinting and others maternal imprinting. This imprint is laid down in the gametes of the parent, and maintained in the zygote. The exception to this imprinting maintenance is in the primordial germ cells of the developing embryo, where the parental imprints are cleared and reset.[6]

There is also evidence from a large range of species, not just mammals, which suggests that cytosine methylation plays a critical role in silencing the large number of retroviruses and other foreign DNA scattered throughout the genome.[7] In fact, some believe this to be the primary role of DNA methylation. Thus, epigenetic modifications to the DNA and the associated chromatin proteins can silence genes, and establishment of the epigenetic states occur in gametogenesis and early development.

Epigenetic Reprogramming Occurs in Early Development

Early development is a critical time for epigenetic reprogramming, with clearing and re-estalishment of epigenetic marks taking place throughout the genome. There is increasing evidence that cytosine methylation plays a crucial role in differentiation and development. After fertilization there is a rapid demethylation of the genome. Fluorescent antibody studies have demonstrated that the paternal genome undergoes a rapid, active demethylation complete within 4 hours of fertilization, while the demethylation of the maternal genome is slower.[8] This maternal demethylation is believed to be the result of a passive loss of DNA methylation due to the absence of the maintenance methyltransferase, DNMT1, leading to progressive loss of methylation with each cell cycle[8] (Fig. 1). Not all sequences undergo this genome-wide clearing, with parentally imprinted genes and some intracisternal-A particle (IAP) retrotransposons escaping this wave of demethylation.[9]

De novo methylation begins in the inner cell mass of the preimplantation blastocyst.[8] There are three major DNA methyltransferases; DNMT1, 3a and 3b. As discussed above, DNMT1 is a maintenance methyltransferase which prefers hemi-methylated DNA as a substrate. DNMT3a and 3b are de novo methyltransferases expressed in early development that can methylate both strands at previously unmethylated cytosines. How DNMT3a and 3b decide which cytosines to methylate, is unknown. Mouse knockouts of *Dnmt3a* and *3b* demonstrate that re-establishment of methylation is crucial for the survival and development of the embryo.[6] Once established, DNA methylation is maintained mitotically through the action of DNMT1.[6] Mouse knockouts of *Dnmt1* are also nonviable. The pattern of DNA methylation differs from tissue to tissue in the adult, suggesting a role for this process in cell differentiation.

In animal models, techniques that involve manipulation of the early embryo such as culturing, nuclear transfer or in vitro fertilization (IVF) can lead to unusual patterns of gene expression and these changes have been attributed to epigenetic events.[10-12] In humans there is a growing concern that techniques such as IVF and intracytoplasmic sperm injection (ICSI) are associated with low birth weight and alterations to parentally imprinted gene expression. This has arisen from the finding that children produced using IVF appear to have an increased risk for imprinting disorders including Beckwith-Wiedemann and Angelman syndrome.[13]

A Possible Role of Epigenetics in Fetal Programming

Parentally imprinted genes could be involved in fetal programming. Imprinted genes such as *Igf2, Peg1, Peg3, H19,* and *Igf2r* are frequently associated, either positively or negatively, with fetal and placental growth. The fact that parentally imprinted genes are often involved in growth of the embryo has led to the parent-offspring conflict theory. This theory aims to explain the function of parental imprinting, stating that in order to successfully propagate their genes into the next generation fathers want as many resources (nutrients) as possible to be available for their offspring, ie an increase in placental and fetal growth. Mothers, however, need to divide their resources across several offspring/litters, and therefore cannot afford to devote too many resources to any one individual. Consistent with this theory is the fact that maternal silencing does often occur at alleles that promote growth, and paternal silencing at those that suppress growth.[14] For example, the imprinted gene, insulin-like growth factor 2 *(Igf2)* is involved in fetal and placental growth by regulating the supply and demand for maternal nutrients and is maternally silenced.[15] IUGR has been seen in *Igf2, Peg1* and *Peg3* mouse knockouts, while overgrowth syndromes are associated with loss of imprinting at maternally expressed genes such as, *H19* and *Igf2r.*[16]

The Influence of Diet on DNA Methylation Levels

Due to the potentially labile nature of DNA methylation there has been interest in the idea of dietary manipulation of these marks. S-adenosylmethionine (SAM) serves as the methyl donor group for DNA methylation of cytosines by DNA methyltranferases. Dietary sources including choline (or its metabolite, betaine), methionine, as well as dietary folates or

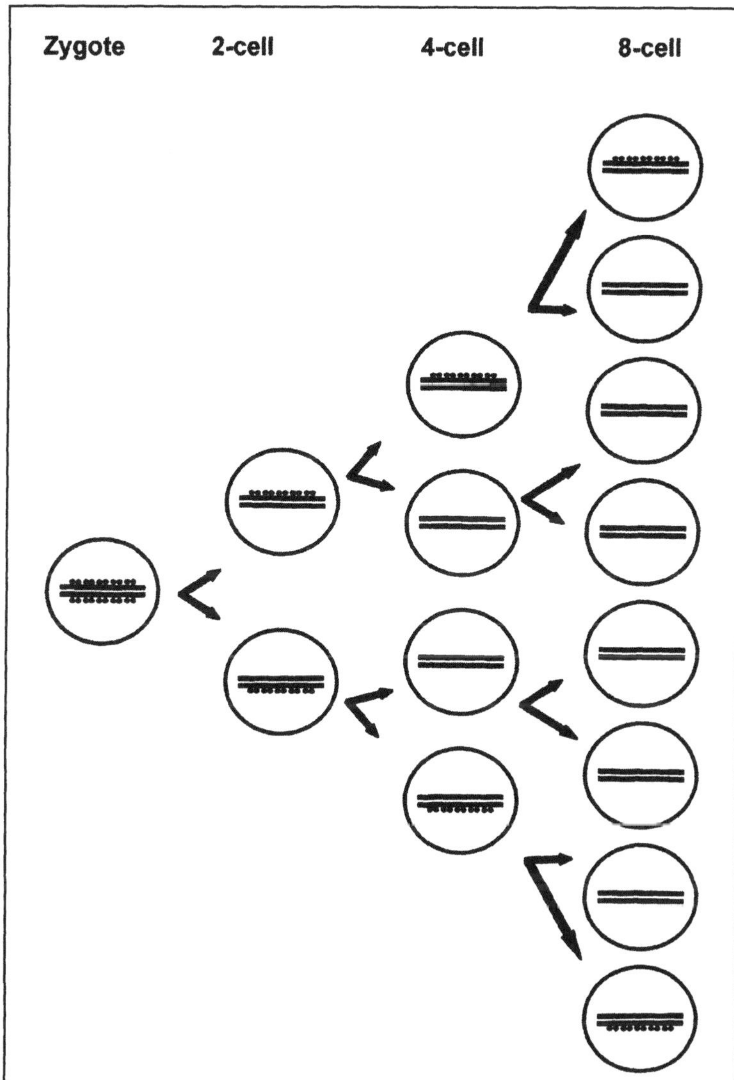

Figure 1. Replication-dependent DNA demethylation occurs on the maternal genome in the zygote. The black lollipops represent methylation at cytosines in the DNA. Exclusion of the maintenance methyltransferase, DNMT1, from the nucleus results in progressive loss of DNA methylation with each round of cell cleavage. At the 8-cell stage there is a transient, (one cell cycle), return of DNMT1 to the nucleus. The relocation of DNMT1 to the nucleus during this time is critical for the maintenance of maternally imprinted marks.[38]

folic acid contribute to the production of SAM in humans[17] (Fig. 2). Hepatic DNA methylation can be altered in rats fed methyl-deficient diets. The changes involve global hypomethylation and altered gene specific hypo- or hypermethylation.[18] However, global DNA hypomethylation is not consistently seen with moderate folate deficiencies, so these results must be interpreted with caution.[18] Reduced DNA methylation in response to diet is not unique to rodents as folate deficiency in humans has also been shown to lower DNA

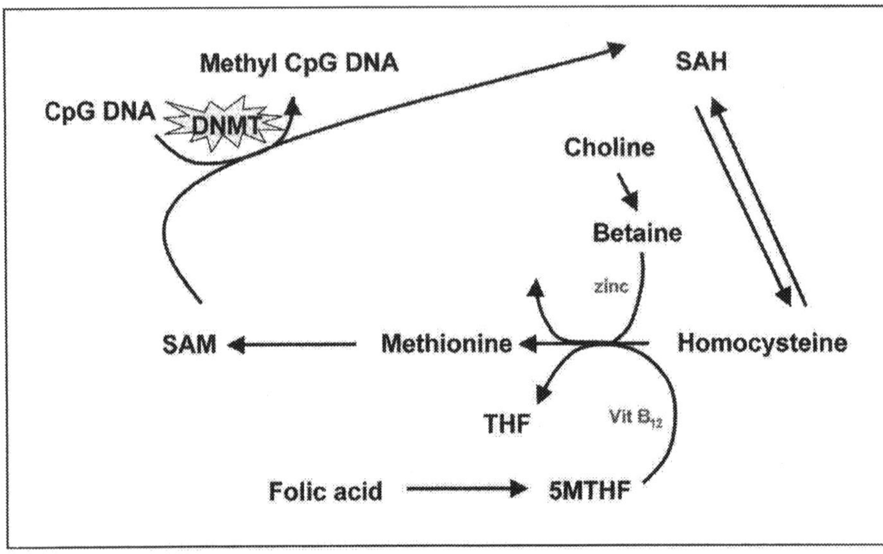

Figure 2. Pathway of methyl metabolism contributing to DNA methylation. Dietary sources of s-adenosylmethionine (SAM) production include choline, or its metabolite betaine, which can convert homocysteine to methionine in the presence of the cofactor zinc. Folates from food sources, or folic acid supplementation, also contribute to the production of methionine, serving as one-carbon unit carriers. The conversion of 5-methyltetrahydrofolate (5MTHF) to tetrahydrofolate (THF) requires the cofactor methylcobalamin, which is derived from vitamin B_{12}. SAM is formed from methionine with the transfer of an adenosyl group from ATP, and serves as a methyl donor for DNA methylation by methyltransferases (DNMTs). Transfer of a methyl group from SAM results in the formation of s-adenosylhomocysteine (SAH), which can be recycled to homocysteine in a reversible reaction. Adapted from reference 19.

methylation levels in lymphocytes.[18] Furthermore, diets deficient in cofactors for methyl metabolism, such as zinc, can alter DNA methylation.[18,19] There is also some evidence that levels of methyl-interacting proteins can be affected by diet.[18-20] Importantly, global and gene-specific methylation changes induced by diet have been implicated in the initiation and progression of cancer.[18]

Maternal Diet Altering Fetal Phenotype

Given the impact that diet can have on DNA methylation in an adult individual, several studies have now been undertaken to see if these effects occur during embryonic development. Rees et al demonstrated that a maternal protein-deficient diet (supplemented with threonine) led to global hypermethylation of DNA in the livers of fetal rats. The authors propose that the hypermethylation seen in the offspring is a result of threonine metabolism competing for enzymes involved in methionine metabolism. The authors suggest that the net result of this competition is elevated homocysteine levels which lead to increased DNA methylation (see Fig. 2).[21] This highlights the difficulty in interpreting dietary studies where overlapping metabolic pathways are involved. Although the long term consequences of hepatic DNA hypermethylation in these embryos was not examined, this type of maternal diet in rats is known to be associated with hypertension and glucose intolerance in the offspring.

Wolff et al[22] demonstrated that the phenotype of offspring could be modulated by micronutrient supplementation of maternal diet using a murine locus, agouti viable yellow, A^{vy}. The A^{vy} allele results from the insertion of an IAP retrotransposon upstream of the agouti gene. A cryptic promoter in the long terminal repeat (LTR) of the IAP can drive constituitive expression of the downstream agouti gene, which is involved in coat colour. The expression state of the cryptic promoter correlates with DNA methylation; an active promoter is hypomethylated, while a silenced promoter is hypermethylated. An inbred population of mice carrying this epiallele display variable expressivity, with some mice having completely yellow coats and others a brown (agouti) coat. As well as the yellow and agouti coats, these isogenic mice can have various degrees of mottling, a trait that stems from variegation of agouti expression from clonally derived patches. Supplementation of maternal diets with methyl donors and cofactors involved in SAM production demonstrated that the proportion of coat colour phenotypes found in the offspring could be shifted from those on a control diet. With maternal micronutrient supplementation, an increased proportion of offspring were found to carry more heavily mottled coats. These effects were found to be influenced by the strain background, suggesting that the differences in the sensitivity to dietary supplements were due to genetic differences in methyl metabolism between the strains.[22]

This finding suggested that methyl supplementation, by providing more substrate, created hypermethylated A^{vy} allele in the offspring, which in turn result in more coat colour mottling. This interpretation was subsequently supported by Waterland and Jirtle, who used bisulfite sequencing to show that increased methylation of the A^{vy} retrotransposon correlated with the shift in coat colours that resulted when dams were fed methyl-group enriched diets.[23] This work not only demonstrates a link between maternal diet and offspring coat colour phenotype, but, has implications on the long-term health of the offspring. Overexpression of agouti, which is a paracrine signalling molecule, results in deleterious health effects in adult mice including obesity, hyperinsulinemia, diabetes, increased susceptibility to cancer and in general a shorter lifespan.[23,24] So, at least in the case of the A^{vy} allele, micronutrient maternal supplementation impacts on the long term health of an individual through epigenetic mechanisms.[23,24]

However, it is worth pointing out that alleles like A^{vy}, which are particularly sensitive to epigenetic state, are not common in the mouse genome, and none have yet been identified in humans. These alleles are now termed metastable epialleles, and to date only a handful have been identified in the mouse.[25] There is now intense interest in the identification of similar alleles in humans, but this is a challenging task due to the outbred nature of our population.

Glucocorticoids and DNA Demethylation

Epigenetic modifications have recently been implicated in the mechanism by which glucocorticoids influence fetal programming.[26] In rats glucocorticoid exposure of dams results in decreased birth weight of the offspring and leads to hypertension and glucose intolerance once these offspring reach adulthood. These effects are similar to those seen in fetal malnutrition. Some people are now suggesting that fetal malnutrition can lead to glucocorticoid overexposure of the fetus as a result of down regulation of 11β-HSD2, a placental enzyme normally involved in inactivation of glucocorticoids.[27] Glucocorticoids, acting through the glucocorticoid receptor (GR), have been shown to mediate epigenetic changes, including DNA demethylation and chromatin remodelling, at the promoters of some genes known to be responsive to glucocorticoids.[28] Interestingly, it has recently been shown that maternal licking of newborns can reset the transcriptional activity of the GR in brains of these offspring. This is associated with hypomethylation of the GR promoter and is retained for the life of the mouse, having long-term behavioural consequences.[29]

Placental Insufficiency as a Cause of Fetal Malnutrition

In developed nations, where maternal diet deprivation is rarely an issue, it has been argued that placental insufficiency is responsible for the majority of fetal malnutrition cases. Placental insufficiency is a condition where the fetus has restricted access to materials through the placenta, often due to vascular conditions.[30] Observations of placental insufficiency in humans is confounded by inheritance of genetic risk factors which may also contribute to the adult chronic diseases.[30] To control for genetic factors involved in this type of fetal programming, rat models for placental insufficiency are used. The placental insufficiency is induced by uterine artery ligation (uteroplacental insufficiency).[31] With this technique Pham et al, demonstrated that uteroplacental insufficiency in rats led to IUGR that was characterized by morphological changes to the nephrons of the offspring. Subsequent molecular analysis of the renal p53 gene revealed promoter region hypomethylation that correlated with increased p53 expression and was associated with altered levels of expression from genes known to be regulated by p53.[32] This indicates that placental insufficiency can lead to permanent changes in patterns of gene expression, which may be the consequence of epigenetic changes.

Other studies using the rat model of uteroplacental insufficiency reveal genome-wide hypomethylation and altered methylation patterns in CpG islands in general. These methylation changes were associated with increased histone H3 acetylation.[31] However, it is dangerous to assume that these changes in patterns of gene expression are the direct consequence of changes to the epigenetic state. There is continuing debate about whether changes in DNA methylation are the causes or the consequences of the changes in transcription.

Transgenerational Effects

Some epidemiological studies indicate that the effects of fetal programming can be passed on to the next generation.[33] Although there is evidence of transgenerational inheritance of low birthweight, cardiovascular and diabetes risk factors in humans, reviewed by Drake and Walker,[33] the most convincing evidence of this phenomena comes from animal models. As discussed previously, rat dams fed protein-deficient diets, produce low birthweight offspring. In one study, these effects were seen to be intergenerational, that is, low birthweight offspring were shown to go on and produce low birthweight offspring of their own.[33] This is despite the fact that they were fed a normal diet. This effect persisted for several generations after a normal protein diet was reintroduced.[33] Transmittance of fetal programming across generations has also been reported for glucose intolerance and hypertension in rats.[33] More recently, intergenerational effects of glucocorticoids have been shown in rats.[26] This study showed that prenatal glucocorticoid overexposure could effect the phenotype of the offspring and their subsequent progeny. Interestingly, in this case, the intergenerational effect could be passed through male individuals, supporting the idea that an epigenetic mechanism is involved. Several studies on intergenerational effects of fetal programming suggest they last approximately three generations before the effect is lost.[26,33]

The inheritance of epigenetic marks between generations has been described in mice.[34] For example, transgenerational epigenetic inheritance has been shown to occur at the previously discussed, A^{vy} allele. The coat colour of the dam influences the range of coat colour in her offspring, with yellow dams producing a higher proportion of yellow offspring than brown dams. The coat colour phenotype of these offspring has been shown to reflect the methylation status of the promoter in the IAP element lying upstream of the agouti gene. At this locus, epigenetic inheritance only occurs through the dam, however it was demonstrated that the coat colour phenotype of offspring was not due to an intrauterine effect.[35] Indeed, at other alleles, paternal epigenetic inheritance has now been described in the mouse.[36] It has been proposed that epigenetic inheritance is a result of incomplete clearing of epigenetic marks in the zygote, leading to a memory of the epigenetic state.[37] This memory leads to nonmendelian inheritance of phenotype, which, at least in some instances, can have health implications in the offspring.[35]

Summary

Although there is more research required in this area, there is a growing body of evidence that maternal diet has the potential to influence epigenetic state in the developing embryo. The impact of epigenetic modifications on phenotype, and the suggestion that they can be affected by maternal diet, aid in explaining the age-old question of how environment can influence phenotype. It is easy to see how, in some cases, fetal programming would be beneficial to an individual if the environmental conditions post-partum were similar to those in utero, and their nongenetic inheritance could offer a mechanism of quick adaptation to the environment. The influence of the maternal environment on offspring phenotype can have profound consequences, as these effects have been shown to be heritable through several generations. In the future, with a better understanding of the mechanisms underlying fetal programming, it is possible that we will be able to modify offspring phenotype through maternal diet. These modifications could have a positive health impact on generations to come.

References

1. Godfrey KM, Barker DJ. Fetal nutrition and adult disease. Am J Clin Nutr 2000; 71(5 Suppl):1344S-1352S.
2. Hales CN, Barker DJ. The thrifty phenotype hypothesis. Br Med Bull 2001; 60:5-20.
3. Schroder HJ. Models of fetal growth restriction. Eur J Obstet Gynecol Reprod Biol 2003; 110(Suppl 1):S29-39.
4. Khan IY, Lakasing L, Poston L et al. Fetal programming for adult disease: Where next? J Matern Fetal Neonatal Med 2003; 13(5):292-299.
5. Waterland RA, Garza C. Potential mechanisms of metabolic imprinting that lead to chronic disease. Am J Clin Nutr 1999; 69(2):179-197.
6. Li E. Chromatin modification and epigenetic reprogramming in mammalian development. Nat Rev Genet 2002; 3(9):662-673.
7. Walsh CP, Chaillet JR, Bestor TH. Transcription of IAP endogenous retroviruses is constrained by cytosine methylation. Nat Genet 1998; 20(2):116-117.
8. Santos F, Hendrich B, Reik W et al. Dynamic reprogramming of DNA methylation in the early mouse embryo. Dev Biol 2002; 241(1):172-182.
9. Dean W, Santos F, Reik W. Epigenetic reprogramming in early mammalian development and following somatic nuclear transfer. Semin Cell Dev Biol 2003; 14(1):93-100.
10. Khosla S, Dean W, Brown D et al. Culture of preimplantation mouse embryos affects fetal development and the expression of imprinted genes. Biol Reprod 2001; 64(3):918-926.
11. Stojanov T, O'Neill C. In vitro fertilization causes epigenetic modifications to the onset of gene expression from the zygotic genome in mice. Biol Reprod 2001; 64(2):696-705.
12. Wrenzycki C, Lucas-Hahn A, Herrmann D et al. In vitro production and nuclear transfer affect dosage compensation of the X-linked gene transcripts G6PD, PGK, and Xist in preimplantation bovine embryos. Biol Reprod 2002; 66(1):127-134.
13. De Rycke M, Liebaers I, Van Steirteghem A. Epigenetic risks related to assisted reproductive technologies: Risk analysis and epigenetic inheritance. Hum Reprod 2002; 17(10):2487-2494.
14. Moore T, Reik W. Genetic conflict in early development: Parental imprinting in normal and abnormal growth. Rev Reprod 1996; 1(2):73-77.
15. Constancia M, Hemberger M, Hughes J et al. Placental-specific IGF-II is a major modulator of placental and fetal growth. Nature 2002; 417(6892):945-948.
16. Reik W, Constancia M, Fowden A et al. Regulation of supply and demand for maternal nutrients in mammals by imprinted genes. J Physiol 2003; 547(Pt 1):35-44.
17. Niculescu MD, Zeisel SH. Diet, methyl donors and DNA methylation: Interactions between dietary folate, methionine and choline. J Nutr 2002; 132(8 Suppl):2333S-2335S.
18. Ross SA. Diet and DNA methylation interactions in cancer prevention. Ann NY Acad Sci 2003; 983:197-207.
19. Van den Veyver IB. Genetic effects of methylation diets. Annu Rev Nutr 2002; 22:255-282.
20. Esfandiari F, Green R, Cotterman RF et al. Methyl deficiency causes reduction of the methyl-CpG-binding protein, MeCP2, in rat liver. Carcinogenesis 2003; 24(12):1935-1940.
21. Rees WD, Hay SM, Brown DS et al. Maternal protein deficiency causes hypermethylation of DNA in the livers of rat fetuses. J Nutr 2000; 130(7):1821-1826.
22. Wolff GL, Kodell RL, Moore SR et al. Maternal epigenetics and methyl supplements affect agouti gene expression in Avy/a mice. FASEB J 1998; 12(11):949-957.
23. Waterland RA, Jirtle RL. Transposable elements: Targets for early nutritional effects on epigenetic gene regulation. Mol Cell Biol 2003; 23(15):5293-5300.
24. Cooney CA, Dave AA, Wolff GL. Maternal methyl supplements in mice affect epigenetic variation and DNA methylation of offspring. J Nutr 2002; 132(8 Suppl):2393S-2400S.
25. Rakyan VK, Blewitt ME, Druker R et al. Metastable epialleles in mammals. Trends Genet 2002; 18(7):348-351.
26. Drake AJ, Walker BR, Seckl JR. Intergenerational consequences of fetal programming by in utero exposure to glucocorticoids in rats. Am J Physiol Regul Integr Comp Physiol 2005; 288(1):R34-38.

27. Holemans K, Aerts L, Van Assche FA. Fetal growth restriction and consequences for the offspring in animal models. J Soc Gynecol Investig 2003; 10(7):392-399.
28. Thomassin H, Flavin M, Espinas ML et al. Glucocorticoid-induced DNA demethylation and gene memory during development. EMBO J 2001; 20(8):1974-1983.
29. Weaver IC, Cervoni N, Champagne FA et al. Epigenetic programming by maternal behavior. Nat Neurosci 2004; 7(8):847-854.
30. Henriksen T, Clausen T. The fetal origins hypothesis: Placental insufficiency and inheritance versus maternal malnutrition in well-nourished populations. Acta Obstet Gynecol Scand 2002; 81(2):112-114.
31. MacLennan NK, James SJ, Melnyk S et al. Uteroplacental insufficiency alters DNA methylation, one-carbon metabolism, and histone acetylation in IUGR rats. Physiol Genomics 2004; 18(1):43-50.
32. Pham TD, MacLennan NK, Chiu CT et al. Uteroplacental insufficiency increases apoptosis and alters p53 gene methylation in the full-term IUGR rat kidney. Am J Physiol Regul Integr Comp Physiol 2003; 285(5):R962-970.
33. Drake AJ, Walker BR. The intergenerational effects of fetal programming: Nongenomic mechanisms for the inheritance of low birth weight and cardiovascular risk. J Endocrinol 2004; 180(1):1-16.
34. Rakyan VK, Preis J, Morgan HD et al. The marks, mechanisms and memory of epigenetic states in mammals. Biochem J 2001; 356(Pt 1):1-10.
35. Morgan HD, Sutherland HG, Martin DI et al. Epigenetic inheritance at the agouti locus in the mouse. Nat Genet 1999; 23(3):314-318.
36. Rakyan VK, Chong S, Champ ME et al. Transgenerational inheritance of epigenetic states at the murine Axin(Fu) allele occurs after maternal and paternal transmission. Proc Natl Acad Sci USA 2003; 100(5):2538-2543.
37. Rakyan V, Whitelaw E. Transgenerational epigenetic inheritance. Curr Biol 2003; 13(1):R6.
38. Reik W, Dean W, Walter J. Epigenetic reprogramming in mammalian development. Science 2001; 293(5532):1089-1093.

CHAPTER 7

Critical Experiments to Determine if Early Nutritional Influences on Epigenetic Mechanisms Cause Metabolic Imprinting in Humans

Robert A. Waterland*

Abstract

Metabolic imprinting occurs when nutritional influences during critical periods of development cause specific metabolic adaptations that persist to adulthood. Epigenetic mechanisms, which regulate the broad diversity of tissue-specific gene expression, are established during development and largely maintained throughout adulthood. Hence, to the extent that nutrition during development affects the establishment of epigenetic gene regulatory mechanisms, metabolic imprinting could occur via this mechanism. This article surveys the growing body of evidence that aberrant epigenetic gene regulation plays an important role in human disease, and recent data from animal models showing that subtle environmental influences during specific ontogenic periods can cause stable alterations in mammalian epigenotype. Experimental approaches are suggested to focus future studies of prenatal and early postnatal nutritional influences on developmental epigenetics in humans.

Introduction

Epigenetics is the study of mitotically or meiotically heritable changes in gene expression that are not caused by changes in DNA sequence.[1] Literally meaning 'above genetics', epigenetics encompasses all the interacting mechanisms that are layered *above* the DNA sequence information to regulate gene expression in a cell type-specific or developmental stage-specific fashion. Just as genetic variability amongst individuals explains differences in susceptibility to disease, it is increasingly clear that so too can inter-individual epigenetic variability.[2-4] However, unlike genetic variation, which is determined by parental inheritance, we are only beginning to understand the factors that determine individual epigenotype.

Five years ago, Waterland and Garza proposed the term 'metabolic imprinting' to encompass a subset of adaptive responses to nutrition during development, characterized by susceptibility limited to a specific ontogenic period early in development and a persistent effect lasting into adulthood.[5] Of the potential mechanisms of metabolic imprinting that were proposed,[5] we have focused on characterizing the circumstances under which nutrition during prenatal and early postnatal development can affect the establishment and maintenance

*Robert A. Waterland—Departments of Pediatrics and Molecular and Human Genetics, Baylor College of Medicine, USDA Children's Nutrition Research Center, 1100 Bates St., Ste. 9070, Houston, Texas 77030, U.S.A. Email: waterland@bcm.tmc.edu

Early Life Origins of Health and Disease, edited by E. Marelyn Wintour and Julie A. Owens.

of epigenetic gene regulatory mechanisms in mammals.[6] This article will briefly review the importance of epigenetics to human disease and recent work in animal models demonstrating that subtle environmental influences during development can change the course of development by permanently altering epigenetic mechanisms. Throughout, we will suggest experimental approaches that will help us understand the role of early nutrition in developmental epigenetics.

Epigenetics and Human Disease

Epigenetic information is conveyed in mammals via a synergistic interaction between mitotically-heritable patterns of DNA methylation, chromatin conformation[7] and autoregulatory DNA binding proteins.[8] (Whitelaw reviews epigenetic gene regulation in this volume). Biological methylation reactions are directly dependent on dietary methyl donors and cofactors including folic acid, vitamin B_{12}, and methionine.[9] Our research, therefore, focuses primarily on the epigenetics of cytosine methylation, which occurs on both strands of palindromic CpG dinucleotides in mammals. Methylation of cytosine to 5-methyl-cytosine affects regional chromatin conformation and gene expression by influencing the affinity of methylation-sensitive DNA binding proteins.[10] Specific patterns of CpG methylation are established during early development and propagated during DNA replication by DNA-methyltransferase 1 (Dnmt1).[11] The mammalian genome is largely depleted of CpG dinucleotides, and those that remain are mostly methylated. However, the promoter regions of about 40-50% of human genes contain small stretches of DNA with a relatively high CpG content,[12] and these 'CpG islands' are normally unmethylated.

Cytosine methylation is critical to mammalian development.[13] Promoter-region methylation does not, however, generally show a clear correlation with gene expression, in that most CpG islands in gene promoters are normally unmethylated regardless of the tissue-specificity of gene expression.[12] Two important exceptions are genomically imprinted genes[14] and transposable elements.[15] (The terms 'tranposable element' and 'transposon' are used here to encompass all parasitic elements in the genome, including retroviruses, retrotransposons and DNA transposons.[15]) Mono-allelic expression of imprinted genes is always associated with differential methylation of the maternal and paternal alleles,[14] and most tranposable elements, which comprise over 45% of the human genome,[16] are constitutively silenced during early development by CpG hypermethylation.[17]

Several recent articles[2-4,18] have reviewed the growing body of evidence that epigenetic dysregulation plays an important role in the etiology of human disease. Aberrant epigenetic regulation is responsible for several devastating developmental diseases including Beckwith-Wiedemann, Angelmann, Prader-Willi, and Rett syndromes.[4] In the last decade a huge body of research has demonstrated that epigenetic dysregulation contributes to many types of human cancer.[4,18,19] Epigenetic mechanisms were first implicated in carcinogenesis by comparing tumor tissue epigenotype with that of surrounding normal tissue.[20] Individual variability in epigenotype at specific loci has now been causally related to cancer susceptibility. For example, epigenetic dysregulation of the gene encoding insulin-like growth factor 2 (*IGF2*) is fairly common among normal adults, and predisposes to colorectal cancer.[21] A heritable epigenetic alteration, or epimutation, in the human DNA mismatch repair gene *MLH1* causes mosaic silencing of *MLH1* which predisposes to various cancers.[22] Several authors[2,23] have made a strong case that, like cancer, many complex diseases are likely to have an epigenetic component. Hence, it is probable that occurrence of the diseases most studied in the context of the 'developmental origins hypothesis', including cardiovascular disease,[24] type 2 diabetes,[25] and obesity[26] is related to epigenetic dysregulation.

The major issue addressed by this paper is illustrated in the causal pathway in Figure 1. Recent data from animal models show clearly that nutrition (and other subtle environmental influences) during development can affect the establishment of gene-specific DNA methylation patterns that are maintained into adulthood. For example, maternal dietary methyl donor

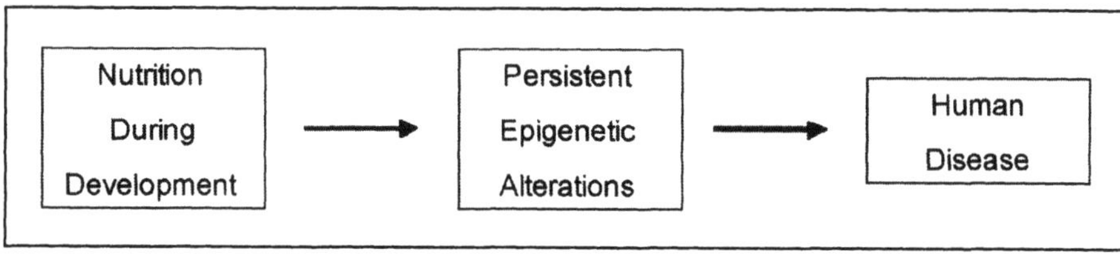

Figure 1. Nutrition during development could influence human disease susceptibility via this causal pathway.

supplementation during pregnancy permanently increases DNA methylation at the *agouti* locus in viable yellow agouti (A^{vy}) mice.[27] Although not a nutritional paradigm, the recent study of Weaver et al[28] is relevant here. They demonstrated that the quality of maternal care received by newborn rats during the first few days of life affects the establishment of DNA methylation of the glucocorticoid receptor gene in the hypothalamus, affecting behavior throughout adulthood. Not only is there is biological precedence for subtle environmental factors to influence developmental epigenetics, but also, as described above, epigenetic dysregulation clearly causes human disease. Therefore, to determine if nutrition during development affects adult chronic disease susceptibility in humans via the pathway shown in Figure 1, the only remaining question is whether nutrition influences the establishment of epigenetic mechanisms during development in humans, as it does in animal models.

What experiments are critically needed to answer this question? Because our understanding of how nutrition influences epigenotype remains rudimentary[23] detailed mechanistic studies in animal models will be required to develop very specific hypotheses to test directly in humans. The entire mammalian genome does not have an equivalent susceptibility to environmentally-induced epigenetic alterations. Rather, there are specific genomic regions whose epigenetic state is especially labile to external influences, and this susceptibility is likely limited to specific ontogenic periods. Once we ascertain the 'genomic signatures' of such regions in appropriate animal models, we can employ comparative genomics to identify candidate human loci at which nutrition may influence developmental epigenetics. Considerable data suggest that specific transposable elements[29] and genomically imprinted genes[6,30] comprise two such classes of epigenetically labile loci.[31]

Epigenetic Lability at Transposable Elements

Except for a brief period of global demethylation in the early mammalian embryo, transposable elements in the genome are generally silenced by hypermethylation.[15] However, these parasitic elements comprise over 45% of the mammalian genome, and their aberrant transcriptional activity can interfere with the expression of neighboring genes.[32] Dysregulation of even a small fraction of human transposable elements could therefore cause substantial genomic instability.[15,29] Notably, transposable elements are transcriptionally activated in *Dnmt1* knockout mice[33] and chimeric mRNAs originating at transposon promoters have been identified in some human tumors.[15,34] Retrotransposition has been directly implicated in human cancer, affecting, for example, the *APC* gene in desmoid tumors and *BRCA2* in breast cancer.[32] Potentially more common than such direct interference, however, is the possibility for transposable elements to affect human genes via epigenetic interference.

As Waterland and Jirtle recently demonstrated in the viable yellow agouti (A^{vy}) mouse,[27] transposable elements in specific genomic regions can lead to epigenetic instability, which renders regional methylation labile to the influence of nutrition (and perhaps other environmental influences). Approximately 5% of human genes contain transposable elements,[35] indicating that a vast number of human genes may be subject to epigenetic interference by neighboring transposons. However, the specific circumstances under which transposable elements can affect epigenetic regulation of neighboring genes are currently unknown.

Rakyan et al[36] introduced the term 'metastable epiallele' to describe loci at which the epigenetic state is established probabilistically early in development, but is stable thereafter. This results in wide variation in individual epigenotype at these sites, even amongst isogenic individuals. It appears that the stochastic nature of establishment of epigenotype at metastable epialleles confers lability to nutritional influences during development.[27] We must determine what types of transposons are associated with epigenetic metastability. Also, we need to know what features of the genomic region into which a transposon inserts are conducive to this phenomenon. For example, the intracisternal A particle (IAP) retrotransposon that causes the A^{vy} mutation is inserted into the *Agouti* promoter region. The IAP that causes a variably-penetrant tail kink phenotype in isogenic $Axin^{Fused}$ heterozygous mice,[37] however, is inserted into an intron of the *Axin* gene. Future studies should identify other animal models in which epigenetic metastability is caused by the insertion of transposon DNA into a specific gene region. Such regions can be contrasted with epigenetically invariant transposon insertion sites to identify genomic characteristics that confer epigenetic metastability. Further, animal models of epigenetic metastability should be tested to determine if epigenetic metastability is a sufficient condition for early nutrition to influence epigenotype.

The pro-methylation supplement used in the A^{vy} studies[27,38] included folic acid, vitamin B_{12}, betaine and choline. In conjunction with the studies described above, we need to determine which specific components of the maternal diet are capable of influencing epigenetic mechanisms in the offspring. Considering the worldwide prevalent supplementation of the food supply with folic acid to combat neural tube defects, it is critical to determine if maternal supplementation before and during pregnancy with folic acid alone can shift offspring epigenotype at specific loci.

Epigenetic Lability at Genomically Imprinted Genes

Genomic imprinting is an epigenetic phenomenon by which certain mammalian genes are expressed preferentially from either the paternally-inherited or maternally-inherited allele. Monoallelic expression of imprinted genes is regulated by allele-specific methylation of specific CpGs.[14] Most imprinted genes are found in clusters, and these imprinted domains are regulated in coordinate fashion via long range mechanisms including antisense RNA interference and methylation-sensitive boundary elements.[14] Such regulatory complexity may render imprinted genes especially sensitive to epigenetic dysregulation.[6,30] This postulate is supported by recent data showing that subtle changes in embryo culture conditions[39] and DNA methyltransferase activity[40] affect the allelic methylation status of imprinted genes in mouse embryos.

Alarmingly, in vitro manipulation of human embryos appears to induce imprinting alterations similar to those characterized in mice. Angelman syndrome is caused by loss of function of imprinted genes including *SNRPN* on chromosome 15q11-13. Recently, there have been several case reports[41,42] of children derived from intracytoplasmic sperm injection (ICSI) who developed Angelman syndrome associated with a loss of methylation in the *SNRPN* region. Assisted reproduction has similarly been linked to an enhanced incidence of Beckwith-Wiedemann syndrome (BWS). BWS is caused by loss of imprinting of a group of genes (including *H19* and *IGF2*) on human chromosome 11p15.[43] Children born after ICSI are at a six-fold higher risk of BWS.[43] These studies taken together provide the first evidence that environmental influences encountered during in vitro manipulation of the early embryo can lead to human disease by inducing epigenetic alterations at imprinted loci.

Research in animal models is now critically needed to determine if, and by what specific mechanisms, early nutritional perturbations can lead to persistent aberrant expression of imprinted genes. For example, does maternal diet before and during pregnancy affect methylation in the differentially methylated regions of imprinted genes in offpring, as it does at the A^{vy} locus?[27,38] If so, which imprinted loci are most labile to such influences? It is logical to focus early investigations on imprinted genes whose methylation status has already been shown, in vitro, to be labile to external influences.[31] The 'early origins' hypothesis is largely predicated on human epidemiologic data relating birth weight to various adult-onset chronic disease outcomes.[5] Given that many genomically imprinted genes are important regulators of fetal growth[44] we should determine if the effects of nutritional exposures known to affect fetal growth (and later physiological outcomes), such as maternal energy or protein restriction during pregnancy, are mediated by epigenetic alterations at imprinted genes.

Another critical issue is identifying the developmental periods when methylation of imprinted genes is most labile to nutritional influences. In the preimplantation embryo the genomic complements derived from the sperm and egg are largely demethylated, and remethylation takes place following implantation.[45] In the germ line, a second wave of demethylation/remethylation occurs later during fetal development, commencing at gestation d 16 in the mouse.[45] The limited ontogenic windows allocated to reestablish appropriate DNA methylation patterns, first in the soma and then in the germ line lineage, suggest potential critical periods when either an excess or deficiency of critical nutrients could affect developmental outcome via epigenetic mechanisms.[6] In the developing germ line, allele-specific methylation at imprinted genes must be set according to sex of the fetus. These primary imprint marks will subsequently escape the wave of demethylation that occurs in the early embryo of the next generation.[45] For these reasons, transgenerational effects of early nutrition could occur via inheritance of induced epigenetic alterations at imprinted genes.[46-48] Research in animal models will enable us to identify the critical windows for early nutritional influences at imprinted genes, and determine if transgenerational effects occur by nutritional influences on imprinted genes during germ line development.

Nutritional Epigenomics

There are almost certainly other classes of loci whose epigenetic states are labile to early nutritional influences. Epigenomic approaches will therefore provide an important complement to targeted studies of transposons and genomically imprinted genes. Several techniques have recently been developed to assess simultaneously the methylation status of thousands of genes.[49-51] All have the potential to identify novel regions of the genome that are labile to early nutritional influences. Future studies will employ these epigenomic approaches to conduct genome-wide comparisons of gene-specific DNA methylation between tissues of animals exposed to different nutritional conditions during development. By identifying the common features of genomic regions that show group differences in DNA methylation, we will advance our understanding of the genomic characteristics that confer epigenetic lability. Epigenomic approaches will also be required to directly test the hypothesis that specific classes of genes, such as genomically imprinted genes, have an enhanced lability to external influences during development.[31,52] For example, one could spot a methylation-specific oligonucleotide array[51] with the differentially methylated regions of 50 genomically imprinted genes, and the analogous regions of 50 nonimprinted genes. By using this array to compare the methylation status of these subsets in animals from different early-diet groups, one could determine if induced methylation changes do indeed occur more readily at genomically imprinted loci.

The Mouse as a Model System for Nutritional Epigenetics

Many epigenetic phenomena, such as genomic imprinting, X-chromosome inactivation, and silencing of transposons by hypermethylation, share much in common between mice and humans. Most genes that are genomically imprinted in humans are also imprinted in the mouse,[53] and the differentially methylated regions that regulate allelic expression of imprinted genes are, for the most part, highly conserved from mouse to human. For example, all four known differentially methylated regions that regulate imprinting of the human gene encoding insulin like growth factor 2 (*IGF2*) also contribute to regulation of *Igf2* expression in the mouse.[54] The eerie precision with which many mouse models of human epigenetic disease recapitulate the presentation of analogously afflicted humans illustrates further that fundamental epigenetic processes are highly conserved between humans and mice. For example, mice with a truncating mutation of the methyl-CpG-binding protein gene (similar to those that cause Rett syndrome in humans) develop a progressive neurological disease sharing many complex Rett syndrome features.[55] The nearly complete availability of DNA sequence data from the human and mouse genomes will simplify comparative genomics approaches based on the results of controlled experiments in mice. Ferguson-Smith et al[53] recently asserted that mice are "the best experimental organism for studying (genomic) imprinting". We agree, and would extend their conclusion to studies of early nutritional influences on epigenetics in general.

Conclusion

For centuries man has recognized that early development is a special time during which subtle environmental influences can dramatically and permanently affect an individual's physiology and behavior. Now, at this exciting time for biology, we are poised to understand the detailed biologic mechanisms mediating this developmental plasticity. It is likely that metabolic imprinting[5] occurs when nutrition during prenatal and early postnatal development induces individual variation in epigenotype that persists to influence adult metabolism and chronic disease susceptibility. An important milestone of research progress in this field will be the identification of human genomic regions that are epigenetically labile to specific dietary influences during development. Carefully-controlled, hypothesis-driven studies in mouse models, combined with genomic and epigenetic comparisons between mouse and human, will yield rapid insights that will facilitate this goal.

Acknowledgments

Supported by NIH grant DK063781 and USDA CRIS #6250-51000-049.

References

1. Bird A. DNA methylation patterns and epigenetic memory. Genes Dev 2002; 16:6-21.
2. Petronis A. Human morbid genetics revisited: Relevance of epigenetics. Trends Genet 2001; 17:142-6.
3. Jiang YH, Bressler J, Beaudet AL. Epigenetics and human disease. Annu Rev Genomics Hum Genet 2004; 5:479-510.
4. Egger G, Liang G, Aparicio A et al. Epigenetics in human disease and prospects for epigenetic therapy. Nature 2004; 429:457-63.
5. Waterland RA, Garza C. Potential mechanisms of metabolic imprinting that lead to chronic disease. Am J Clin Nutr 1999; 69:179-197.
6. Waterland R, Garza CG. Potential for metabolic imprinting by nutritional perturbation of epigenetic gene regulation. In: Black R, Michaelson KF, eds. Public Health Issues in Infant and Child Nutrition. Vol. 48. New York: Lippincott Williams and Wilkins, 2002:317-333.
7. Rakyan VK, Preis J, Morgan HD et al. The marks, mechanisms and memory of epigenetic states in mammals. Biochem J 2001; 356:1-10.
8. Riggs AD, Porter TN. Overview of epigenetic mechanisms. In: Russo VE, Martienssen RA, Riggs AD, eds. Epigenetic Mechanisms of Gene Regulation. Plainview: Cold Spring Harbor Laboratory Press, 1996:29-46.
9. Van den Veyver I. Genetic effects of methylation diets. Annu Rev Nutr 2002; 22:255-82.
10. Jaenisch R, Bird A. Epigenetic regulation of gene expression: How the genome integrates intrinsic and environmental signals. Nat Genet 2003; 33(Suppl):245-54.
11. Jones PA. The DNA methylation paradox. Trends Genet 1999; 15:34-7.
12. Antequera F, Bird A. CpG islands as genomic footprints of promoters that are associated with replication origins. Curr Biol 1999; 9:R661-7.
13. Li E, Bestor TH, Jaenisch R. Targeted mutation of the DNA methyltransferase gene results in embryonic lethality. Cell 1992; 69:915-926.
14. Reik W, Walter J. Genomic imprinting: Parental influence on the genome. Nat Rev Genet 2001; 2:21-32.
15. Yoder JA, Walsh CP, Bestor TH. Cytosine methylation and the ecology of intragenomic parasites. Trends Genet 1997; 13:335-40.
16. Bestor TH. Cytosine methylation mediates sexual conflict. Trends Genet 2003; 19:185-90.

17. Walsh CP, Chaillet JR, Bestor TH. Transcription of IAP endogenous retroviruses is constrained by cytosine methylation. Nat Genet 1998; 20:116-7.
18. Jones PA, Baylin SB. The fundamental role of epigenetic events in cancer. Nat Rev Genet 2002; 3:415-28.
19. Issa JP, Baylin SB. Epigenetics and human disease. Nat Med 1996; 2:281-2.
20. Warnecke PM, Bestor TH. Cytosine methylation and human cancer. Curr Opin Oncol 2000; 12:68-73.
21. Cui H, Cruz-Correa M, Giardiello FM et al. Loss of IGF2 imprinting: A potential marker of colorectal cancer risk. Science 2003; 299:1753-5.
22. Suter CM, Martin DI, Ward RL. Germline epimutation of MLH1 in individuals with multiple cancers. Nat Genet 2004; 36:497-501.
23. Beaudet AL. Is medical genetics neglecting epigenetics? Genet Med 2002; 4:399-402.
24. Hiltunen MO, Yla-Herttuala S. DNA methylation, smooth muscle cells, and atherogenesis. Arterioscler Thromb Vasc Biol 2003; 23:1750-3.
25. Maier S, Olek A. Diabetes: A candidate disease for efficient DNA methylation profiling. J Nutr 2002; 132:2440S-2443S.
26. Levin BE. The obesity epidemic: Metabolic imprinting on genetically susceptible neural circuits. Obes Res 2000; 8:342-7.
27. Waterland RA, Jirtle RL. Transposable elements: Targets for early nutritional effects on epigenetic gene regulation. Mol Cell Biol 2003; 23:5293-300.
28. Weaver IC, Cervoni N, Champagne FA et al. Epigenetic programming by maternal behavior. Nat Neurosci 2004.
29. Whitelaw E, Martin DI. Retrotransposons as epigenetic mediators of phenotypic variation in mammals. Nat Genet 2001; 27:361-5.
30. Pembrey M. Imprinting and transgenerational modulation of gene expression; human growth as a model. Acta Genet Med Gemellol (Roma) 1996; 45:111-125.
31. Waterland RA, Jirtle RL. Early nutrition, epigenetic changes at transposons and imprinted genes, and enhanced susceptibility to adult chronic diseases. Nutrition 2004; 20:63-68.
32. Kazazian Jr HH, Moran JV. The impact of L1 retrotransposons on the human genome. Nat Genet 1998; 19:19-24.
33. Jones PA, Laird PW. Cancer epigenetics comes of age. Nat Genet 1999; 21:163-7.
34. Nigumann P, Redik K, Matlik K et al. Many human genes are transcribed from the antisense promoter of L1 retrotransposon. Genomics 2002; 79:628-34.
35. Nekrutenko A, Li WH. Transposable elements are found in a large number of human protein-coding genes. Trends Genet 2001; 17:619-21.
36. Rakyan VK, Blewitt ME, Druker R et al. Metastable epialleles in mammals. Trends Genet 2002; 18:348-51.
37. Vasicek TJ, Zeng L, Guan XJ et al. Two dominant mutations in the mouse fused gene are the result of transposon insertions. Genetics 1997; 147:777-86.
38. Wolff GL, Kodell RL, Moore SR et al. Maternal epigenetics and methyl supplements affect agouti gene expression in Avy/a mice. FASEB J 1998; 12:949-957.
39. Khosla S, Dean W, Brown D et al. Culture of preimplantation mouse embryos affects fetal development and the expression of imprinted genes. Biol Reprod 2001; 64:918-26.
40. Biniszkiewicz D, Gribnau J, Ramsahoye B et al. Dnmt1 overexpression causes genomic hypermethylation, loss of imprinting, and embryonic lethality. Mol Cell Biol 2002; 22:2124-35.
41. Cox GF, Burger J, Lip V et al. Intracytoplasmic sperm injection may increase the risk of imprinting defects. Am J Hum Genet 2002; 71:162-4.
42. Orstavik KH, Eiklid K, van der Hagen CB et al. Another case of imprinting defect in a girl with Angelman syndrome who was conceived by intracytoplasmic semen injection. Am J Hum Genet 2003; 72:218-9.
43. DeBaun MR, Niemitz EL, Feinberg AP. Association of in vitro fertilization with Beckwith-Wiedemann syndrome and epigenetic alterations of LIT1 and H19. Am J Hum Genet 2003; 72:156-60.
44. Tilghman SM. The sins of the fathers and mothers: Genomic imprinting in mammalian development. Cell 1999; 96:185-193.
45. Reik W, Dean W, Walter J. Epigenetic reprogramming in mammalian development. Science 2001; 293:1089-93.
46. Pembrey ME. Time to take epigenetic inheritance seriously. Eur J Hum Genet 2002; 10:669-71.
47. Kaati G, Bygren LO, Edvinsson S. Cardiovascular and diabetes mortality determined by nutrition during parents' and grandparents' slow growth period. Eur J Hum Genet 2002; 10:682-8.
48. Holliday R. The inheritance of epigenetic defects. Science 1987; 238:163-170.

49. Huang TH, Perry MR, Laux DE. Methylation profiling of CpG islands in human breast cancer cells. Hum Mol Genet 1999; 8:459-70.
50. Costello JF, Plass C, Cavenee WK. Restriction landmark genome scanning. Methods Mol Biol 2002; 200:53-70.
51. Shi H, Maier S, Nimmrich I et al. Oligonucleotide-based microarray for DNA methylation analysis: Principles and applications. J Cell Biochem 2003; 88:138-43.
52. Thompson SL, Konfortova G, Gregory RI et al. Environmental effects on genomic imprinting in mammals. Toxicol Lett 2001; 120:143-50.
53. Ferguson-Smith A, Lin SP, Tsai CE et al. Genomic imprinting-insights from studies in mice. Semin Cell Dev Biol 2003; 14:43-9.
54. Sullivan MJ, Taniguchi T, Jhee A et al. Relaxation of IGF2 imprinting in Wilms tumours associated with specific changes in IGF2 methylation. Oncogene 1999; 18:7527-34.
55. Shahbazian M, Young J, Yuva-Paylor L et al. Mice with truncated MeCP2 recapitulate many Rett syndrome features and display hyperacetylation of histone H3. Neuron 2002; 35:243-54.

CHAPTER 8

Manipulation of the Maternal Diet in Rat Pregnancy:

Different Approaches to the Demonstration of the Programming Principle

Simon C. Langley-Evans,* Leanne Bellinger, Dean Sculley, Alison Langley-Evans and Sarah McMullen

Abstract

Animal studies of nutritional programming confirm the biological principle underpinning the "Barker Hypothesis". Most studies have modelled the hypothesis in its simplest form, seeking to test the proposal that low birthweight predicts hypertension and, indeed, growth restricted offspring in many species do exhibit raised blood pressure as adults. A growing body of work with rodents has considered the programming effects of restricting single nutrients, including low protein feeding, high fat feeding and micronutrient restriction. Quite subtle shifts in the composition of the diet in pregnancy appear to produce potent effects, with hypertension, glucose intolerance, impaired immunity and reduced longevity noted with restriction of maternal protein, iron, sodium or calcium intakes. Although the nature and severity of the insults applied vary greatly between models, the general finding is that either balanced undernutrition or restriction of specific nutrients promotes metabolic and physiological disturbance and also relative adiposity in adult life. Intrauterine influences upon feeding, metabolism and the deposition of adipose tissue may well be mediated at the level of the hypothalamus. Microarray studies of the offspring of protein-restricted pregnant rats, which exhibit a preference for a high-fat food, indicate altered hypothalamic expression of a number of genes relating to signal transduction and homeostatic functions. The common outcomes of a range of nutrient manipulations in pregnancy suggest that a small number of common mechanisms may operate to reset the structure and long-term functions of most tissues. Timing and duration of the insult appears to be a more important determinant of long-term disease outcomes than the nature of the nutrient challenge.

Introduction

The epidemiological evidence indicating an early life programming influence upon the development of major human disease states has been subject to heavy criticism, largely on the grounds of inconsistency between cohorts, and the application of inadequate statistical methodologies.[1-3] These criticisms are however largely irrelevant given the range of literature now available from animal studies that confirm the biological principle underpinning the "Barker

*Corresponding Author: Simon C. Langley-Evans—Centre for Reproduction and Early Life, School of Biosciences, University of Nottingham, Sutton Bonington, Loughborough, Leicestershire, LE12 5RD, U.K. Email: Simon.Langley-Evans@Nottingham.ac.uk

Early Life Origins of Health and Disease, edited by E. Marelyn Wintour and Julie A. Owens.

Hypothesis."[4,5] The purpose of this chapter is to provide an overview of some of the animal models currently in use and to explore the nature of the maternal nutrients that are likely to have the greatest impact upon the long-term health of the developing fetus.

Most animal models in this field were developed with the intention of modelling the Barker Hypothesis in its simplest form, namely that low birthweight predicts hypertension. Good examples of this approach are the uterine ligation models used in both rats and guinea pigs.[6,7] Here surgical restriction of the blood flow to one horn of the two-lobed uterus results in unilateral undernutrition and growth restriction of the fetuses, which can then be compared to littermates in the contralateral horn. Growth restricted offspring exhibit raised blood pressure as adults, in keeping with the programming hypothesis.[6] Similarly the global food restriction model of Woodall et al, in which very severe maternal food restriction limits fetal growth,[8] has a number of physiological and metabolic effects, including hypertension and obesity.[8-10]

Models of this nature may, however, be criticised as nonphysiological given that they depend upon either surgical intervention or extremely severe dietary restriction. In the case of the global restriction model of Woodall et al there are major detrimental effects upon maternal weight gain, and presumably body composition, in pregnancy.[8] In developing suitable animal models of nutritional programming it has been argued that the approach that is most representative of human undernutrition is this form of balanced, or global, restriction of nutrient intakes.[11] However, most of the studies of disease programming in humans have demonstrated associations between early life and disease in more affluent populations for whom this form of undernutrition is relatively rare.[12-15] In the light of the type of diet typically consumed by westernised populations, models in which specific nutrients are restricted, or where certain nutrients are provided in excess may be more relevant.

A large number of nutritional manipulations have been imposed upon pregnant rodents, including less severe food restriction (50-70% of *ab libitum*), low protein diets, high saturated fat diets, iron and zinc depletion and high sodium and calcium diets. The most interesting outcome of these studies is that, in general, the same spectrum of physiological and metabolic adaptations occurs in the fetal tissues, becoming permanently fixed in place and promoting adverse functional consequences in later life. This strongly suggests that a relatively small number of common pathways provide the mechanism linking poor nutritional *quality* to long-term disease. Importantly in many of these studies the programming of tissue functions occurs without evidence of fetal growth restriction, indicating that the continued restriction of human studies to exploring associations between disease and infant birth anthropometry is unlikely to be a fruitful activity.

Whilst nutritional programming studies of rodents are very informative as proof of principle and in developing initial hypotheses regarding programming mechanisms, it is important to recognise the constraints on drawing parallels between animals and humans. One specific drawback of rodent studies is that species like the rat are resistant to the development of coronary heart disease. In terms of blood pressure and the systems that regulate it however, the rat is a useful model for the study of nutritional programming.[16]

Nutritional Programming of Disease—How Does It Happen?

For diverse nutritional insults to produce a similar profile of metabolic and physiological outcomes, simple common adaptive mechanisms in the fetus must drive the programming response that leads to disease. There have many suggestions for such a common pathway, including a primary role for glucocorticoids[17-20] and modulation of DNA methylation patterns.[21-23]

The ultimate disease end-points associated with nutritional programming in adult animals appear to relate generally to a reduction in the number of functional units (e.g., number of nephrons, or number of islet cells)[24,25] within an organ, or modification of the profile of cell types present within a tissue,[26] or to defects in hormone-receptor interactions (over-expression of certain receptors promoting a hypersensitivity to normal concentrations of the hormones).[27,28] The latter may easily relate to the former cases, as the profile of receptors borne by cells is

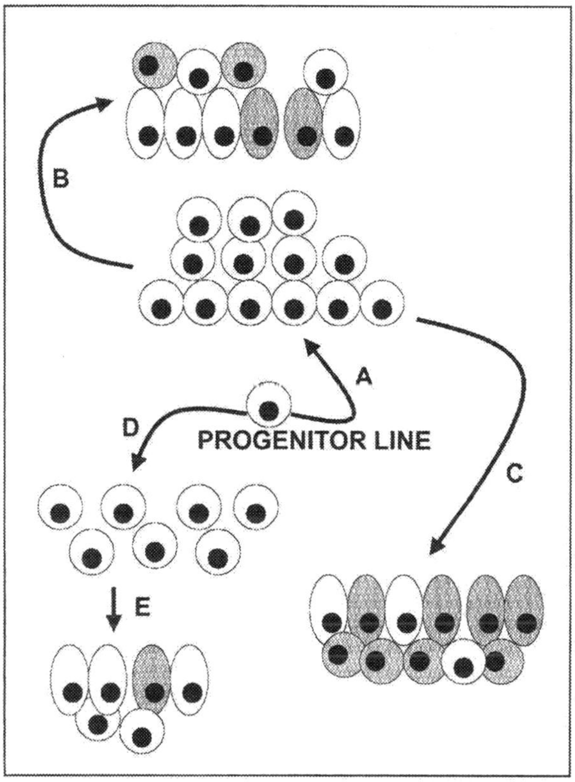

Figure 1. Schematic representation of tissue development from a progenitor cell line in the embryo. A) Normal development involves the proliferation of the progenitor cell line followed by: B) Differentiation of the early cell lineage to form mature functional cell types. C) Abnormal development may involve dysregulated differentiation generating a mature tissue that is functionally abnormal. Overall tissue size is unchanged. D) Abnormal development may involve impairment of early proliferation of the cell line. This reduces the pool of cells available for differentiation into mature lines, leading to: E) A smaller tissue with fewer functional units.

related to cell type and function, and the overall sensitivity of a tissue to a hormone will be determined by the number of cells bearing receptors.

These observations suggest that at the simplest level, programming of physiology is a matter of the fetus or infant employing adaptive survival strategies which impact upon the differentiation and proliferation of the early cells lines that will eventually form the mature organ systems. Exposures that inhibit cell proliferation are likely to result in mature tissues that contain fewer cells. Treatments that induce earlier cell differentiation are likely to reduce the numbers of key functional cells, possibly without a change in organ size, due to the presence of more immature, undifferentiated cells or changes in the proportions of the different cell types present. In some tissues (e.g., muscle) there may be a degree of compensatory growth, such that after the inhibitor is removed, undifferentiated cells either proliferate more rapidly to catch-up, or they undergo hypertrophy, thus leading to no overall effect on organ/tissue size and weight (Fig. 1).

There is ample evidence of these processes occurring in rats exposed to undernutrition in utero. In the kidney for example, we have shown that low protein diets in pregnancy reduce the number of functional units (nephrons) present, with no gross change in organ size.[24,29,30] In the pancreas, the same dietary approach, both reduces organ size and the number of islets present[25]. In the brain, protein restriction reduces the volume of key hypothalamic nuclei involved in the control of feeding and alters the density of neurones present and the types of neuropeptides they produce or respond to.[31] Similarly in soleus muscle from rats subject to protein restriction in utero, the number of secondary fibres present is decreased, suggesting that proliferation of myoblasts is impaired in fetal life, or that myoblast differentiation occurs early (Langley-Evans, Brameld and Hurley, unpublished data). In all of these examples, it can be argued that the basic process leading to metabolic consequences or disease is a reduction in the numbers of active functional units for the organ or system, subsequent to impairments of early cell proliferation and differentiation.

Organ Systems and Disease States Shown to Be Programmed in Animal Models

Blood Pressure and Renal Programming

A number of rodent models have been developed to explore the impact of global nutrient restriction upon long-term vascular functions, and in particular blood pressure. Studies of uterine ligation[6,7] and severe maternal food restriction[8] have already been noted earlier in this chapter. Whilst these particular global restriction models appear to support the nutritional programming hypothesis in general terms, less severe global nutrient restriction studies have less clear-cut effects. In rats, 50% reduction of food intake in the second half of pregnancy did not increase blood pressure in the resulting offspring. However, although these offspring were not hypertensive they exhibited altered vascular responsiveness to nitric oxide, indicating subtle programming of vascular reactivity and function.[32] Rats allowed to consume 70% of ad libitum intake produced pups that were hypertensive relative to control animals from 13 weeks of age.[33] Similarly a very mild global nutrient restriction in guinea pig pregnancy (85% of ad libitum intake) also programmed later blood pressure.[34] At around 14 weeks of age guinea pig pups exposed to undernutrition in utero had blood pressures that were 9% higher than in control animals. Studies in which rodents are subject to global nutrient restriction thus appear to simply reproduce the central assertion of the original Barker Hypothesis. Nutrient restriction produces lower birth weight. The growth-retarded offspring develop high blood pressure and/or altered vascular function in later life. A common feature of all of these studies is that the cardiovascular effects of fetal undernutrition are small and are subject to some delay before their appearance in postnatal life. In contrast the offspring of rats subject to specific nutrient restriction in pregnancy tend to manifest a greater degree of hypertension, with an earlier onset (Table 1).

Table 1. A comparison of offspring blood pressure responses to maternal nutritional manipulations in small animal species

Dietary Manipulation	Species	Age of Onset (Weeks)	Maximal SBP ↑	References
Global	Rat	14-30	6-22 mmHg	8,9
Restriction	Guinea pig	12	9%	34
Low protein	Rat	4	30 mmHg	36-39,41
	Mouse	12	14 mmHg	92
Low iron	Rat	6-10	10-20mmHg	53,54
Low calcium	Rat	52	12 mmHg	56
High calcium	Rat	52	7.5 mmHg	56
Low sodium	Rat	12	8 mmHg	55
↑ saturated fat	Rat	4-25	8-20 mmHg	52,58

Many nutrients in the human diet are likely to be limiting in terms of fetal growth and development. Much of the work with animal models that has considered the programming effects of restricting single nutrients has utilized low protein feeding. This is of considerable relevance in the study of human health as at the global level there is wide variation in protein intakes. Whilst women from affluent nations generally consume 60-80 g protein per day, those from poorer countries may consume 40-50 g per day and have marginal protein intakes. Even within the more affluent countries, protein intakes may be lower in pregnant women from particular sectors of the population, including the lower socio-economic classes[35] (Table 2).

The offspring of rats fed a maternal low protein diet in pregnancy (MLP diet) exhibit elevated systolic blood pressure from an early age, despite being of normal weight at birth.[36] Among animals aged 3-4 weeks pressures are increased by 15-30 mmHg relative to control animals.[37-40] The effect of prenatal protein restriction appears to be permanent and blood pressures remain elevated well into adult life (Fig. 2). The programming hypothesis suggests that exposure to insults during critical periods of development exerts permanent physiological or metabolic effects and this has been demonstrated with the MLP model. Rat pregnancy may be conveniently divided into 3 critical phases; embryogenesis; tissue differentiation and; tissue maturation. Exposure to MLP during any of these phases appears to produce significant effects upon the vasculature but the greatest elevations of blood pressure were associated with either restriction over the rapid fetal growth phase (days 15-22), or with restriction throughout pregnancy[41,42](Fig. 3).

Elevation of blood pressure in the MLP-exposed rat is associated with reduction of nephron number.[24] This association has often been regarded as a causal relationship as it is argued that to maintain renal haemodynamic functions in the face of a nephron deficit, local blood pressure increases are necessary.[43] Rising pressures lead to further nephron loss and hence to still greater increases in pressure to maintain function further. We have recently, however, been able to demonstrate effects of maternal nutritional manipulations upon blood pressure, without adverse effects on the kidney, suggesting either that other mechanisms may drive the development of hypertension, or that reduction of the number of functional filtration units is not the sole renal mechanism.[44] Much of our recent work has focussed on the expression of angiotensin II receptors in the kidney. Our data suggest that, in contrast to AT1R protein,[27] the long-term expression of AT1R mRNA is unaffected by MLP feeding.[28] This finding coupled to measurements of the responses of MLP-exposed offspring to specific AT1R

Table 2. Protein intake in pregnant women

Socio-Economic Group	Trimester 1 Protein Intake (g/Day)	Trimester 1 Protein Intake (% Energy)	Trimester 3 Protein Intake (g/Day)	Trimester 3 Protein Intake (% Energy)
I	75.7 ± 13.8	14.1 ± 2.1	72.6 ± 14.3	13.8 ± 2.2
II	69.0 ± 14.0	13.8 ± 1.9	73.0± 14.7	15.3 ± 6.2
III M III NM	71.9 ± 14.8	14.1 ± 2.0	70.8 ± 13.2	13.5 ± 2.2
IV	68.7 ± 15.9 *	13.4 ± 2.0	67.6 ± 13.8 *	13.9 ± 2.3
V	64.8 ± 22.8 *	13.8 ± 2.5	63.6 ± 15.6 *	14.0 ± 3.1

Women from Northampton, UK, were studied in the first (n = 220) and third (n = 172) trimesters of pregnancy respectively. Daily protein intake was estimated using 5-day food records. Women were grouped on the basis of their partners occupation. * indicates significantly different to social class I (P<0.05). Data are mean ± SD. Social class I represents professional occupations, class IV represents unskilled labour and V is unemployed.

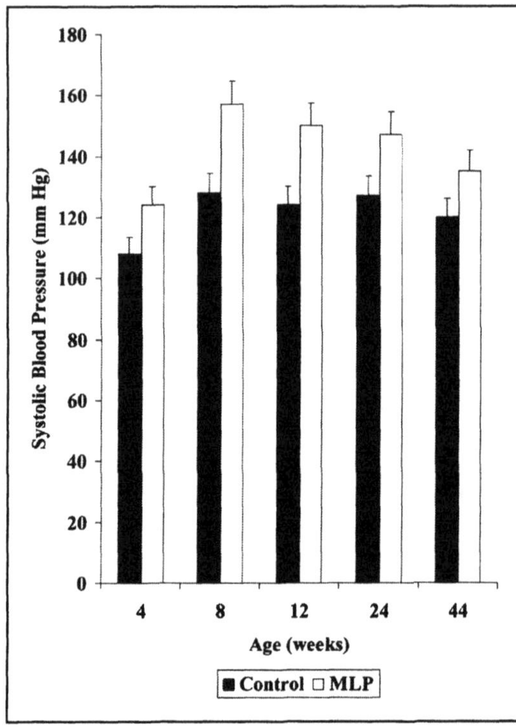

Figure 2. Blood pressure in rats exposed to MLP diet in utero. Data are shown as mean ± SEM. $P < 0.05$ for MLP vs control at all ages shown.

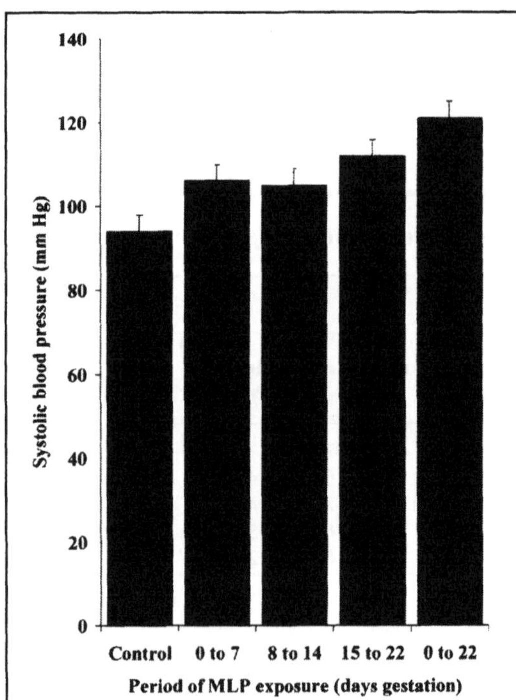

Figure 3. Blood pressure in rats exposed to MLP diet in utero during specific phases of development. Data are shown as mean ±SEM. $P < 0.05$ for MLP vs control at all developmental stages shown.

Table 3. Summary of data from microarray studies of renal tissue

Up- or Down-Regulation	Functional Classes	Genes with Altered Expression
Up-	Transcription Factors	*Pax4* (3.72),
	Intermediary metabolism	*Hal* (3.45), *Smp-2* (3.42), *2,3-oxidosqualene cyclase* (2.59), *Dihydrofolate reductase* (2.23), *17βHSD6* (3.88).
	Receptors/signal transduction	*Dynorphin* (3.26), *Argbp2* (2.89) *5HTR* (2.86), *Go-vn13c* (2.73)
	Structural proteins	*Colq* (3.35), *FAK* (3.02), *Crydg* (2.74), *Ugt8* (2.18)
Down-	Ion channels	*Cacn1d* (0.38), *Slc5a3* (0.50),
	Structural proteins	*Myosin heavy chain 6* (0.32), *Myocilin* (0.29)
	Receptors/signal transduction	*Npr3* (0.34)
	Hormones/growth factors	*Nov* (0.35), *prlpf* (0.33), *Smpd3* (0.48)
	Intermediary metabolism	*Tpi1* (0.46), *Acsl4* (0.33), *20αHSD* (0.48)

23 genes were up-regulated by MLP exposure in utero (2-fold or greater increase in expression) in kidney. 13 genes were down-regulated (2-fold or greater decrease in expression). Data shows genes grouped into general functional classes. Figures in parentheses show fold-change in expression. Kidney were samples from 4-week-old male rats. Data were generated in a 2-slide dye-swap experiment using pooled mRNA from 6-9 rats per group.

antagonists suggests that there may be some programming effects upon the post-transcriptional regulation of this receptor. Expression of AT2R mRNA is down-regulated in the kidneys of rats exposed to MLP diet in utero.[28] As this receptor appears to be involved in the attenuation of responses to angiotensin II, this may provide a direct mechanism for blood pressure elevation. The balance of AT1R and AT2R receptors may also determine the development of the fetal kidney and later in life control blood pressure responses to this important peptide hormone.[45,46]

Interestingly recent microarray studies of renal tissue from MLP-exposed offspring indicate that a significant proportion of the genes whose expression is up- or down-regulated encode ion channels involved in the excretion and reuptake of sodium, calcium and potassium (Table 3). This may disturb electrolyte homeostasis and hence the normal regulation of fluid and sodium balance, either as a consequence of, or by provoking, changes in the angiotensin II-aldosterone axis.

There has been considerable interest in whether the effects of MLP feeding in rat pregnancy upon the offspring may be related to deficiencies of specific amino acids. Rees et al suggested that reductions in the threonine concentration of fetal serum during MLP feeding were related to the use of threonine to generate glycine, which is conditionally essential in pregnancy.[47] Experiments in which MLP diet is supplemented with glycine show that this specifically reverses the hypertensive effect of the diet upon the offspring.[48] It has been proposed that the importance of glycine may lie in its use in the metabolism of methionine and homocysteine. The ratio of S-adenosyl methionione and S-adenosyl homocysteine, intermediates in the methionine-homocysteine cycle dictates that capacity for DNA methylation, which may provide a mechanism for nutritional programming.[49] These observations have prompted further interest in the potential role of folate, which donates methyl groups to the methionine-homocysteine pathway. There is evidence from rat studies that both folate deficiency and excess may programme vascular function in utero.[50,51]

Feeding a low protein diet in rodent pregnancy is not the only approach that has been used to study nutritional programming of the vasculature. Quite subtle shifts in the composition of the diet in pregnancy can produce equally potent effects. For example, the substitution of polyunsaturated fat-rich oil with a saturated fat rich oil in the diet of pregnant rats produces an elevation of blood pressure in their offspring.[52] Low intakes of micronutrients in pregnancy also appear to impact upon long-term cardiovascular health in laboratory rats. Low iron diets sufficient to produce anaemia in pregnant rats, produce an initial hypotension in the offspring that is the precursor to an elevation of blood pressure later in life.[53,54] Low sodium diets in rat pregnancy also elevated blood pressure by 7-8 mmHg in the offspring from around 6 weeks of age, through a mechanism that appears dependent upon renin-angiotensin system components.[55] Calcium has also been implicated in cardiovascular programming. High consumption of calcium in rat pregnancy resulted in a small elevation of blood pressure in the offspring (4.3 mmHg at 1 year of age), whilst in the same study the offspring of calcium deficient mothers had blood pressures 12 mmHg above the control value.[56]

The excess of fat in the typical diet of humans in the developed world has promoted interest in the potential role for fat as a programming agent. Holemans et al reported that feeding a high saturated fat diet to the pregnant offspring of diabetic rats altered vascular sensitivity to acetylcholine and noradrenaline, but no assessment was made of their offspring.[57] More recently it has been shown that offspring from rats fed a lard-rich diet had systolic blood pressures that were elevated by between 8 and 13 mmHg and blunted endothelium-dependent relaxation responses to acetycholine.[58]

The clear implication of this body of work is that variation in maternal nutritional status can exert an important influence on the cardiovascular function and health of the resulting offspring. The ideal composition of the control diet used for comparative purposes in such experiments remains an unanswered question for researchers in this area. The overall food matrix, against which intakes of particular nutrients may vary, could be a critical element in nutritional programming. This is illustrated well by consideration of the many variants of low protein feeding protocols in the literature, most of which have little commonality in terms of exact diet composition.[25,29,37,59-62]

Whilst most studies show that low protein diets in pregnancy programme later hypertension in the developing fetus, this is not true of all. Work by Hoet et al in Louvain[25,63] and by Hales et al in Cambridge,[61] has examined the nutritional programming of type II diabetes by low protein feeding in rat pregnancy and lactation. The low protein diet used in these studies contained 8% protein (provided as casein), with carbohydrate provided mainly in the form of glucose and fat as soybean oil. This diet has a clear impact on pancreatic structure and function, promoting glucose intolerance and insulin resistance. However, this low protein diet does not programme blood pressure changes.[64] The MLP diet used in our studies provides a similar source and quantity of protein, but provides fat as corn oil (at a greater concentration) and carbohydrate as a starch:sucrose mixture. It is clear, therefore, that variation in the source of fat and/or carbohydrate may explain the discrepancies. This example highlights the need for the full composition of diets used in studies of programming to be published in papers. All low protein diets are not the same and without awareness of the possible nutrient interactions occurring within the undernourished mother, it is not possible to assess of the metabolic adjustments and nutritional demands necessary to deal with sub-optimal nutrition or frank malnutrition.

This is extremely important in the context of criticism of the methodology used in many rat studies of programmed hypertension (see Chapter 9 by Denton et al). Tail-cuff plethysmography has been the basis for almost all studies of this kind, but is prone to artefacts associated with restraint stress. Tonkiss and colleagues reported that in rats exposed to maternal low protein diets in utero, the only abnormal components of cardiovascular function were stress induced blood pressure increases[60] and this is erroneously interpreted as a confounder of studies using similar models. However, this was demonstrated using ammonia inhalation, rather than restraint, as the stressor and the low protein diet used in pregnancy bore little relation to that used in other studies.[39,48,59]

Appetite and Obesity

A number of experimental approaches have been used to assess the impact of early life nutritional exposures on long-term feeding behaviour and obesity. Although the nature and severity of the insults applied vary greatly, the general finding is that either balanced undernutrition or restriction of specific nutrients promotes increased food intake and relative adiposity in adult life.[9,65-67] We have noted that mild protein restriction in rat pregnancy leads to increased deposition of abdominal fat in male offspring, in keeping with other reports of increased adiposity following mild or severe maternal food restriction.[68] Prenatal high protein diets, which are generally considered to be a risk factor for low birth weight in human pregnancy also have programming effects on fat deposition. Daenzer et al have demonstrated that feeding a high protein (40% by weight) diet in rat pregnancy resulted in a greater fat mass in the offspring at 9 weeks of age.[69]

As described in the chapter by Breier and Vickers (see Chapter 12), prenatal undernutrition followed by hypercaloric feeding in postnatal life is associated with a greater degree of diet-induced obesity than in animals subject to an uncompromised intrauterine environment. It is suggested that this presusceptibility to obesity is a consequence of increased appetite and reduced physical activity. Our own studies of rats exposed to low protein diets in utero suggests that the prenatal environment may also have a profound effect on food choice. Offspring of protein-restricted rats exhibit an increased preference for a high fat food when allowed to self-select from fat, carbohydrate and protein-rich sources (Fig. 4). Over time, this would be expected to induce obesity.[68]

Intrauterine influences upon feeding and the deposition of adipose tissue may well be mediated at the level of the hypothalamus. Plagemann et al have reported that protein restriction targeted at both pregnancy and lactation in rats programmes both the density of neurones in key areas of the hypothalamus that regulate food intake and the production of regulatory neuropeptides.[31] Microarray studies of the offspring of protein-restricted pregnant rats, which exhibit a preference for a high-fat food, indicate altered hypothalamic expression of genes encoding important olfactory and taste receptors, for example up-regulation of the olfactory receptor *olfr41* and the pheremone receptor *vn6*, alongside down-regulation of the taste receptor *rt2rg*, which is involved in sensing bitter components (Table 4). Together these findings appear to support the hypothesis that most nutritional stressors initiate a common pathway of adaptive responses that include modulation of cell proliferation and differentiation, and hence the final profile of cells and associated functions present within a tissue.

Immune Function

As described above, undernutrition may exert programming effects through reductions in cell numbers within a tissue, or changes in the cell types present. With regard to the immune system it seems likely that undernutrition may determine the development of the thymus and lymphoid tissues and hence their long-term functions.[70-72] We have recently noted that the feeding of a maternal low protein (MLP) diet in rat pregnancy, as in human IUGR, reduces the size of the thymus in the offspring, particularly when undernutrition is targeted to the early phase of pregnancy.[73] Whilst this indicates changes to gross structure, it is likely that more subtle changes to the thymic microenvironment may also occur and that these will dictate specialised immune functions in the long-term.

Beach et al first reported that the feeding of a zinc-depleted diet in mouse pregnancy compromised the immune functions of the resulting offspring. Mice that were zinc depleted in utero exhibited impaired production of IgM, and lower plaque-forming cell responses to sheep red blood cell inoculation.[74] Interestingly these effects persisted for three generations after the initial dietary insult, suggesting that programming of these functions could be transmitted

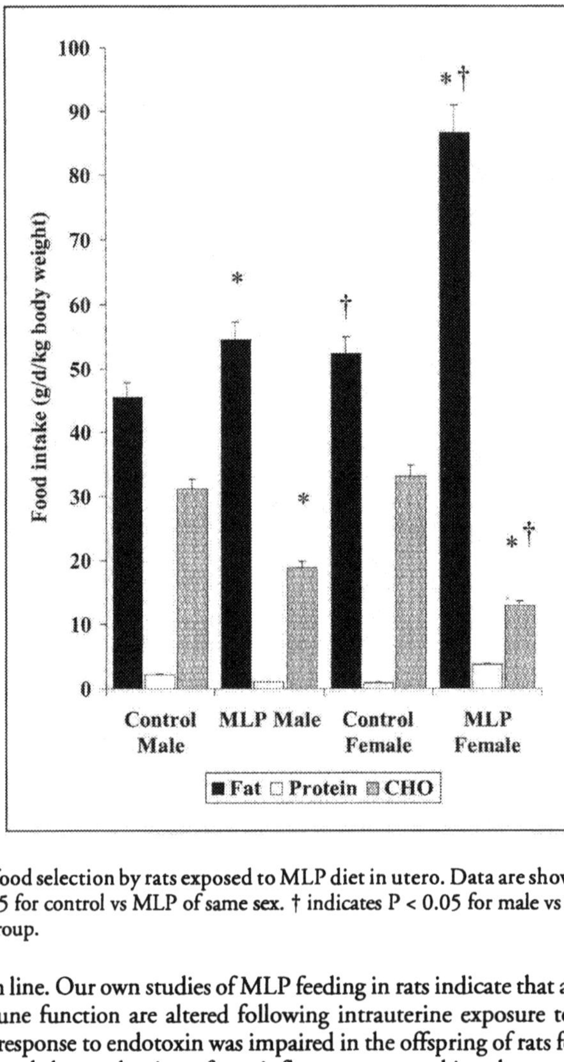

Figure 4. Altered food selection by rats exposed to MLP diet in utero. Data are shown as mean ± SEM. * indicates P < 0.05 for control vs MLP of same sex. † indicates P < 0.05 for male vs female within same maternal dietary group.

through the germ line. Our own studies of MLP feeding in rats indicate that a number of crude markers of immune function are altered following intrauterine exposure to undernutrition. The acute phase response to endotoxin was impaired in the offspring of rats fed MLP throughout pregnancy[75] and the production of pro-inflammatory cytokines by macrophages isolated from the male animals was also reduced.[76] Calder and Yaqoob have more recently reported that altering the level of protein and type of fat in the diet during rat pregnancy, modulates the proliferative responses and NKC activities of lymphocytes isolated from neonatal and weanling offspring.[77]

We have shown that targeting MLP feeding to specific periods early in rat pregnancy results in a reduction of thymic size and an increase in the circulating numbers of lymphocytes. Interestingly these offspring of rats fed MLP diet in early pregnancy appear more resistant to infection by the gut parasite *Nippostrongylus brasiliensis*. When MLP feeding continues throughout pregnancy the resulting offspring have a smaller spleen than control animals and fewer lymphocytes.[78] These discrepancies suggest that exposure to undernutrition during the separate phases of embryonic development, organogenesis and the rapid fetal growth spurt associated with the last few days of rat development, may have specific effects upon the developing immune tissues. It is important therefore to study the effects of undernutrition during these periods in isolation and identify the most critical time points for immune system programming.

Table 4. Summary of data from microarray studies of hypothalamus

Up- or Down-Regulation	Functional Classes	Genes with Altered Expression
Up-	Receptors/signal transduction	*Hrh1* (3.28), *Gaba-aR4* (3.12), *2mu* (3.18), *snap23* (3.05), *inpp4a* (2.94), *vn2* (2.95), *prlr* (2.93), *ptgerep4* (2.42), *grk4a* (2.13)
	Structural proteins	*Myelin basic protein* (3.58), *dspp* (2.42)
	Transcription factors	*Tmf1* (3.64)
	Apoptosis	*αbadα* (2.23)
	Endopeptidase inhibitors	*Aplp2* (3.23), *nexin* (2.12)
Down-	Receptors/signal transduction	*Syt8* (0.43), *Gal2r* (0.41), *Adra1b* (0.41),*Chrna1* (0.21)
	Transcription factors	*Nf1-b1* (0.14)
	Ion channels	*Kcnmb1* (0.36), mrp14 (0.40), *Slc7a9* (0.32), nckx1 (0.36)
	Immune function	*IFNα* (0.47)
	Steroid metabolism	*Cyp17* (0.41)
	Cell cycle regulators	*c-mos* (0.42)

33 genes were up-regulated by MLP exposure in utero (2-fold or greater increase in expression) in hypothalamus. 69 genes were down-regulated (2-fold or greater decrease in expression). Data shows genes grouped into general functional classes (not all genes shown). Figures in parentheses show fold-change in expression. Hypothalamus samples were from 13-week-old male rats. Data were generated in a 2-slide dye-swap experiment using pooled mRNA from 6-9 rats per group. LP diet was fed only during the first 7 days gestation.

Ageing and Longevity

Studies of rodents subject to intrauterine insult suggest that lifespan may be shortened and degenerative processes associated with ageing may be enhanced. Studies of rats and mice exposed to maternal low protein diets indicate that lifespan may be shortened by as much as 15%.[79-81] This is of considerable interest as the impact of prenatal undernutrition appears to be the opposite of postnatal restriction of energy intake. It is well-established that postnatal caloric restriction increases longevity and provides resistance to cancer in a range of species from *Drosophila* to primates.[82]

The degenerative processes of ageing, including progressive loss of tissue and organ function, are a consequence of cellular apoptosis. This may be triggered through a mechanism termed replicative senescence, which involves the progressive loss of DNA from the chromosome ends (telomeres) with successive cell divisions. Shortening of telomeres to critical levels results in the activation of the p53 tumor suppressor gene which leads to a cascade of events that promote cell death.[83] There is some evidence of this in the kidneys of aged male rats exposed to low protein diets in utero and throughout lactation.[79]

Apoptosis is also triggered by oxidative injury to cells.[84] Progressive accumulation of damaged proteins within tissues also has the capacity to activate p53. The balance of oxidant-antioxidant processes is governed by both the dietary provision of antioxidant nutrients and the expression of antioxidant enzyme proteins within cells. In *Drosophila* over-expression of the genes for key antioxidant enzymes increases lifespan and administration of antioxidants limits age-related lipid peroxidation in rats.[85,86] We have been studying the impact of prenatal protein restriction upon the oxidant-antioxidant balance in an ageing colony of rats. As shown in Figure 5, maternal protein restriction results in an increase in hepatic protein oxidation as early as 4 weeks of age, and this may be related to changes in substrate flux through the gamma-glutamyl cycle.[75,87-89]

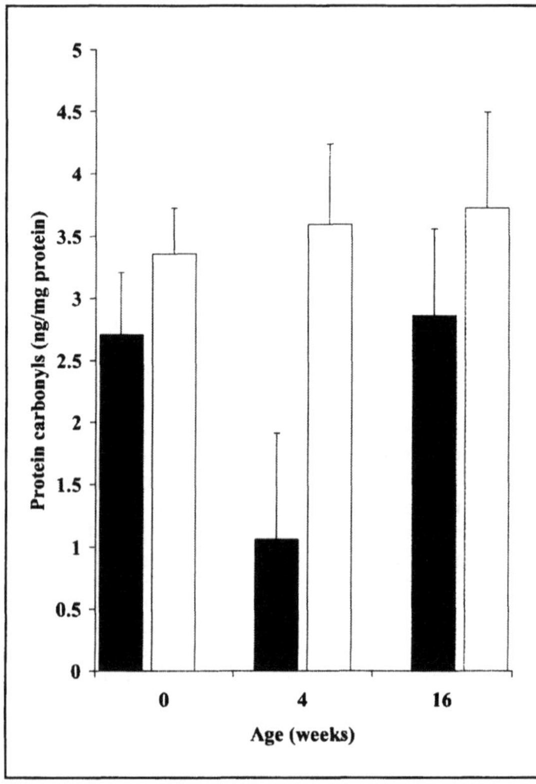

Figure 5. Oxidative injury in the livers of rats exposed to MLP diet in utero. Data are mean ± SEM. $P < 0.05$ at 4 weeks of age.

These studies of ageing and tissue degeneration in rodents provide an early first step towards exploring programming influences on the ageing process in human populations. As described throughout this book, epidemiological studies in man clearly show that chronic life-threatening diseases associated with ageing are related to fetal and infant growth indices, but it should also be noted that other less serious conditions that detract from quality of life (osteoporosis, cataracts, deafness, reduction in muscle mass) also appear to be related to intrauterine growth.[90,91]

Conclusions

Studies of animal models of nutritional programming demonstrate very clearly that most tissues and organ systems are vulnerable to the effects of undernutrition during critical periods of development. Restriction of food intake, or specific nutrient intake and the provision of excess nutrients have all been shown to have the capacity to modify long-term interactions between the genome and the environment and hence determine disease risk. These effects are appear to occur largely (though not wholly) irrespective of the nutrient manipulation applied, but do depend to a large extent on the timing of insults within the fetal and/or lactation period of the animal.

At the present time our knowledge of the nutrients within the human diet that may play a key role in determining disease risk is poor and extrapolation from studies of rats may be of limited value. The rodent models have however indicated that, in general, the long-term effects of a mild-to-moderate restriction of a specific nutrient are greater than the effects of an equivalent balanced restriction of food intake. This may relate to the metabolic consequences of processing other nutrients that are effectively in excess in the absence of essential cofactors. In the future, epidemiologists will need to focus their efforts upon detailed assessment of nutrition in pregnancy, with long-term follow-up of infants. This is the only way in which effective intervention and disease prevention strategies can begin to be formulated.

References

1. Huxley R, Neil A, Collins R. Unravelling the fetal origins hypothesis: Is there really an inverse association between birth weight and subsequent blood pressure? Lancet 2002; 360:659-665.
2. Kramer MS, Joseph KS. Enigma of fetal/infant-origins hypothesis. Lancet 1996; 348:1254-1255.
3. Lucas A, Fewtrell MS, Cole TJ. Fetal origins of adult disease-the hypothesis revisited. BMJ 1999; 319:245-9.
4. Langley-Evans SC. Fetal programming of cardiovascular function through exposure to maternal undernutrition. Proc Nutr Soc 2001; 60:505-513.
5. Hoet JJ, Hanson MA. Intrauterine nutrition: Its importance during critical periods for cardiovascular and endocrine development. J Physiol 1999; 514:617-27.
6. Persson E, Jansson T. Low birth weight is associated with elevated adult blood pressure in the chronically catheterized guinea-pig. Acta Physiol Scand 1992; 115:195-196.
7. Jansson T, Lambert GW. Effect of intrauterine growth restriction on blood pressure, glucose tolerance and sympathetic nervous system activity in the rat at 3-4 months of age. J Hypertens 1999; 17:1239-48.
8. Woodall SM, Johnston BM, Breier BH et al. Chronic maternal undernutrition in the rat leads to delayed postnatal growth and elevated blood pressure of offspring. Ped Res 1996; 40:438-443.
9. Vickers MH, Breier BH, Cutfield WS et al. Fetal origins of hyperphagia, obesity, and hypertension and postnatal amplification by hypercaloric nutrition. Am J Physiol Endocrinol Metab 2000; 279:E83-7.
10. Woodall SM, Breier BH, Johnston BM et al. A model of intrauterine growth retardation caused by chronic maternal undernutrition in the rat: Effects on the somatotrophic axis and postnatal growth. J Endocrinol 1996; 150:231-42.
11. Bertram CE, Hanson MA. Animal models and programming of the metabolic syndrome. Br Med Bull 2001; 60:103-121.
12. Fall CHD, Vijayakumar M, Barker DJP et al. Weight in infancy and prevalence of coronary heart disease in adult life. BMJ 1995; 310:17-19.
13. Curhan GC, Chertow GM, Willett WC et al. Birth weight and adult hypertension and obesity in women. Circulation 1996; 94:1310-1315.
14. Moore VM, Miller AG, Boulton TJC et al. Placental weight, birth measurements and blood pressure at age 8 years. Arch Dis Child 1996; 74:538-541.
15. Forsen T, Eriksson JG, Tuomilehto J et al. Growth in utero and during childhood among women who develop coronary heart disease: Longitudinal study. BMJ 1999; 319:1403-1407.
16. Langley-Evans SC. Nutritional programming and the development of hypertension. In: McCarty R, Blizzard DA, Chevalier RL, eds. Development of the Hypertensive Phenotype: Basic and Clinical Studies. Handbook of Hypertension. Vol. 19. Amsterdam: Elsevier, 1999:539-574.
17. Edwards CRW, Benediktsson R, Lindsay RS et al. Dysfunction of placental glucocorticoid barrier: Link between fetal environment and adult hypertension. Lancet 1993; 341:355-357.
18. Langley-Evans SC. Intrauterine programming of hypertension by glucocorticoids. Life Sci 1997; 60:1213-1221.
19. Langley-Evans SC, Phillips GJ, Benediktsson R et al. Protein intake in pregnancy, placental glucocorticoid metabolism and the programming of hypertension. Placenta 1996; 17:169-172.
20. Gardner DS, Jackson AA, Langley-Evans SC. Maintenance of maternal diet-induced hypertension in the rat is dependent upon glucocorticoids. Hypertension 1997; 30:1525-1530.
21. Young LE. Imprinting of genes and the Barker hypothesis. Twin Research 2001; 4:307-17.
22. Bird A. DNA methylation patterns and epigenetic memory. Genes Dev 2002; 16:6-21.
23. Petrie L, Duthie SJ, Rees WD et al. Serum concentration sof homocysteine are elevated during early pregnancy in rodent models of fetal programming. Br J Nutr 2002; 88:471-77.
24. Langley-Evans SC, Welham SJM, Jackson AA. Fetal exposure to a maternal low protein diet impairs nephrogenesis and promotes hypertension in the rat. Life Sci 1999; 64:965-974.
25. Snoeck A, Remacle C, Reussens B et al. Effect of a low protein diet during pregnancy on the fetal rat endocrine pancreas. Biol Neonate 1990; 57:107-118.

26. Bayol S, Jones D, Goldspink G et al. Influence of maternal nutrition on postnatal skeletal muscle growth. Arch Animal Breeding 2003; 46:158-159.
27. Sahajpal V, Ashton N. Renal function and AT1 receptor expression in young rats following intrauterine exposure to a maternal low protein diet. Clin Sci 2003; 104:607-614.
28. McMullen S, Gardner DS, Langley-Evans SC. Prenatal programming of angiotensin II type 2 receptor expression in the rat. Br J Nutr 2004; 91:133-140.
29. Vehaskari VM, Aviles DH, Manning J. Prenatal programming of adult hypertension in the rat. Kid Int 2001; 59:238-245.
30. Zimanyi MA, Bertram JF, Black JM. Nephron number in the offspring of rats fed a low protein diet during pregnancy. Image Anal Stereol 2000; 19:219-222.
31. Plagemann A, Harder T, Rake A et al. Hypothalamic nuclei are malformed in weanling offspring of low protein malnourished rat dams. J Nutr 2000; 130:2582-2590.
32. Holemans K, Gerber R, Meurrens K et al. Maternal food restriction in the second half of pregnancy affects vascular function but not blood pressure of rat female offspring. Br J Nutr 1999; 81:73-79.
33. Ozaki T, Nishina H, Hanson MA et al. Dietary restriction in pregnant rats causes gender-related hypertension and vascular dysfunction in offspring. J Physiol 2001; 530:141-152.
34. Kind KL, Simonetta G, Clifton PM et al. Effect of maternal feed restriction on blood pressure in the adult guinea pig. Exp Physiol 2002; 87:469-477.
35. Langley-Evans SC, Langley-Evans AJ, Marchand MC. Nutritional programming of blood pressure and renal morphology. Arch Physiol Biochem 2003; 111:8-16.
36. Langley-Evans SC, Gardner DS, Jackson AA. Association of disproportionate growth of fetal rats in late gestation with raised systolic blood pressure in later life. J Reprod Fert 1996; 106:307-312.
37. Langley SC, Jackson AA. Increased systolic blood pressure in adult rats induced by fetal exposure to maternal low protein diet. Clin Sci 1994; 86:217-222.
38. Langley-Evans SC, Gardner DS, Jackson AA. Evidence of programming of the hypothalamic-pituitary-adrenal axis by maternal protein restriction during pregnancy. J Nutr 1996; 126:1578-1585.
39. Langley-Evans SC, Phillips GJ, Jackson AA. In utero exposure to maternal low protein diets induces hypertension in weanling rats, independently of maternal blood pressure changes. Clin Nutr 1994; 13:319-324.
40. Langley-Evans SC, Jackson AA. Captopril normalises systolic blood pressure in rats with hypertension induced by fetal exposure to maternal low protein diets. Comp Biochem Physiol 1995; 110A:223-228.
41. Langley-Evans SC, Welham SJM, Sherman RC et al. Weanling rats exposed to maternal low protein diets during discrete periods of gestation exhibit differing severity of hypertension. Clin Sci 1996; 91:607-615.
42. Kwong WY, Wild AE, Roberts P et al. Maternal undernutrition during the preimplantation period of rat development causes blastocyst abnormalities and programming of postnatal hypertension. Development 2002; 127:4195-4202.
43. Mackenzie HS, Lawler EV, Brenner BM. Congenital oligonephropathy: The fetal flaw in essential hypertension? Kid Int 1996; 49(suppl 55):S30-S34.
44. Marchand MC, Langley-Evans SC. Intrauterine programming of nephron number: The fetal flaw revisited. J Nephrology 2001; 14:327-331.
45. Ray PE, Bruggeman LA, Horikoshi S et al. Angiotensin II stimulates human fetal mesangial cell proliferation and fibronectin biosynthesis by binding to AT1 receptors. Kid Int 1994; 45:177-184.
46. Siragy HM. Angiotensin receptor blockers: How important is selectivity? AJH 2002; 15:1006-1014.
47. Rees WD, Hay SM, Buchan V et al. The effects of maternal protein restriction on the growth of the rat fetus and its amino acid supply. Br J Nutr 1999; 81:243-250.
48. Jackson AA, Dunn RL, Marchand MC et al. Increased systolic blood pressure in rats induced by maternal low protein diet is reversed by dietary supplementation with glycine. Clin Sci 2002; 103:633-639.
49. Rees WD. Manipulating the sulfur amino acid content of the early diet and its implications for long-term health. Proc Nutr Soc 2002; 61:71-7.
50. Dunn RL, Burdge GC, Jackson AA. Folic acid reduces blood pressure in rat offspring from maternal low protein diet but increases blood pressure in offspring of the maternal control diet. Ped Res 2003; 53:2A.
51. Dance CS, Brawley L, Dunn RL et al. Folate supplementation of a protein restricted diet during pregnancy: Restoration of vascular dysfunction in small mesenteric arteries of female adult rat offspring. Ped Res 2003; 53:19A.
52. Langley-Evans SC. Intrauterine programming of hypertension: Nutrient interactions. Comp Biochem Physiol 1996; 114A:327-333.

53. Crowe C, Dandekar P, Fox M et al. The effects of anaemia on heart, placenta and body weight, and blood pressure in fetal and neonatal rats. J Physiol (London) 1995; 488:515-519.
54. Gambling L, Dunford S, Wallace DI et al. Iron deficiency during pregnancy affects postnatal blood pressure in the rat. J Physiol 2003; 552:603-10.
55. Battista M-C, Oligny LL, St-Louis J et al. Intrauterine growth restriction in rats is associated with hypertension and renal dysfunction in adulthood. Am J Physiol 2002; 283:E124-E131.
56. Bergel E, Belizan JM. A deficient maternal calcium intake during pregnancy increases blood pressure of the offspring in adult rats. Br J Obstet Gynaecol 2002; 109:540-545.
57. Holemans K, Gerber R, O'Brien-Coker I et al. Raised saturated-fat intake worsens vascular function in virgin and pregnant offspring of streptozocin-diabetic rats. Br J Nutr 2000; 84:285-296.
58. Khan IY, Taylor PD, Dekou V et al. Gender-linked hypertension in offspring of lard-fed pregnant rats. Hypertension 2003; 41:168-75.
59. Woods LL, Ingelfinger JR, Nyengaard JR et al. Maternal protein restriction suppresses the newborn renin-angiotensin system and programs adult hypertension in rats. Pediatr Res 2001; 49:460-7.
60. Tonkiss J, Trzcinska M, Galler JR et al. Prenatal malnutrition-induced changes in blood pressure - Dissociation of stress and nonstress responses using radiotelemetry. Hypertension 1998; 32:108-114.
61. Ozanne SE, Martensz ND, Petry CJ et al. Maternal low protein diet in rats programmes fatty acid desaturase activities in the offspring. Diabetologia 1998; 41:1337-42.
62. Merlet-Benichou C, Gilbert T, Muffat-Joly M et al. Intrauterine growth retardation leads to a permanent nephron deficit in the rat. Pediatr Nephrol 1994; 8:175-180.
63. Sparre T, Reusens B, Cherif H et al. Intrauterine programming of fetal islet gene expression in rats—effects of maternal protein restriction during gestation revealed by proteome analysis. Diabetologia 2003; 46:1497-511.
64. Langley-Evans SC. Critical differences between two low protein diet protocols in the programming of hypertension in the rat. Int J Food Sci Nutr 2000; 51:11-17.
65. Anguita RM, Sigulem DM, Sawaya AL. Intrauterine food restriction is associated with obesity in young rats. J Nutr 1993; 123:1421-1428.
66. Jones AP, Friedman MI. Obesity and adipocyte abnormalities in offspring of rats undernourished during pregnancy. Science 1982; 215:1518-1519.
67. Jones AP, Simson EL, Friedman MI. Gestational undernutrition and the development of obesity in rats. J Nutr 1983; 114:1482-1484.
68. Bellinger L, Lilley C, Langley-Evans SC. Prenatal exposure to a low protein diet programmes a preference for high fat foods in the rat. Ped Res 2003; 53:P603.
69. Daenzer M, Ortmann S, Klaus S et al. Prenatal high protein exposure decreases energy expenditure and increases adiposity in young rats. J Nutr 2002; 132:142-144.
70. McDade TW, Kuzawa CW, Adair LS et al. Prenatal and early postnatal environments are significant predictors of total immunoglobulin E concentration in Filipino adolescents. Clin Exp Allergy 2004; 34:44-50.
71. Benn CS, Jeppesen DL, Hasselbalch H et al. Thymus size and head circumference at birth and the development of allergic diseases. Clin Exp Allergy 2001; 31:1862-6.
72. Moore SE, Cole TJ, Collinson AC et al. Prenatal or early postnatal events predict infectious deaths in young adulthood in rural Africa. Int J Epidemiol 1999; 28:1088-95.
73. Langley-Evans SC, Buttery PJ, Wakelin D. Fetal exposure to a maternal low protein diet and the immune system. Proc Nutr Soc 2002; 61:121A.
74. Beach RS, Gershwin ME, Hurley LS. Gestational zinc deprivation in mice: Persistence of immunodeficiency for three generations. Science 1982; 218:469-471.
75. Langley SC, Seakins M, Grimble RF et al. The acute phase response of adult rats is altered by in utero exposure to maternal low protein diets. J Nutr 1994; 124:1588-1596.
76. Tappia PS, McCarthy HD, Langley-Evans SC et al. Prenatal nutritional adequacy and gender influence the ability of adult rats to produce interleukins 1, 6 and tumour necrosis alpha. Proc Nut Soc 1994; 53:182A.
77. Calder PC, Yaqoob P. The level of protein and type of fat in the diet of pregnant rats both affect lymphocyte function in the offspring. Nut Res 2000; 20:995-1005.
78. Langley-Evans SC, Wakelin D, Buttery PJ. Undernutrition during fetal life programmes immune function in the rat. Ped Res 2003; 53:P104.
79. Jennings BJ, Ozanne SE, Dorling MW et al. Early growth determines longevity in male rats and may be related to telomere shortening in the kidney. FEBS Lett 1999; 448:4-8.
80. Sayer AA, Dunn RL, Langley-Evans SC et al. Intrauterine exposure to a maternal low protein diet shortens lifespan in rats. Gerontology 2001; 47:9-14.
81. Ozanne SE, Hales CN. Lifespan: Catch-up growth and obesity in male mice. Nature 2004; 427:411-2.

82. Merry BJ. Effect of dietary restriction on lifespan. Rev Clin Gerontol 1991; 1:203-213.
83. Jennings BJ, Ozanne SE, Hales CN. Nutrition, oxidative damage, telomere shortening and cellular senescence: Individual or connected agents of aging? Mol Gen Metab 2000; 71:32-42.
84. Kaufmann JA, Bickford PC, Taglialatela G. Oxidative-stress-dependent up-regulation of Bcl-2 expression in the central nervous system of aged Fischer-344 rats. J Neurochem 2001; 76:1099-1108.
85. Sohal RS, Agarwal S, Orr WC. Simultaneous overexpression of copper-containing and zinc-containing superoxide dismutase and catalase retards age-related oxidative damage and increases metabolic potential in Drosophila melanogaster. J Biol Chem 1995; 270:15671-15674.
86. Arivazhagan P, Juliet P, Panneerselvam C. Effect of DL-alpha lipoic acid on the status of lipid peroxidation and antioxidants in aged rats. Pharm Res 2000; 41:299-303.
87. Langley-Evans SC, Sculley DV, McMullen S. Increased oxidative injury in the liver of newborn rats exposed to intrauterine undernutrition is associated with reduced activity of superoxide dismutase. Ped Res 2003; 53:P103.
88. Langley-Evans SC, Wood S, Jackson AA. Enzymes of the gamma-glutamyl cycle are programmed in utero by maternal nutrition. Ann Nutr Metab 1995; 39:28-35.
89. Langley-Evans SC, Phillips GJ, Jackson AA. Fetal exposure to a maternal low protein diet alters the susceptibility of the young adult rat to sulphur dioxide-induced lung injury. J Nutr 1997; 127:202-209.
90. Evans JR, Rauf A, Sayer AA et al. Age-related nuclear lens opacities are associated with reduced growth before 1 year of age. Inv.Opthal.Vis.Sci 1998; 39:1740-1744.
91. Sayer AA, Cooper C, Evans JR et al. Are rates of ageing determined in utero? Age and Ageing 1998; 27:579-583.
92. Dunn RL, Langley-Evans SC, Jackson AA et al. Hypertension in the mouse following intrauterine exposure to a low protein diet. Proc Nutr Soc 2001; 60:51A.

CHAPTER 9

Programming Hypertension—Animal Models: Causes and Mechanisms

Kate M. Denton,* Michelle M. Kett and Miodrag Dodic

Abstract

Hypertension can be programmed by experimental manipulation of the intrauterine environment. Studies to date suggest that, at least in some models, common pathways such as glucocorticoids or the renin-angiotensin system cause programming of arterial pressure. How mechanisms involved in controlling "normal" arterial pressure have been altered, remains a largely unanswered question, though the process may include the programming of the major organs and endocrine/neural systems involved in long-term blood pressure regulation. Clear evidence demonstrates a prominent role for the programming of the kidney in the development of hypertension. The major mechanisms examined to date include a reduced nephron endowment and alterations to the function of renal renin-angiotensin system. These studies do not preclude a role for other major cardiovascular organ systems (brain, vasculature, heart) in the programming of hypertension. Several studies have identified sex-specific differences in the programming of hypertension, which may relate to fetal sex-specific rates of placental gene expression and/or sex-specific timing of fetal development. Future studies should be directed towards examining the integrative control of blood pressure in prehypertensive animals to differentiate between the primary initiating programming events and events secondary to the development of hypertension. Understanding the mechanisms involved will be essential for devising preventative and/or treatment strategies.

Introduction

High blood pressure affects 20% of adults and is a major risk factor for cardiovascular diseases such as stroke, myocardial infarction, peripheral vascular disease and chronic renal failure.[1,2] In the majority of cases, the cause of the hypertension is unknown, with less than 10% of cases accounted for by secondary (i.e., renal artery stenosis, adrenal tumour) or genetic factors. Recently, attention has shifted to the idea that adult hypertension can be programmed in utero.[3] It is hypothesised that an adverse intrauterine environment during critical stages of development permanently alters, or 'programmes' the development of fetal tissues, which enables the fetus to survive, but with adverse consequences in postnatal life.[3] The mechanisms by which an altered intrauterine environment might exert these effects may involve epigenetic effects in the embryo/fetus (discussed elsewhere in this book, Chs. 6, 7).

Here we will briefly outline animal models of adverse intrauterine environments that have been demonstrated to lead to adult hypertension. However, our primary focus will be to explore, where evidence is available, the organs and physiological systems that may be affected and thus underlie the development of hypertension (Fig. 1). A clearer understanding of these

*Corresponding Author: Kate M. Denton—Department of Physiology, Monash University, Victoria, Australia 3800. Email: kate.denton@med.monash.edu.au

Early Life Origins of Health and Disease, edited by E. Marelyn Wintour and Julie A. Owens.

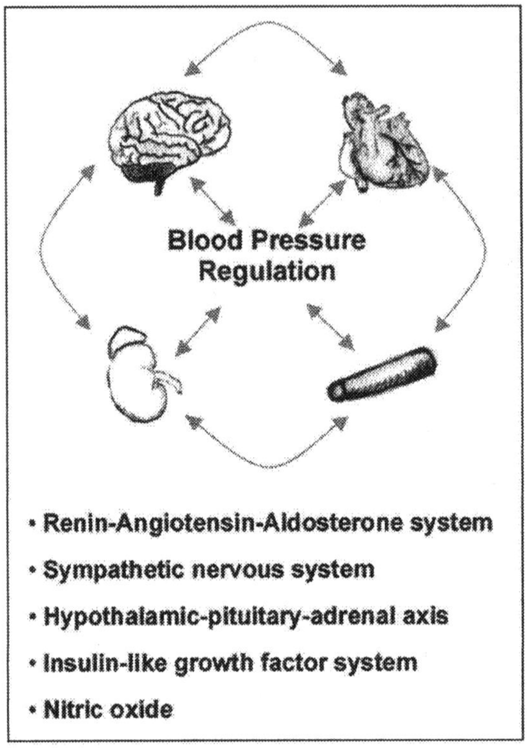

Figure 1. Schema suggesting possible targets for the programming of hypertension.

mechanisms and the delineation of possible common pathways, such as glucocorticoids for which there is strong evidence,[4,5] may ultimately lead to potential treatments or strategies for prevention of programmed hypertension.

Models of Arterial Pressure Programming

There is now compelling evidence to support the hypothesis that events occurring during fetal life can have life-long consequences for the health of the adult. The first models centred on producing low birth weight via maternal nutrient restriction, in line with the original hypothesis that low birth weight was associated with high blood pressure.[6,7] With an increasing understanding of the mechanisms of fetal programming, models have become more specific, examining the impact of micro-nutrient deficiencies, hormones, and conditions that are common in human pregnancy, such as anaemia and hypertension. Attention has also begun to focus on critical windows during development when different organs have a greater susceptibility to programming.[8]

Arterial Blood Pressure

In considering the topic of programming of blood pressure, it is timely to evaluate the methodologies associated with its measurement.[9,10] The most significant factor is whether blood pressure is measured directly, that is via an indwelling arterial catheter, or indirectly via tail-cuff. This is an important consideration for two reasons: (1) the degree of stress associated with each method and (2) the length of time over which the measurement is made varies considerably. Thus, whilst the tail-cuff method can provide reliable measurements and is the most frequently used method in rats (see Table 1), for reasons that will be discussed, direct measurement of blood pressure, preferably by telemetry, is the gold standard.

Table 1. Different methodologies used to measure blood pressure in models of "programmed" hypertension

Anaesthetised	Intra-arterial			Rat [20,28,29,31-33,143]
				Sheep [30,143]
Conscious	In-direct	Restrained	[1]Single period	Rat [16-21,24,25,42,43,45,77,78,110]
			Intermittent	Rat [22,23,27,36-38,44,56,60,105,106]
Conscious	Intra-arterial	Restrained	[2]"Awake"	Rat [35,92,130,131,139,140]
			[3]Single period	Rat [74,93,95]
			or intermittent	Rabbit [53]
			(recovered)	Sheep [15,40,80,128,135,144]
			24-hour	Sheep [76,90,113,129]
Conscious	Intra-arterial	Unrestrained	Intermittent	–
		(telemetry)	24-hour	Rat [11-14]

Conscious unrestrained 24-hour telemetry recording is the gold standard. [1]Single period (tail-cuff)—blood pressure measurements taken at a single time-point. [2]"Awake"—refers to those studies that did not allow the animals to fully recover from the surgical implantation of catheters before blood pressure was measured (minimum of 3 days). [3]Single period or intermittent (intra-arterial)—blood pressure recorded for a short period (hours) on a single day or over days to weeks, (minimum 3 days recovery from surgery). The [numbers] are references cited in this review using each method

Telemetry offers long-term, 24 hour intra-arterial blood pressure recording in conscious unrestrained, and thus unstressed, animals. Further, it allows for analysis of day versus night pressures and, due to the sensitivity of the technique, small differences in blood pressure can be detected (~5 mmHg). Unfortunately, the high cost associated with telemetry means that for long-term studies, such measurements are not always practicable. Indeed, whilst telemetry is common in the field of hypertension in general, only a handful of studies to date have used telemetry to examine the in utero programming of hypertension.[11-14] Chronic indwelling catheters in the carotid and femoral arteries also provide quality measures of blood pressure in rodents and larger animals. An advantage of this technique is that blood sampling can be performed in addition to measurement of blood pressure and supplemented with a venous catheter, allows for concomitant infusion of agents such as antihypertensives. However, the presence of externalised catheters does add an element of restraint stress. Further, the practicality of this technique for long-term studies is limited by the ability to maintain catheter patency for longer than a few weeks. Due to the invasive nature of both these techniques it is critical that the animals are given the appropriate length of time to recover from surgery before measurements of blood pressure begin.[9] Unfortunately, studies in animals equipped with indwelling catheters often fail to take full advantage of the benefits conferred, still only measuring blood pressure for short periods. For example, sheep from undernourished mothers demonstrated elevated morning blood pressure prior to, but not after feeding. The question remains as to whether these animals had significant hypertension or not; 24-hour recordings of blood pressure would have given a more accurate picture.[15]

Tail-cuff plethysmography allows blood pressure of rodents to be followed long-term within animal, but only measures single-time point systolic blood pressure accurately and the animals are subject to the stress of restraint. The element of restraint stress can be minimised in rats with training; however, as mice fail to show significant training, tail-cuff measurements in mice are questionable.[10] Another important consideration is that an adverse intrauterine environment may not alter blood pressure per se, but rather the blood pressure response to stresses such as restraint. Therefore if elevations in blood pressure detected by tail-cuff occur in the absence of left ventricular hypertrophy, an indicator of increased after-load with elevated blood pressure, blood pressure should be confirmed intra-arterially (preferably by telemetry) before concluding the stimulus has programmed hypertension. There are considerable disadvantages with measurements, such as those obtained by tail-cuff, that are of only a single time-point, or collection of direct, intra-arterial measurements for only short periods each day. These measurements, taken predominantly during the day, can be affected by factors such as feeding, the particular time of the day in relation to diurnal rhythm, and presence or absence of other activity in the room during recording. This is especially significant for rodents that are nocturnal as blood pressure during the day is considerably lower, and thus small differences in blood pressure that may be evident during night-time might not be detectable during the day-time. In the clinic it is the widely accepted practice not to make judgments about the significance of raised blood pressure until at least three measurements have been taken over a period of weeks, since anxiety, stress or discomfort can temporarily increase blood pressure of people who do not have significant hypertension. Yet, the majority of animal studies examining the impact of an adverse intrauterine environment on adult blood pressure, particularly those in rodent models, utilised tail-cuff plethysmography to determine blood pressure often only on a single day (Table 1).

Tonkiss and colleagues, who used telemetry to measure blood pressure in offspring of dams malnourished during pregnancy, presented a telling example of these drawbacks.[14] Previous studies in this model, based on indirect tail-cuff blood pressure measurements, demonstrated increases in systolic blood pressure of greater than 20 mmHg in the offspring of malnourished rats.[16-27] However, Tonkiss et al demonstrated a much more modest increase in blood pressure (+4 mmHg in diastolic pressure during the night) and provided evidence that the responsiveness to stress was augmented in prenatally malnourished rats.[14] Indeed, this study strongly suggests that the stress associated with the tail-cuff procedure, contributed to the large elevations in blood pressure seen previously in this model. However, these differences may also reflect the importance of protein contect and overall composition of a diet to programming of hypertension (see Chapter by Langley-Evans).

Finally, whilst differences in conscious blood pressure between animal groups can be reflected in anaesthetised measurements,[28-33] albeit at lower pressures in general, anaesthetised blood pressure is a poor indicator of conscious blood pressure since anaesthetic depth can be arbitrarily set. Thus the limitations of each technique with each animal model must be taken into consideration to prevent false positives and false negatives in the hypertensive programming effect of particular intrauterine stimuli.

Nutrition

Maternal dietary manipulation has been demonstrated in many animal studies to programme arterial pressure. Perturbations such as maternal under-nutrition (total calorie), restriction in specific dietary components (protein, vitamins, minerals), or restricting placental function (decreased uterine blood flow reducing both nutrient and oxygen availability) lead to elevated blood pressure in progeny across many species (see Chapter by Langley-Evans). In models of under-nutrition, it has been suggested that the programming of hypertension is mediated by glucocorticoid-induced endocrine changes.[5,8,34] Over-nutrition (lard, sodium) has also been reported to programme hypertension.[11,35,36] In some cases it is apparent that maternal diets both low or high in a particular nutrient (calcium,[37] sodium[35,38]) can programme adult hypertension. It is interesting to speculate whether these nutrients act by stimulation or suppression of the same pathway or whether they are acting independently via alternate mechanisms. Importantly, increasing evidence demonstrates that hypertension can occur without impaired fetal growth,[21,39] conversely intrauterine growth restriction does not always result in high adult blood pressure.[40]

Anaemia during Pregnancy

Of particular clinical import are studies examining the influence of diets low in iron. A physiological drop in haemagloblin (to ~100 g/l) occurs in normal pregnancy, due to the increase in plasma volume. However, it has been shown that iron deficiency (70-100 g/l) occurs in ~20% of pregnancies in 'first-world' countries, and up to 75% of pregnancies in developing countries.[41] Three groups have now demonstrated that iron deficiency induced prior to, and continued throughout, pregnancy in rats leads to intrauterine growth restriction and elevated arterial pressures in the offspring.[42-45] By cross-fostering all pups onto to control fed dams at birth, Gambling et al[44] confirmed that the elevated blood pressure was the result of the iron deficiency in utero and not due to continued iron deficiency during lactation. Interestingly, these elevated adult pressures were preceded by relative hypotension in the early post-weaning period, particularly in females.[42,44] The mechanisms by which intrauterine iron deficiency translates to adult hypertension are as yet unclear, however a reduced nephron endowment has been implicated.[45] It is yet to be determined whether the programming of hypertension by iron deficiency in utero is independent of a generalised effect on intrauterine growth retardation.

Hypertension during Pregnancy

Another condition common during pregnancy is hypertension.[46] It has been predicted that the incidence of chronic hypertension will increase from 1 to 5 in 100 pregnancies over the next decade.[47] This is due to the shift to an older child bearing age in women and the increased risk of hypertension in this older population.[48] However, few studies have followed the children of mothers with hypertension into adulthood,[49-51] though both low-birth weight and macrosomic babies have been linked with mild maternal hypertension.[50,52] Thus, the question of whether chronic hypertension during pregnancy exposes the fetus to an increased risk of developing hypertension and cardiovascular disease later in life is an important one.

Several animal studies have examined the influence of chronic hypertension on fetal development and adult blood pressure. Denton et al[53] published the first study to demonstrate that maternal secondary hypertension could programme hypertension in offspring. In a rabbit model of chronic maternal hypertension, induced using a two-kidney, one-wrapped model of perinephritic hypertension, it was demonstrated that offspring were hypertensive as adults (Fig. 2). The increase in blood pressure only occurred in adult female offspring, though the variation in

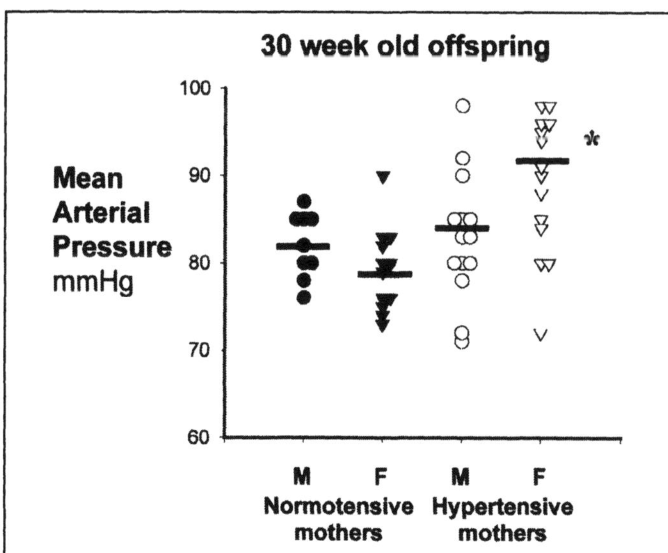

Figure 2. Conscious mean arterial pressure measured at 30 weeks of age in offspring of hypertensive and normotensive rabbit mothers. Individual data presented for male (M, circles) and female (F, triangles). The bars represent the group average. Hypertensive mothers: open symbols; n = 6 mothers; 14 male, 14 female offspring. Normotensive mothers: solid symbols; n = 6; mothers; 9 male, 12 female offspring. * $P < 0.05$ compared to normotensive control.

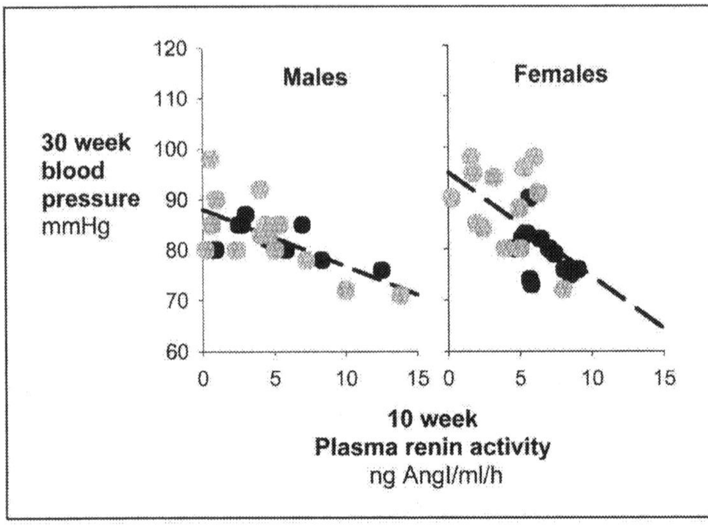

Figure 3. Relationship between plasma renin activity at 10 weeks of age (adolescent) and arterial blood pressure at 30 weeks of age (adult) measured in conscious individual male and female rabbits born of hypertensive (grey) or normotensive (black) mothers. Male offspring R^2 = 0.48, $P < 0.001$; Female offspring $R^2 = 0.31$, $P < 0.005$.

blood pressure in male offspring was increased (Fig. 2). The renin-angiotensin system (RAS) was implicated in this model with significantly lower plasma renin activities in the female offspring of hypertensive mothers at 10 weeks of age (adolescence), prior to any rise in blood pressure. Indeed, plasma renin activity at 10 weeks of age was found to directly correlate with adult blood pressure at thirty weeks of age (Fig. 3). It has been suggested that low plasma renin activity may reflect a reduction in nephron number, which is linked to the development of hypertension.[54,55]

In agreement with the study by Denton and collegues,[53] male offspring of one-kidney, one-clip hypertensive dams also showed no rise in arterial pressure.[56] Interestingly, these male pups were found to be more susceptible to DOCA-salt treatment.[56] These studies suggest that chronic hypertension during pregnancy differentially influences programming in the sexes, an effect documented previously in other models.[38,57-59] In contrast, no effect of increased maternal blood pressure on offspring was demonstrated when blood pressure was increased by central administration of aldosterone,[60] leading to the suggestion that it may not be maternal arterial pressure per se that is responsible for the programming of hypertension in offspring. There is little doubt that the changes in the maternal environment during hypertension, of whatever cause, are complex and thus the stimuli impacting on the fetus may be multifactorial.

In the rabbit model of maternal hypertension discussed above, there a number of possible maternal stimuli that might affect fetal development. Not only was arterial pressure increased but plasma renin activity was also elevated,[53] suggesting that both angiotensin II (AngII) and aldosterone levels in the mothers were elevated during pregnancy.[61,62] Aldosterone can cross the placenta and may possibly have a direct effect on fetal development.[63] Maintenance and growth of the placenta is essential for the normal growth and wellbeing of the developing fetus. The uteroplacental circulation has a local renin-angiotensin system (RAS) that plays important roles in placental angiogenesis and in modulating placental production of cytokines, growth factors and vasoactive substances, which also influence fetal development.[64] Chronic infusion of AngII to pregnant rabbits[65] and ewes[66] has been shown to decrease uterine blood flow and evidence suggests that uteroplacental perfusion is reduced in humans and animal models with chronic hypertension.[49,67,68] Normally, during pregnancy the uterine artery is particularly insensitive to AngII due to the predominance of angiotensin type 2 receptors (AT_2R),[69] however, uterine artery AT_2R density decreases with chronic AngII.[66] Thus, uterine blood flow may be reduced via this mechanism in pregnancy when maternal plasma AngII levels are increased, affecting placental nutrient transfer. In cultured human placental cells AngII has been shown to decrease 11-beta-hydroxysteriod dehydrogenase type-2 (11β-HSD_2).[70] If 11β-HSD_2 is decreased in mild chronic hypertension, maternal glucocorticoids may cross the placenta and influence organs/systems in the fetus. Another possible contributor to fetal programming of hypertension in this model is maternal renal function which may be compromised,[61,62,71] possibly altering maternal plasma levels of sodium, potassium or urea which may influence fetal development.[72]

Glucocorticoids

Glucocorticoids are potent regulators of fetal growth and development. Mechanisms that tightly regulate fetal glucocorticoid exposure are of considerable importance, as certain organs (kidney, brain) are adversely affected by excess glucocorticoids. Placental 11β-HSD_2 reduces trans-placental passage of maternal glucocorticoids to the fetus, thus protecting the fetus from the deleterious effects of maternal glucocorticoids. Many studies have observed the effect of glucocorticoids to programme high blood pressure in sheep and rat models using either prenatal exposure to stress (e.g., restraint) or infusions of cortisol, corticosterone, ACTH or dexamethasone.[34,73] Prenatal glucocorticoid exposure, induced by blocking placental inactivation of endogenous glucocorticoids, also leads to high blood pressure in adult rats.[19,74] Importantly, it has been shown reproducibly in sheep, that elevated arterial pressure in adults can be programmed in both female and male adult offspring by as little as 2 days of exposure to glucocorticoids at days 26-28 of the 150 day gestation.[75,76] A similar critical window, during the earliest stages of metanephric development, has also been demonstrated in rats.[39,77-79] Glucocorticoid exposure at this critical stage in kidney development also causes high blood pressure in adult rat progeny without affecting birth weight.[39,78] In contrast, glucocorticoid treatment late in gestation does not result in subsequent hypertension.[78,80]

Possible Mechanisms Leading to Adult Hypertension

The cardiovascular system regulates blood pressure to maintain an adequate perfusion to meet the needs of each tissue (Figs. 1, 4). "Normal" blood pressure is regulated by a number of organs and physiological systems, exerting both short (reflex) and long-term effects. Mechanisms integrating the control of arterial blood pressure are outlined and possible adaptations in the development of components of the cardiovascular system resulting in alterations in function and the programming of hypertension have been summarised in Figure 4. A caveat that should be considered when examining the mechanisms underlying the programming of hypertension is whether such changes are present before the onset of hypertension or occur as a consequence of the hypertension. Thus ideally, the mechanisms controlling blood pressure should be examined prior to the establishment of chronic hypertension since compensatory mechanisms might confound interpretation of the results once hypertension has developed.

Long-term blood pressure regulation is inextricably linked to renal excretory function,[81,82] and there is also strong evidence linking the renal actions of the RAS and the sympathetic nervous system to adult hypertension.[83-85] However, the initial stimulus for hypertension to develop need not originate in the kidney. Thus a stimulus from other organs or systems involved in cardiovascular homeostasis, such as altered central sympathetic out-flow, myocardial function or vascular reactivity, may trigger a shift in renal function and an increase in arterial pressure. Thus while the kidney has received the bulk of attention, these other organs and systems need also to be considered in the effort to determine the mechanisms behind developmental programming of hypertension (Figs. 1, 4).

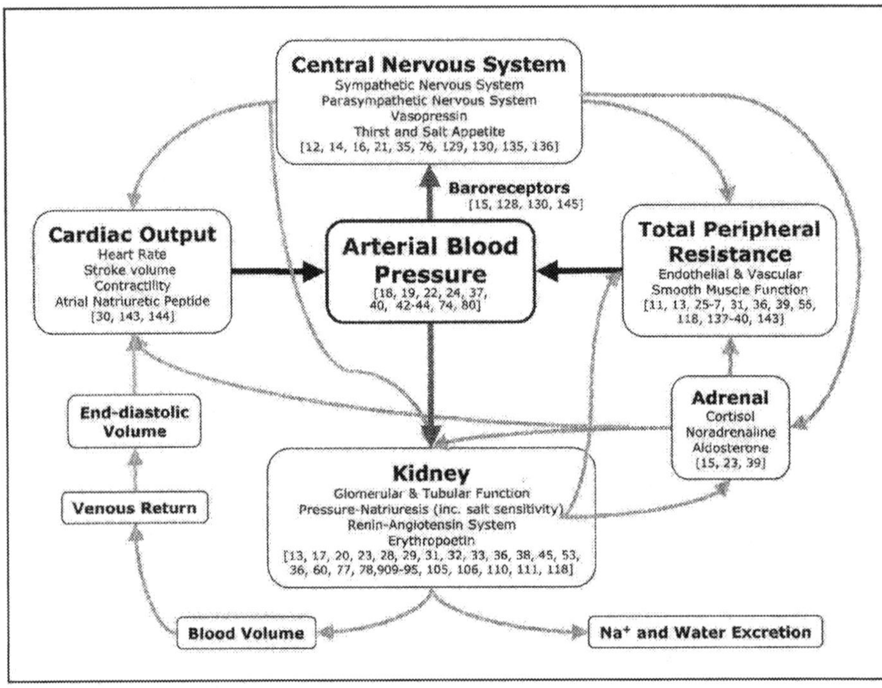

Figure 4. Diagram showing the basic mechanisms controlling blood pressure. Arterial pressure is the product of cardiac output and total peripheral resistance, changes in blood pressure are sensed primarily by the central nervous system and kidneys (red arrows) and mechanisms are activated (grey arrows) that restore blood pressure to "normal". Programming in utero may affect blood pressure by altering the cardiovascular system at any point in this loop. Possible mechanisms are suggested in italics and studies that have examined some of these are cited.

Kidney

The kidney, as stated above, plays a very important role in the control of blood pressure due to its influence on salt and water excretion and thus plasma volume.[81,82] Considerable attention has therefore been directed towards the kidney to look for changes in fetal kidney development and alterations in adult renal structure and function. Disruption of kidney development, due to a programming effect, may permanently alter normal function. Compensatory mechanisms may ensue resulting in hypertension. Where available, the evidence implicating these mechanisms as possible contributors to programming adult hypertension are discussed.

Reduced Nephron Number in Models of Programmed Hypertension

Investigations have focused on nephron development due to the hypothesis advanced by Brenner and colleagues that nephron endowment at birth is inversely related to the risk of developing essential hypertension in later life.[54,55,86,87] In brief, it is postulated that a low nephron number at birth signifies a reduction in total kidney filtration surface area, which is not adequate to meet the demands of the growing animal, with resultant sodium retention.[54,55] Further to this, compensatory mechanisms cause arterial and glomerular hypertension leading to hyperfiltration, a vicious cycle then ensues as the increased work load placed on each nephron causes glomerular sclerosis and further loss of nephron function.[54,55] Reduced nephron number has been documented in a number of animal models of programmed hypertension, including food restriction, uterine artery ligation, low protein diet, iron deficiency, and glucocorticoid treatment (see ref. 86). Glomerular number has yet to be measured in a number of models of programmed hypertension (chronic hypertension[53]) and blood pressure has yet to be measured in some models known to programme reduced nephron endowment (vitamin A,[88] hyperglycemia[89]). Studies examining the mechanisms whereby a reduction in nephron number may be programmed in utero are discussed elsewhere in greater detail (see Moritz & Cullen-McEwen).

Studies in models with reduced nephron number in which glomerular filtration rate has been measured all show evidence that the remaining nephrons are hyperfiltering.[17,20,29,33,78,90-92] Altered expression of components of the RAS have also been documented in the adult as well as the fetus in these models, suggesting that not only has the developmental role of the RAS been altered, but that the functionality of the system in the adult may have been reset.[17,76,90,93-95] It has been speculated that failure to suppress intrarenal AngII activity during chronic salt loading may lead to salt-sensitive hypertension.[96] Certainly, salt-sensitive hypertension has been demonstrated in offspring of mothers fed a low protein diet, which had previously been shown to have reduced renin expression and indeed fewer nephrons.[93,95]

As a result of this hypothesis, attention has also centred on alterations in the expression of components of the RAS due to this systems prominent role in renal development and its importance in regulating blood pressure in the adult.[81,97,98] Impetus for this direction of research has also been fuelled by the clinical correlate that growth retarded infants, which are prone to later hypertension,[99] have particularly small kidneys, have elevated cord blood renin and AngII concentrations[100,101] as well as elevated renin gene expression in the kidney,[102] suggesting that intra-renal RAS activity may be elevated.

When taken in context with other studies in humans or in experimental animal models, in which hypertension resulted when nephrogenesis was impaired, it is highly suggestive that a kidney abnormality is an essential part of the etiology of the subsequent hypertension.[103,104] However, there is also evidence to suggest that reduced nephron endowment and hypertension may be coincident.[18,87,105] A study has shown that dietary supplements given in combination with a low protein diet can prevent low nephron number without affecting the development of hypertension in the adult.[18,87] Furthermore, reduced nephron number has been documented in the absence of hypertension[105] and programmed hypertension has been demonstrated in the absence of changes in nephron number.[106] Perhaps programmed hypertension is more than reduced nephron number, and compensatory changes in tubular function and/or renal hormonal systems must occur concomitantly for hypertension to develop. These are important questions, awaiting confirmation in future studies.

Tubular Epithelial Sodium Co-Transporters and Hypertension

Programming of epithelial sodium transporters in the renal tubules offer another mechanism by which sodium retention may cause adult hypertension. These transporters are localised to specific segments of the nephron and mediate sodium entry across the apical membrane. These include the Na/H exchanger of the proximal tubule, the Na/K/2Cl co-transporter of the thick ascending limb of Henle, the Na/Cl co-transporter of the distal convoluted tubule, and the α, β, γ-subunits of the epithelial sodium channel (ENaC) of the distal tubule and the collecting duct.[107] Whilst the bulk of the reabsorption of sodium is carried out in the proximal tubule of the nephron, the fine control of sodium reabsorption is carried out in the distal nephron and collecting duct.[108,109] Gene-targeted studies in mice have lead to the suggestion that it is in these later segments of the tubule, downstream to the macula densa, in which sodium delivery is not monitored, that changes in sodium transport play a key role in controlling sodium balance and blood pressure.[108,109] For example, while an increase in the Na/H exchanger in the proximal tubule can be compensated for by other later segments of the tubule, an increase in ENaC activity in the collecting duct, as found in Liddle's syndrome, results in excess sodium reabsorption and hypertension.[107-109]

Few studies have directly examined renal sodium transporters in models of programmed hypertension. One study in 4 week old rats exposed to a low protein diet throughout the second half of gestation, demonstrated up-regulation (at both the mRNA and protein level) of Na/K/2Cl cotransporter of the thick ascending limb of Henle (302% compared to controls) and the Na/Cl cotransporter of the distal convoluted tubule (160% compared to controls).[110] Thus in this model, before the hypertension becomes manifest, the fetal kidney was programmed to inappropriately retain sodium, a finding consistent with the hypothesis that sodium retention might directly contribute to the development of hypertension. Subsequently, the same authors showed that sodium transporters were not down-regulated after hypertension became manifest at 8 weeks of age.[110] This finding is particularly important since it is known that down-regulation of the Na/Cl cotransporter of the distal convoluted tubule is an important component of the pressure-natriuresis response, the crucial mechanism in long-term blood pressure control.[111] Another study, examining the effects of maternal hypercholesterolemia on offspring, demonstrated an increase in Na^+/K^+-ATPase activity in the outer medulla associated with reduced creatinine clearance (estimate of glomerular filtration rate) but not hypertension, though blood pressure was measured in anaesthetised animals and needs to be confirmed.[33] Further studies are required to examine the possibility that prenatal programming of renal epithelial sodium cotransporters can lead to hypertension in the adult.

Other Renal Mechanisms

There are other renal mechanisms controlling blood pressure that should also be considered when examining possible mechanisms leading to programming of adult hypertension. For example, it has been proposed that hypertension may be caused by structural changes that narrow intrarenal blood vessels, increasing preglomerular vascular resistance and the aortic-glomerular capillary pressure gradient.[112] Such a situation present in spontaneously hypertensive rats, and analogous to renal artery stenosis, would result in a cascade of events, including activation of the RAS, leading to hypertension.[112] Whilst pro-hypertensive vascular structural changes have not been investigated specifically in models of programmed hypertension, in sheep exposed to dexamethasone during early gestation accumulation of collagen in the tubular interstitium and peri-adventitia of renal cortical vessels has been demonstrated.[113]

Programming of the sympathetic nervous system has been demonstrated[114] and there is strong evidence implicating increased renal sympathetic activity in the pathogenesis of essential hypertension.[84] Developmentally, growth of the renal nerves is closely linked to the fetal RAS, specifically the timing of renal innervation of the vessels is concomitant with the regression of renin expression along the vasculature.[115] Nerve growth factors are expressed in the fetal kidney and are inducers of differentiation and survival of nerves,[116] thus altered expression of these factors may lead to hyper-innervation of the renal vasculature, an affect which is pro-hypertensive.[84] A few studies have demonstrated alterations in sympathetic function in models of sub-optimal maternal environments. In chick embryos, chronic moderate hypoxia leads to hyper-innervation of the arterial vasculature.[117] In a model of uterine artery ligation increased sympathetic nervous system activity was observed in female rats at 3 months of age, though this was not associated with hypertension.[12] In a model of prenatal stress in rats, adrenoreceptor responses were altered in renal, but not femoral, mesenteric or saphenous arteries.[118] Additional tests led to the conclusion that the enhanced responsiveness to phenylephrine was due to alterations in signal transduction not increased nerve or receptor densities.[118]

No one to date has examined the intrinsic renal mechanisms that maintain glomerular filtration rate, the first step in sodium excretion, constant: tubulo-glomerular feedback, the myogenic response or the phenomenon of pressure-natriuresis. Resetting of these mechanisms due to alterations in hormone sensitivity be it due to increased receptor density, increased hormone availability or up-regulation of second messenger systems has yet to be studied. Interestingly, however, human data has suggested that the pressure-natriuresis relationship is influenced by birth weight.[119] In the future, attention should also focus on sex-related programming effects on renal structure and function given the striking differences in renal function previously reported for healthy males and females.[59,120,121]

Brain

The central nervous system also plays a major role in maintaining body fluid homeostasis via sympathetic stimulation, vasopressin release and increase in salt and water appetite. For example, the hypothalamus is involved in fluid balance through salt and water intake and control of sympathetic drive,[122,123] while the medulla oblongata affects cardiovascular function, mainly through the control of peripheral sympathetic drive, including baroreflexes.[123,124] The lamina terminalis is situated in the anterior wall of the third ventricle and consists of the median preoptic nucleus and the circumventricular organs; the subfornical organ and the organum vasculosum. This region of the brain has a crucial role in osmoregulatory vasopressin secretion and thirst.[125] There is a local brain RAS and hyperactivity of this system has been implicated in the development and maintenance of hypertension. Confirmation of the role of the central RAS and its effects on blood pressure and fluid balance has been obtained using transgenic mouse models that selectively over-express components of the RAS within the brain.[126,127] Other signalling systems (i.e., noradrenergic or glutaminergic) in brain regions involved in cardiovascular control may also be implicated in the fetal programming of adult hypertension, but have yet to be considered (see ref. 123).

Evidence of Altered Brain RAS in Models of Programmed Hypertension

To date, there are only a few studies in the literature that suggest a link between altered brain RAS, as a result of exposure to a sub-optimal intrauterine environment, and adult hypertension.[35,128,129] Studies have demonstrated an up-regulation of AT_1 receptors in the medulla oblongata and higher expression of angiotensinogen in the hypothalamus of late gestational fetuses, previously exposed to dexamethasone at the end of the first month of pregnancy.[129] This increase in AT_1 receptors expression of the medulla oblongata persisted in adult sheep measured at 7 years of age, when high blood pressure was clearly evident.[129] A recent study of 1 year-old lambs exposed to maternal under-nutrition (50% of daily intake) from day 1-30 of gestation, demonstrated blunted baroreflex sensitivity during AngII infusion.[128] Similarly, rats of low-protein fed mothers had increased blood pressure and demonstrated altered baroreflex function.[130] The hypertension of these offspring was significantly attenuated by intracerebroventicular administration of an AngII antagonist.[130] Further, Swenson et al showed that the hypertension of 30-day old rats subjected to a high-salt diet throughout gestation and the post-natal period, was partly due to increased brain AT_1 receptor activation.[35] Taken together, these studies suggest that increased AngII action within cardiovascular control centres in the brain contribute to programmed hypertension.[35,128-130] it is important to bear in mind that resetting of the baroreflex is found, commonly, as a consequence of developed hypertension.[131] However, in some strains of rats (spontaneously hypertensive rats, Dahl salt-sensitive rats) abnormal baroreflex function precedes the development of hypertension and may very well be the cause rather than the consequence of hypertension.[132,133]

Research also supports the hypothesis that salt appetite and thirst can be programmed in utero (see ref. 134). In a study in sheep, maternal dehydration during late gestation, has been demonstrated to programme hypertension.[135] Further, this study demonstrated that the offspring of water-restricted ewes had increased plasma osmolality, hematocrit and threshold for AVP secretion.[135] In another study, in which extracellular dehydration and exaggerated sodium appetite was produced in pregnant rats by polyethylene glycol treatment, salt appetite of offspring was increased.[136]

Heart and Vasculature

Adaptations in the cardiovascular system are linked to the development and maintenance of systemic hypertension. Alterations in myocardial, conduit and resistance artery geometry and reactivity will have a direct impact on cardiac output and total peripheral resistance, the primary determinants of arterial pressure (see Fig. 4). However, little is known of the role of the cardiovascular system in the translation of intrautero insults into a chronic elevation of arterial pressure in the adult. Of the limited studies available, vascular dysfunction of both the conduit and/or resistance vasculature is a common feature, though the responses vary from study to study due to differences in vessel size (conduit or resistance), vascular bed, gender, intrauterine insult, and age of offspring at time of examination.

Vasculature

The primary vascular defect identified in studies to date, appears to be an impaired endothelium-dependent relaxation and this has been demonstrated in offspring with hypertension induced by various adverse intra-uterine events including undernutrition,[26,137] protein restriction,[25] high fat intake[11,138] and placental insufficiency.[139,140] The precise nature of the defect underlying the reduced endothelium-dependent dilation is of considerable conjecture but includes impaired synthesis of NO,[26,139] and/or impaired response of the vascular smooth muscle cells to NO,[27,140] Whilst not a consistent finding, some studies have demonstrated an increased responsiveness to vasoconstrictors, though the effect is not universal to constrictors in general, even within the same study.[11,26,137-139] This increased responsiveness is likely the result of impaired buffering by endothelial factors,[139] however the contribution of increased numbers of specific receptor types mediating vasoconstriction cannot be rule out (see Poston).

Heart

Intrauterine insults such as anaemia and hypoxemia have been shown to have significant effects on the fetal heart.[30,141,142] However, few studies have examined the consequences of these stimuli during fetal life on the adult. In one interesting study, in a model of perinatal anaemia, evidence of coronary vascular remodelling has been described in adult sheep, in which maximal coronary conductance and reserve increased, providing a physiological advantage.[30,143] Whilst maternal dexamethasone exposure led to hypertension and increased cardiac output in 7 year old offspring, associated ventricular hypertrophy and reduced cardiac functional reserve, these changes are likely due to secondary effects of the hypertension.[144,145] Further studies performing detailed analysis of the structure and function of hearts in juvenile and adult offspring, subjected to an adverse intrauterine environment, are required.

Conclusions

Strong evidence in both human and in animal studies supports the hypothesis that hypertension can be programmed in utero. Future studies should encompass the following: (1) Prehypertensive animals should be studied to differentiate between the primary initiating programming events and events secondary to the consequent development of hypertension. (2) Best practice methods should be employed to determine arterial blood pressure; single time-point measures are open to misinterpretation particularly in young restrained animals. (3) It is unlikely that the interventions (i.e., under-nutrition, glucocorticoids) used to programme hypertension affect single organs but will rather affect multiple organs or systems (i.e., programming of RAS may alter brain, heart and kidney function), unless adverse stimuli are restricted to narrow windows in the timing of development. Thus, the reductionist approach of examining single organs or systems will not provide a complete picture of the physiological adaptations that have taken place. (4) Furthermore, accumulating evidence demonstrating sexually dimorphic programming in response to an adverse maternal environment highlights the need to consider male and female offspring separately. Understanding the mechanisms involved in the programming of hypertension will be essential for devising preventative and/or treatment strategies.

References

1. Burt VL, Whelton P, Roccella EJ et al. Prevalence of hypertension in the US adult population. Results from the Third National Health and Nutrition Examination Survey, 1988-1991. Hypertension 1995; 25(3):305-313.
2. MacMahon S, Peto R, Cutler J et al. Blood pressure, stroke, and coronary heart disease. Part 1, Prolonged differences in blood pressure: Prospective observational studies corrected for the regression dilution bias. Lancet 1990; 335(8692):765-774.
3. Barker DJ. The fetal origins of adult hypertension. J Hypertens Suppl 1992; 10(7):S39-44.
4. Dodic M, Peers A, Coghlan JP et al. Can excess glucocorticoid, predispose to cardiovascular and metabolic disease in middle age? Trends Endocrinol Metab 1999; 10(3):86-91.
5. Fowden AL, Forhead AJ. Endocrine mechanisms of intrauterine programming. Reproduction 2004; 127(5):515-526.
6. Barker DJ, Osmond C, Golding J et al. Growth in utero, blood pressure in childhood and adult life, and mortality from cardiovascular disease. BMJ 1989; 298(6673):564-567.
7. Gennser G, Rymark P, Isberg PE. Low birth weight and risk of high blood pressure in adulthood. BMJ 1988; 296(6635):1498-1500.
8. Wintour EM, Johnson K, Koukoulas I et al. Programming the cardiovascular system, kidney and the brain—a review. Placenta 2003; 24(Suppl A):S65-71.
9. Pelaez LI, Manriquez MC, Nath KA et al. Low-dose angiotensin II enhances pressor responses without causing sustained hypertension. Hypertension 2003; 42(4):798-801.
10. Lorenz JN. A practical guide to evaluating cardiovascular, renal, and pulmonary function in mice. Am J Physiol Regul Integr Comp Physiol 2002; 282(6):R1565-1582.
11. Khan IY, Taylor PD, Dekou V et al. Gender-linked hypertension in offspring of lard-fed pregnant rats. Hypertension 2003; 41(1):168-175.
12. Jansson T, Lambert GW. Effect of intrauterine growth restriction on blood pressure, glucose tolerance and sympathetic nervous system activity in the rat at 3-4 months of age. J Hypertens 1999; 17(9):1239-1248.
13. Contreras RJ, Wong DL, Henderson R et al. High dietary NaCl early in development enhances mean arterial pressure of adult rats. Physiol Behav 2000; 71(1-2):173-181.
14. Tonkiss J, Trzcinska M, Galler JR et al. Prenatal malnutrition-induced changes in blood pressure: Dissociation of stress and nonstress responses using radiotelemetry. Hypertension 1998; 32(1):108-114.
15. Gopalakrishnan GS, Gardner DS, Rhind SM et al. Programming of adult cardiovascular function after early maternal undernutrition in sheep. Am J Physiol Regul Integr Comp Physiol 2004; 287(1):R12-20.
16. Langley-Evans SC. Hypertension induced by foetal exposure to a maternal low-protein diet, in the rat, is prevented by pharmacological blockade of maternal glucocorticoid synthesis. J Hypertens 1997; 15(5):537-544.
17. Nwagwu MO, Cook A, Langley-Evans SC. Evidence of progressive deterioration of renal function in rats exposed to a maternal low-protein diet in utero. Br J Nutr 2000; 83(1):79-85.
18. Jackson AA, Dunn RL, Marchand MC et al. Increased systolic blood pressure in rats induced by a maternal low-protein diet is reversed by dietary supplementation with glycine. Clin Sci (Lond) 2002; 103(6):633-639.
19. Langley-Evans SC. Maternal carbenoxolone treatment lowers birthweight and induces hypertension in the offspring of rats fed a protein-replete diet. Clin Sci (Colch) 1997; 93(5):423-429.
20. Langley-Evans SC, Welham SJ, Jackson AA. Fetal exposure to a maternal low protein diet impairs nephrogenesis and promotes hypertension in the rat. Life Sci 1999; 64(11):965-974.
21. Langley-Evans SC, Welham SJ, Sherman RC et al. Weanling rats exposed to maternal low-protein diets during discrete periods of gestation exhibit differing severity of hypertension. Clin Sci (Colch) 1996; 91(5):607-615.
22. Langley SC, Jackson AA. Increased systolic blood pressure in adult rats induced by fetal exposure to maternal low protein diets. Clin Sci (Colch) 1994; 86(2):217-222.
23. Vehaskari VM, Aviles DH, Manning J. Prenatal programming of adult hypertension in the rat. Kidney Int 2001; 59(1):238-245.
24. Langley-Evans SC, Phillips GJ, Jackson AA. In utero exposure to maternal low protein diets induces hypertension in weanling rats, independently of maternal blood pressure changes. Clin Nutr 1994; 13:319-324.
25. Brawley L, Itoh S, Torrens C et al. Dietary protein restriction in pregnancy induces hypertension and vascular defects in rat male offspring. Pediatr Res 2003; 54(1):83-90.

26. Franco MDP, Arruda R, Dantas APV et al. Intrauterine undernutrition: Expression and activity of the endothelial nitric oxide synthase in male and female adult offspring. Cardiovasc Res 2002; 56(1):145-153.
27. Lamireau D, Nuyt AM, Hou X et al. Altered vascular function in fetal programming of hypertension. Stroke 2002; 33(12):2992-2998.
28. Sanders MW, Fazzi GE, Janssen GM et al. Reduced uteroplacental blood flow alters renal arterial reactivity and glomerular properties in the rat offspring. Hypertension 2004; 43(6):1283-1289.
29. Sahajpal V, Ashton N. Renal function and angiotensin AT1 receptor expression in young rats following intrauterine exposure to a maternal low-protein diet. Clin Sci (Lond) 2003; 104(6):607-614.
30. Broberg CS, Giraud GD, Schultz JM et al. Fetal anemia leads to augmented contractile response to hypoxic stress in adulthood. Am J Physiol Regul Integr Comp Physiol 2003; 285(3):R649-655.
31. Arguelles J, Lopez-Sela P, Brime JI et al. Changes of blood pressure responsiveness in rats exposed in utero and perinatally to a high-salt environment. Regul Pept 1996; 66(1-2):113-115.
32. De Assis SM, Seguro AC, Helou CM. Effects of maternal hypercholesterolemia on pregnancy and development of offspring. Pediatr Nephrol 2003; 18(4):328-334.
33. Celsi G, Kistner A, Aizman R et al. Prenatal dexamethasone causes oligonephronia, sodium retention, and higher blood pressure in the offspring. Pediatr Res 1998; 44(3):317-322.
34. Dodic M, Moritz K, Wintour EM. Prenatal exposure to glucocorticoids and adult disease. Arch Physiol Biochem 2003; 111(1):61-69.
35. Swenson SJ, Speth RC, Porter JP. Effect of a perinatal high-salt diet on blood pressure control mechanisms in young Sprague-Dawley rats. Am J Physiol Regul Integr Comp Physiol 2004; 286(4):R764-770.
36. Contreras RJ. Differences in perinatal NaCl exposure alters blood pressure levels of adult rats. Am J Physiol 1989; 256(1 Pt 2):R70-77.
37. Bergel E, Belizan JM. A deficient maternal calcium intake during pregnancy increases blood pressure of the offspring in adult rats. BJOG 2002; 109(5):540-545.
38. Battista MC, Oligny LL, St-Louis J et al. Intrauterine growth restriction in rats is associated with hypertension and renal dysfunction in adulthood. Am J Physiol Endocrinol Metab 2002; 283(1):E124-131.
39. Dodic M, May CN, Wintour EM et al. An early prenatal exposure to excess glucocorticoid leads to hypertensive offspring in sheep. Clin Sci (Colch) 1998; 94(2):149-155.
40. Louey S, Cock ML, Stevenson KM et al. Placental insufficiency and fetal growth restriction lead to postnatal hypotension and altered postnatal growth in sheep. Pediatr Res 2000; 48(6):808-814.
41. van den Broek N. Anaemia and micronutrient deficiencies. Br Med Bull 2003; 67:149-160.
42. Crowe C, Dandekar P, Fox M et al. The effects of anaemia on heart, placenta and body weight, and blood pressure in fetal and neonatal rats. J Physiol 1995; 488(Pt 2):515-519.
43. Lewis RM, Petry CJ, Ozanne SE et al. Effects of maternal iron restriction in the rat on blood pressure, glucose tolerance, and serum lipids in the 3-month-old offspring. Metabolism 2001; 50(5):562-567.
44. Gambling L, Dunford S, Wallace DI et al. Iron deficiency during pregnancy affects postnatal blood pressure in the rat. J Physiol 2003; 552(Pt 2):603-610.
45. Lisle SJ, Lewis RM, Petry CJ et al. Effect of maternal iron restriction during pregnancy on renal morphology in the adult rat offspring. Br J Nutr 2003; 90(1):33-39.
46. Lindheimer MD, Akbari A. Hypertension in pregnant women. In: Oparil S, Weber MJ, eds. Hypertension: A companion to Brenner and Rector's The Kidney. 1st ed. Philadelphia: WB. Saunders Company, 1996:688-701.
47. Roberts JM, Pearson G, Cutler J et al. Summary of the NHLBI working group on research on hypertension during pregnancy. Hypertension 2003; 41(3):437-445.
48. Martin JA, Park MM, Sutton PD. Births: Preliminary data for 2001. National Vital Statisitics Reports 2002; 50(10).
49. Shah DM. Perinatal implications of maternal hypertension. Semin Pediatr Neurol 2001; 8(2):108-119.
50. Ferrer RL, Sibai BM, Mulrow CD et al. Management of mild chronic hypertension during pregnancy: A review. Obstet Gynecol 2000; 96(5 Pt 2):849-860.
51. Sibai BM. Chronic hypertension in pregnancy. Obstet Gynecol 2002; 100(2):369-377.
52. Boulet SL, Alexander GR, Salihu HM et al. Macrosomic births in the united states: Determinants, outcomes, and proposed grades of risk. Am J Obstet Gynecol 2003; 188(5):1372-1378.
53. Denton KM, Flower RL, Stevenson KM et al. Adult rabbit offspring of mothers with secondary hypertension have increased blood pressure. Hypertension 2003; 41(3Pt 2):634-639.

54. Mackenzie HS, Lawler EV, Brenner BM. Congenital oligonephropathy: The fetal flaw in essential hypertension? Kidney Int 1996; 55:S30-34.
55. Brenner BM, Garcia DL, Anderson S. Glomeruli and blood pressure. Less of one, more the other? Am J Hypertens 1988; 1(4 Pt 1):335-347.
56. Jimenez AE, Passmore JC. Cardiovascular correlates of predisposition to hypertension in pups of one kidney: One clip renal hypertensive dams. Clin Exp Hypertens A 1990; 12(2):227-241.
57. Dahlgren J, Nilsson C, Jennische E et al. Prenatal cytokine exposure results in obesity and gender-specific programming. Am J Physiol Endocrinol Metab 2001; 281(2):E326-334.
58. Ingelfinger JR, Woods LL. Perinatal programming, renal development, and adult renal function. Am J Hypertens 2002; 15(2 Pt 2):46S-49S.
59. Ingemarsson I. Gender aspects of preterm birth. BJOG 2003; 110(Suppl 20):34-38.
60. Gomez-Sanchez EP, Gomez-Sanchez CE. Maternal hypertension and progeny blood pressure: Role of aldosterone and 11beta-HSD. Hypertension 1999; 33(6):1369-1373.
61. DeForrest JM, Davis JO, Freeman RH et al. Circadian changes in plasma renin activity and plasma aldosterone concentration in two-kidney hypertension rats. Hypertension 1979; 1(2):142-149.
62. DeForrest JM, Scalese RJ, Oehl RS et al. Perinephritis hypertension in Macaca fascicularis (cynomolgus monkey): Studies of the renin-angiotensin-aldosterone axis and renal hemodynamic function. J Hypertens 1989; 7(9):763-767.
63. Wintour EM, Coghlan JP, Hardy KJ et al. Placental transfer of aldosterone in the sheep. J Endocrinol 1980; 86(2):305-310.
64. Poisner AM. The human placental renin-angiotensin system. Front Neuroendocrinol 1998; 19(3):232-252.
65. Binder ND, Laird MR, Faber JJ. Interrelationships between the renin angiotensin system and uteroplacental blood flow—a recent perspective. Reprod Fertil Dev 1995; 7(6):1437-1442.
66. McMullen JR, Gibson KJ, Lumbers ER et al. Selective down-regulation of AT2 receptors in uterine arteries from pregnant ewes given 24-h intravenous infusions of angiotensin II. Regul Pept 2001; 99(2-3):119-129.
67. Lumbers ER, Burrell JH, Stevens AD et al. Effects of one-clip, one-kidney hypertension in chronically catheterized pregnant ewes. Clin Exp Pharmacol Physiol 1997; 24(5):336-343.
68. Karlsson K, Ljungblad U, Lundgren Y. Blood flow of the reproductive system in renal hypertensive rats during pregnancy. Am J Obstet Gynecol 1982; 142(8):1039-1044.
69. Rosenfeld CR. Mechanisms regulating angiotensin II responsiveness by the uteroplacental circulation. Am J Physiol Regul Integr Comp Physiol 2001; 281(4):R1025-1040.
70. Lanz B, Kadereit B, Ernst S et al. Angiotensin II regulates 11B-hydroxysteroid dehydrogenase type 2 via AT2 receptors. Kidney Int 2003; 64:970-977.
71. Denton KM, Anderson WP. Role of angiotensin II in renal wrap hypertension. Hypertension 1985; 7(6 Pt 1):893-898.
72. Feig DI, Nakagawa T, Karumanchi SA et al. Hypothesis: Uric acid, nephron number, and the pathogenesis of essential hypertension. Kidney Int 2004; 66(1):281-287.
73. Welberg LA, Seckl JR. Prenatal stress, glucocorticoids and the programming of the brain. J Neuroendocrinol 2001; 13(2):113-128.
74. Lindsay RS, Lindsay RM, Edwards CR et al. Inhibition of 11-beta-hydroxysteroid dehydrogenase in pregnant rats and the programming of blood pressure in the offspring. Hypertension 1996; 27(6):1200-1204.
75. Dodic M, Baird R, Hantzis V et al. Organs/systems potentially involved in one model of programmed hypertension in sheep. Clin Exp Pharmacol Physiol 2001; 28(11):952-956.
76. Dodic M, Hantzis V, Duncan J et al. Programming effects of short prenatal exposure to cortisol. FASEB J 2002; 16(9):1017-1026.
77. Ortiz LA, Quan A, Zarzar F et al. Prenatal dexamethasone programs hypertension and renal injury in the rat. Hypertension 2003; 41(2):328-334.
78. Ortiz LA, Quan A, Weinberg A et al. Effect of prenatal dexamethasone on rat renal development. Kidney Int 2001; 59(5):1663-1669.
79. Benediktsson R, Lindsay RS, Noble J et al. Glucocorticoid exposure in utero: New model for adult hypertension. Lancet 1993; 341(8841):339-341.
80. Moss TJ, Sloboda DM, Gurrin LC et al. Programming effects in sheep of prenatal growth restriction and glucocorticoid exposure. Am J Physiol 2001; 281(3):R960-970.
81. Hall JE. Control of blood pressure by the renin-angiotensin-aldosterone system. Clinical Cardiology 1991; 14(8 Suppl 4):IV6-21.
82. Guyton AC, Coleman TG, Cowley Jr AV et al. Arterial pressure regulation. Overriding dominance of the kidneys in long-term regulation and in hypertension. Am J Med 1972; 52(5):584-594.

83. Ritz E, Adamczak M, Zeier M. Kidney and hypertension-causes. Herz 2003; 28(8):663-667.
84. Grisk O, Rettig R. Interactions between the sympathetic nervous system and the kidneys in arterial hypertension. Cardiovasc Res 2004; 61(2):238-246.
85. Lohmeier TE. The sympathetic nervous system and long-term blood pressure regulation. Am J Hypertens 2001; 14(6 Pt 2):147S-154S.
86. Kett MM, Bertram JF. Nephron endowment and blood pressure: What do we really know? Curr Hypertens Rep 2004; 6(2):133-139.
87. Langley-Evans SC, Langley-Evans AJ, Marchand MC. Nutritional programming of blood pressure and renal morphology. Arch Physiol Biochem 2003; 111(1):8-16.
88. LelievrePegorier M, Vilar J, Ferrier ML et al. Mild vitamin A deficiency leads to inborn nephron deficit in the rat. Kidney Int 1998; 54(5):1455-1462.
89. Amiri F, Garcia R. Differential regulation of renal glomerular and preglomerular vascular angiotensin II receptors. American Journal of Physiology 1996; 270(5 Pt 1):E810-815.
90. Moritz KM, Wintour EM, Dodic M. Fetal uninephrectomy leads to postnatal hypertension and compromised renal function. Hypertension 2002; 39(6):1071-1076.
91. Woods LL. Neonatal uninephrectomy causes hypertension in adult rats. Am J Physiol 1999; 276(4 Pt 2):R974-978.
92. Alexander BT. Placental insufficiency leads to development of hypertension in growth-restricted offspring. Hypertension 2003; 41(3):457-462.
93. Woods LL, Weeks DA, Rasch R. Programming of adult blood pressure by maternal protein restriction: Role of nephrogenesis. Kidney Int 2004; 65(4):1339-1348.
94. Moritz KM, Johnson K, Douglas-Denton R et al. Maternal glucocorticoid treatment programs alterations in the renin-angiotensin system of the ovine fetal kidney. Endocrinology 2002; 143(11):4455-4463.
95. Woods LL, Ingelfinger JR, Nyengaard JR et al. Maternal protein restriction suppresses the newborn renin-angiotensin system and programs adult hypertension in rats. Pediatr Res 2001; 49(4):460-467.
96. Brooks VL. Interactions between angiotensin II and the sympathetic nervous system in the the long-term control of arterial pressure. Clin Exp Pharmacol Physiol 1997; 24(1):83-90.
97. Wolf G. Angiotensin II and tubular development. Nephrol Dial Transplant 2002; 17(Suppl 9):48-51.
98. Sequeira Lopez ML, Gomez RA. The role of angiotensin II in kidney embryogenesis and kidney abnormalities. Curr Opin Nephrol Hypertens 2004; 13(1):117-122.
99. Huxley RR, Shiell AW, Law CM. The role of size at birth and postnatal catch-up growth in determining systolic blood pressure: A systematic review of the literature. J Hypertens 2000; 18(7):815-831.
100. Kingdom JC, McQueen J, Connell JM et al. Fetal angiotensin II levels and vascular (type I) angiotensin receptors in pregnancies complicated by intrauterine growth retardation. Br J Obstet Gynaecol 1993; 100(5):476-482.
101. Konje JC, Bell SC, Morton JJ et al. Human fetal kidney morphometry during gestation and the relationship between weight, kidney morphometry and plasma active renin concentration at birth. Clin Sci (Colch) 1996; 91(2):169-175.
102. Kingdom JC, Hayes M, McQueen J et al. Intrauterine growth restriction is associated with persistent juxtamedullary expression of renin in the fetal kidney. Kidney International 1999; 55(2):424-429.
103. Dodic M, Moritz K, Koukoulas I et al. Programmed hypertension: Kidney, brain or both? Trends Endocrinol Metab 2002; 13(9):403-408.
104. Moritz KM, Dodic M, Wintour EM. Kidney development and the fetal programming of adult disease. Bioessays 2003; 25(3):212-220.
105. Zimanyi MA, Bertram JF, Black MJ. Nephron number and blood pressure in rat offspring with maternal high-protein diet. Pediatr Nephrol 2002; 17(12):1000-1004.
106. da Silva AA, de Noronha IL, de Oliveira IB et al. Renin-angiotensin system function and blood pressure in adult rats after perinatal salt overload. Nutr Metab Cardiovasc Dis 2003; 13(3):133-139.
107. Su YR, Menon AG. Epithelial sodium channels and hypertension. Drug Metab Dispos 2001; 29(4 Pt 2):553-556.
108. Schnermann J. Sodium transport deficiency and sodium balance in gen-targeted mice. Acta Physiol Scand 2001; 173:59-66.
109. Schnermann J. NaCl transport deficiencies—hemodynamics to the rescue. Pflugers Arch 2000; 439(6):682-690.
110. Manning J, Beutler K, Knepper MA et al. Upregulation of renal BSC1 and TSC in prenatally programmed hypertension. Am J Physiol Renal Physiol 2002; 283(1):F202-206.

111. Wang XY, Masilamani S, Nielsen J et al. The renal thiazide-sensitive Na-Cl cotransporter as mediator of the aldosterone-escape phenomenon. J Clin Invest 2001; 108(2):215-222.
112. Anderson WP, Kett MM, Stevenson KM et al. Renovascular hypertension : Structural changes in the renal vasculature. Hypertension 2000; 36(4):648-652.
113. Wintour EM, Moritz KM, Johnson K et al. Reduced nephron number in adult sheep, hypertensive as a result of prenatal glucocorticoid treatment. J Physiol 2003; 549(Pt 3):929-935.
114. Young JB. Programming of sympathoadrenal function. Trends Endocrinol Metab 2002; 13(9):381-385.
115. Pupilli C, Gomez RA, Tuttle JB et al. Spatial association of renin-containing cells and nerve fibers in developing rat kidney. Pediatr Nephrol 1991; 5(6):690-695.
116. Karavanov A, Sainio K, Palgi J et al. Neurotrophin 3 rescues neuronal precursors from apoptosis and promotes neuronal differentiation in the embryonic metanephric kidney. Proc Natl Acad Sci USA 1995; 92(24):11279-11283.
117. Ruijtenbeek K, le Noble FA, Janssen GM et al. Chronic hypoxia stimulates periarterial sympathetic nerve development in chicken embryo. Circulation 2000; 102(23):2892-2897.
118. Sanders M, Fazzi G, Janssen G et al. Prenatal stress changes rat arterial adrenergic reactivity in a regionally selective manner. Eur J Pharmacol 2004; 488(1-3):147-155.
119. Lurbe E, Redon J, Tacons J et al. Current and birth weights exert independent influences on nocturnal pressurenatriuresis relationships in normotensive children. Hypertension 1998; 31(2):546-551.
120. Miller JA, Anacta LA, Cattran DC. Impact of gender on the renal response to angiotensin II. Kidney Int 1999; 55(1):278-285.
121. Evans RG, Stevenson KM, Bergstrom G et al. Sex differences in pressure diuresis/natriuresis in rabbits. Acta Physiol Scand 2000; 169(4):309-316.
122. McKinley MJ, Johnson AK. The physiological regulation of thirst and fluid intake. News Physiol Sci 2004; 19:1-6.
123. Dampney RA. Functional organization of central pathways regulating the cardiovascular system. Physiol Rev 1994; 74(2):323-364.
124. Head GA, Saigusa T, Mayorov DN. Angiotensin and baroreflex control of the circulation. Braz J Med Biol Res 2002; 35(9):1047-1059.
125. McKinley MJ, Mathai ML, McAllen RM et al. Vasopressin secretion: Osmotic and hormonal regulation by the lamina terminalis. J Neuroendocrinol 2004; 16(4):340-347.
126. Lazartigues E, Dunlay SM, Loihl AK et al. Brain-selective overexpression of angiotensin (AT1) receptors causes enhanced cardiovascular sensitivity in transgenic mice. Circ Res 2002; 90(5):617-624.
127. Morimoto S, Sigmund CD. Angiotensin mutant mice: A focus on the brain renin-angiotensin system. Neuropeptides 2002; 36(2-3):194-200.
128. Gardner DS, Pearce S, Dandrea J et al. Peri-implantation undernutrition programs blunted angiotensin II evoked baroreflex responses in young adult sheep. Hypertension 2004; 43(6):1290-1296.
129. Dodic M, Abouantoun T, O'Connor A et al. Programming effects of short prenatal exposure to dexamethasone in sheep. Hypertension 2002; 40(5):729-734.
130. Pladys P, Lahaie I, Cambonie G et al. Role of brain and peripheral angiotensin II in hypertension and altered arterial baroreflex programmed during fetal life in rat. Pediatr Res 2004; 55(6):1042-1049.
131. Head GA. Cardiac baroreflexes and hypertension. Clin Exp Pharmacol Physiol 1994; 21(10):791-802.
132. Gordon FJ, Matsuguchi H, Mark AL. Abnormal baroreflex control of heart rate in prehypertensive and hypertensive Dahl genetically salt-sensitive rats. Hypertension 1981; 3(3Pt 2):I135-141.
133. Howe PR, Rogers PF, Head GA. Limited baroreflex control of heart rate in young stroke-prone spontaneously hypertensive rats. J Hypertens 1989; 7(1):69-75.
134. El-Haddad MA, Desai M, Gayle D et al. In utero development of fetal thirst and appetite: Potential for programming. J Soc Gynecol Investig 2004; 11:123-130.
135. Desai M, Guerra C, Wang S et al. Programming of hypertonicity in neonatal lambs: Resetting of the threshold for vasopressin secretion. Endocrinology 2003; 144(10):4332-4337.
136. Nicolaidis S, Galaverna O, Metzler CH. Extracellular dehydration during pregnancy increases salt appetite of offspring. Am J Physiol 1990; 258(1 Pt 2):R281-283.
137. Ozaki T, Nishina H, Hanson MA et al. Dietary restriction in pregnant rats causes gender-related hypertension and vascular dysfunction in offspring. J Physiol 2001; 530(Pt 1):141-152.
138. Koukkou E, Ghosh P, Lowy C et al. Offspring of normal and diabetic rats fed saturated fat in pregnancy demonstrate vascular dysfunction. Circulation 1998; 98(25):2899-2904.
139. Payne JA, Alexander BT, Khalil RA. Reduced endothelial vascular relaxation in growth-restricted offspring of pregnant rats with reduced uterine perfusion. Hypertension 2003; 42(4):768-774.

140. Payne JA, Alexander BT, Khalil RA. Decreased endothelium-dependent NO-cGMP vascular relaxation and hypertension in growth-restricted rats on a high-salt diet. Hypertension 2004; 43(2):420-427.
141. Murotsuki J, Challis JR, Han VK et al. Chronic fetal placental embolization and hypoxemia cause hypertension and myocardial hypertrophy in fetal sheep. Am J Physiol 1997; 272(1 Pt 2):R201-207.
142. Jonker S, Davis LE, van der Bilt JD et al. Anaemia stimulates aquaporin 1 expression in the fetal sheep heart. Exp Physiol 2003; 88(6):691-698.
143. Davis L, Roullet JB, Thornburg KL et al. Augmentation of coronary conductance in adult sheep made anaemic during fetal life. J Physiol 2003; 547(Pt 1):53-59.
144. Dodic M, Samuel C, Moritz K et al. Impaired cardiac functional reserve and left ventricular hypertrophy in adult sheep after prenatal dexamethasone exposure. Circ Res 2001; 89(7):623-629.
145. Dodic M, Peers A, Coghlan JP et al. Altered cardiovascular haemodynamics and baroreceptor-heart rate reflex in adult sheep after prenatal exposure to dexamethasone. Clin Sci (Colch) 1999; 97(1):103-109.

CHAPTER 10

Developmental Programming of Cardiovascular Dysfunction

Lucilla Poston,* James A. Armitage and Paul D. Taylor

Abstract

Population based studies of developmental programming of adulthood cardiovascular disease have implied associations between intrauterine growth restriction and a range of adulthood indices of cardiovascular dysfunction and risk. Whilst the emphasis has been on the programming of hypertension, there is also evidence for an impact of the early life environment on later development of vascular endothelial dilator dysfunction and associated risk factors including inflammatory and thrombogenic bio-markers, dyslipidaemia and vascular compliance. In animal models, researchers have been more circumspect in the cardiovascular parameters studied and it is not always possible to draw parallels with the human situation. There is, nonetheless, strong evidence for developmental programming of reduced endothelium dependent dilatation in a variety of models of maternal nutritional imbalance which share similarity with the human data and may imply an important role in the aetiology of developmentally induced cardiovascular risk. Studies of inflammatory bio-markers, lipid profiles and compliance in animal models are too few to allow comparison. Increasing evidence for altered sympathetic activity in man and animals provides an important channel for future research effort.

Introduction

Investigations in man and in animal models have frequently reported associations between perturbations of the in utero and early life environment and elevation of adulthood blood pressure. These have been reviewed in Chapters 2, 8, 9, and 11 in this volume. The elevation of blood pressure documented in population based studies in man appears to be one of a constellation of disorders centred on insulin resistance which together contribute to the metabolic syndrome. In animal studies there has been considerable emphasis on the measurement of blood pressure whilst other aspects of cardiovascular function, including those which may provide mechanistic insight into hypertension and insulin resistance, have been less extensively studied. This chapter briefly reviews those studies which investigate the relationship between peturbations of the early life environment and adulthood disturbances of cardiovascular function other than hypertension.

*Corresponding Author: Lucilla Poston—Maternal and Fetal Research Unit, KCL Division of Reproductive Health, Endocrinology and Development, St. Thomas' Hospital, London SE1 7EH, U.K. Email: lucilla.poston@kcl.ac.uk

Early Life Origins of Health and Disease, edited by E. Marelyn Wintour and Julie A. Owens.

Endothelial Function

The pivotal role of the endothelial cell layer in cardiovascular homeostasis has provided the impetus to determine whether in utero growth restriction or adverse nutritional influences during pregnancy and early postnatal life may have permanent and adverse influences on endothelial function in later life. The endothelium, which forms a lining to all blood vessels was once considered to be a passive barrier which limited transport of water and molecules to the extravascular compartment. Whilst undoubtedly playing a crucial role in vascular permeability, endothelial cells are now recognised to be a major determinant of cardiovascular function and haemostasis. The endothelium, which shows clear heterogeneity between vascular beds, express a wide range of vasoactive factors which influence the tone of the underlying vascular smooth muscle, control platelet and leucocyte adhesion and function and directly affect the thrombogenetic potential of the blood.[1,2] These endothelial factors perform an essential physiological role and contribute to the inflammatory response. Endothelial dysfunction involving abnormal expression of these proteins has been directly implicated in the aetiology of disease, particularly atherosclerosis[3] and insulin resistance.[4]

Endothelium Dependent Dilatation

The endothelium provides local control of underlying vascular smooth muscle tone through synthesis and release of vasoactive agents, most notably the nitrogen radical, nitric oxide (NO) which leads to reduction of vascular smooth muscle calcium, and thus tone, through elevation of cyclic GMP.[5] Other endothelial derived dilators include prostacyclin, evoking vasodilation through elevation of cyclic AMP, and endothelial derived hyperpolarising factor, (EDHF, which is likely to be several different molecules) achieving vasodilatation through hyperpolarization of the vascular smooth muscle membrane and reduction of cell calcium entry through voltage gated calcium channels.[6] The endothelium also synthesises vasoconstrictor factors including endothelin and thromboxane. Synthesis and release of these vasoactive agents is under control of local humoral and mechanical influences. Thus NO synthesis and release may occur through stimulation by vasodilatory agents such as histamine or Calcitonin Gene Related Peptide CGRP), but is also tonically simulated by the shear stress created as the blood flows past the vessel wall. Blunted endothelium dependent dilatation in man is an important risk factor for subsequent cardiovascular disease, particularly atherosclerosis.[7] Reduced synthesis of NO and increased expression of cell adhesion molecules provides a proatherogenic stimulus. Current theory suggests that leucocyte adhesion and monocyte migration into the intima leads to a local inflammatory response and ultimately to generation of an atherosclerotic plaque.[8] Poor endothelial function has also been implicated in insulin resistance since impairment of endothelium dependent dilatation is a frequent accompaniment to insulin resistance and type 2 diabetes.[4]

Developmental Programming of Endothelial Function

Studies in Human Populations

Relatively few studies have probed the relationship between birth weight or early catch up growth and endothelial function in later life. Endothelial dilator function may be evaluated noninvasively in man by estimation of flow mediated dilatation in the brachial artery. A rapid increment in flow is achieved by application of a cuff on the lower arm. Inflation and subsequent deflation leads to a hyperaemic response and increased flow in the brachial artery. This evokes a shear mediated dilator response which is assessed by measurement of the diameter of the vessel by high resolution ultrasound. Impairment of endothelial function in the brachial artery correlates with that in the carotid artery, a vessel prone to development of atherosclerotic plaques.[9] Leeson et al[10] were the first to suggest that low birth weight could herald the later development of blunted endothelial dependent dilatation. In a study of 333, 9-11 year old British children these authors found a significant, graded and positive association of flow-mediated dilation with birth weight which was unaffected by adjustment for potential confounding variables such as body build, cardiovascular risk factors and socio-economic status. An inverse relationship with HDL cholesterol concentrations was also found. In a subsequent study of 3125 adults, low birthweight was associated with reduced flow-mediated dilatation. However increasing levels of acquired risk factors with age overwhelmed this association i.e., it became nonsignificant after correction.[11] Another group showed reduced endothelium dependent dilatation in fit 19-20 year olds of low birthweight.[12] Responses to vasodilatation in skin (to local application of the endothelium dependent dilator acetylcholine) and to local heating to 44°C in 9 year old children have also been assessed. The mean skin perfusion in response to acetylcholine increased by 240% in low birthweight children, compared with 650% in normal birthweight children.[13] Others have recently shown an association between exaggerated catch up growth in the first two weeks of life and reduced brachial artery endothelium dependent dilatation in adolescents aged between 13-16years who were born prematurely. Interestingly, these children were also insulin resistant.[14,15]

Animal Models of Developmental Programming

Models in rodents designed to assess 'programming' effects of maternal nutrient restriction have uniformly shown blunted responses to endothelium dependent dilators in isolated arteries from adult offspring. Thus these animal models share this characteristic with patients at risk of development of cardiovascular disease. Brawley et al[16,17] and Torrens et al[18] have evaluated responses to the endothelium dependent dilators, bradykinin and acetylcholine in arteries from virgin and pregnant adult offspring, respectively, of protein restricted (50% reduction) rat dams. Whilst responses in mesenteric small arteries were impaired, dilator function in the aorta was unaffected.[17] The same group have also shown that maternal glycine supplementation prevents the development of this defect in the offspring, implicating a mechanistic role for altered folate metabolism in cardiovascular dysfunction in this model.[17] This accords with another study demonstrating that maternal glycine supplementation reversed the hypertension in the adult offspring.[19] Others have noted a reduction in endothelium dependent dilatation in microvessels in the cerebral circulation from offspring of protein restricted (50% reduction) dams.[20] The reduction of endothelium dependent dilatation in these models is accompanied by an increase in vascular smooth muscle sensitivity to NO which may in part compensate for failure of synthesis or bioavailability. This may offer some protection against cardiovascular dysfunction.

Similar abnormalities have been described in offspring of calorific deprived animals. Studies from our group in collaboration with Professor Hanson (Southampton University) in small femoral arteries from fetal sheep subjected to maternal dietary deprivation (50%) have shown blunted endothelium dependent dilatation and suggest that the fetal circulation may be compromised in late gestation (0.9).[21] In collaboration with Dr Holemans (Leuven University), we have also shown that 120 day old offspring of rat dams fed 50% of the normal diet from mid gestation and through lactation show reduced endothelium dependent dilatation in small mesenteric arteries.[22] In contrast to the offspring of protein restricted dams the defect seems to include the larger vessels as Franco et al[23] have shown blunted relaxation in the aorta of offspring from dams subjected to 50% dietary restriction in pregnancy. The same group[24] have also documented in the small mesenteric vessels an increase in superoxide synthesis by NAD(P)H oxidase. Superoxide may reduce the bioavailability of NO through the formation of peroxynitrite and so contribute to poor endothelium dependent dilatation. A similar increase in vascular synthesis of superoxide through NAD(P)H oxidase has been described in an animal model of hypertension, the stroke prone spontaneously hypertensive rat[25] and there has recently been focus on the potentially important role of this enzyme in cardiovascular disease in man.[26] In a study from our group in which rat dams were subjected to a more moderate 30% reduction in dietary intake, no defect in endothelium dependent dilatation in small arteries from the femoral circulation was found in weanlings or in 100 or 200 day old offspring.[27] These data together may suggest that a relatively severe dietary insult in utero is required for the development of a sustained defect in offspring endothelium dependent dilatation.

Dietary restriction models in rodents have relevance to times of famine in the developing world and may reflect the fetal response to intrauterine nutrient restriction arising from placental disease. The protein content of the Western diet has fallen in recent years and the protein deprivation model may also provide a model for this downward trend. In our laboratory we have focussed on a model of dietary excess which has relevance to the diet consumed by many Western populations. Pregnant rats are fed a diet rich in animal lard during pregnancy and lactation. Male and female adult offspring demonstrate marked impairment of endothelium dependent dilatation to acetycholine in small mesenteric arteries[28] whereas females alone are hypertensive. An investigation of the NO, prostacyclin and EDHF components of relaxation demonstrated that the EDHF component, which is generally the principal contributor to endothelium dependent dilatation in small arteries, was much reduced and accounted for the failure of relaxation.[29] Further studies suggested that exposure to the diet during lactation is particularly deleterious to later evolution of poor endothelium dependent dilatation[30] but that if the offspring themselves consumed the same diet as the dams, that the defect was prevented.[31] This is one of the few examples of a study in which endothelial function has been assessed in animals habituated to the diet experienced by the dam, and is an example of a 'predictive adaptive response.'[32] The majority of studies of this kind wean the offspring onto a normal chow diet and do not attempt to assess whether the in utero experience has prepared the offspring for an adult diet similar to that of their dam. Endothelial function has infrequently been assessed in animal models in which maternal glucocorticoid administration provides the programming 'stimulus' although Molnar et al[33] have shown that five month old lambs exposed prenatally to dexamethasone (in late gestation) have enhanced sensitivity to endothelin in isolated femoral arteries. This was attributed to blunted endothelin $(ET)_B$ receptor mediated stimulation of endothelial nitric oxide.

Markers of Endothelial Cell Activation and the Inflammatory Response

Reduced endothelium dependent dilatation is only one of the facets of endothelial dysfunction which contributes to cardiovascular risk in man. The endothelium plays an important role in the inflammatory response and increased expression of cell adhesion molecules e.g., vascular cell adhesion molecule 1 (VCAM-1) and intercellular adhesion molecule (ICAM-1) and prothrombotic factors e.g., plasminogen activator inhibitor (PAI-1) and tissue factor (TF) are linked to disease. Likely stimuli include oxidised lipids, cytokines and C reactive protein (CRP), a marker of hepatic origin and a significant cardiovascular risk factor.[34,35] Few investigators have attempted assessment of circulating concentrations of endothelial or inflammatory markers in human studies. Although demonstrating a relationship between birthweight and endothelium dependent dilatation, Goodfellow et al found no evidence of a correlation between birthweight and serum concentrations of von Willebrand factor, a marker of endothelial cell activation in fit young adults[12] whereas in a smaller study McAllister et al found higher concentrations in low birth weight subjects.[36] A report of adolescents born prematurely has shown that those fed formula feed in neonatal life have higher concentrations of C reactive protein than those fed banked breast milk[37] indicating that fat rich formula feed in early life may confer cardiovascular risk. Recently, a study of 1663 individuals taking part in the MIDPSPAN family study has reported a negative association between birthweight and adulthood CRP after adjusting for potential confounders, whereby a 1-kg increase in birth weight was associated with a 10.7% decrease in CRP.[38] An inverse association between low birth and plasma concentrations of fibrinogen,[39] a thrombogenic factor has also been observed, although a recent study comparing monozygotic and dizygotic twins[40] has suggested that this relationship may depend entirely on genetic influences. Another report has documented an association between low birth weight and the endothelial maker PAI-1 in middle aged Swedish men.[41]

Animal Studies

There are very few reports of inflammatory mediators in offspring of animals subjected to nutritional imbalance in pregnancy or any other programming stimulus. This may partly reflect the relative resistance of rat endothelium to activation stimuli but requires further study given the obvious parallel between human and animal studies in relation to endothelium dependent dilator function. Although we are unaware of any baseline investigation of markers of the inflammatory response, one study has assessed acute inflammatory responses to a pleurisy inducing agent (carrageenan) in offspring of protein restricted rat dams and has shown that these animals mount a reduced inflammatory response, as assessed by pulmonary oedema, neutrophil migration and ICAM expression.[42]

Dyslipidaemia

Studies in Human Populations

The suggestion that low birth weight may contribute to later development of metabolic syndrome and attendant cardiovascular sequelae (including endothelial dysfunction) has been supported by studies documenting an association of birth weight and or catch-up growth and an altered adult lipid profile,[43] although relationships are not always found.[41,44] In one of the earliest studies, amongst children aged between 7-11 years involved in the Bogalusa Heart Study, an association between low birthweight and raised serum triglycerides was found, although no correlation with cholesterol was apparent.[45] In a study from Finland nearly half of a cohort of 12 year old small for gestational age children were in the highest quartile for serum total cholesterol of the appropriate for gestational age children, but *poor* catch up growth predicted higher cholesterol.[46] Associations between birth length, birth weight, ponderal index and total serum cholesterol were also examined in 545 Danish men and women (31- 51 years). No correlations were found in women, but in men a negative association was found between birth weight and serum total cholesterol, with a fall in mean serum total cholesterol from 6.03 mmol/l for birth weight below 3300 g to 5.64 mmol/l for birth weight above 4000.[47] In a Filipino population, mothers with low energy status during pregnancy, as assessed by maternal arm fat area, gave birth to male offspring who had a high CVD risk in adolescence, as indicated by lipid profiles.[48] Very recently, in the largest cohort study to-date of 25,843 men and women from the UK, lower birth weight in men was associated with higher adult total cholesterol levels (a 0.07 mmol/l reduction in total cholesterol for each 1-kg increase in birth weight), whereas no association was observed in women. The authors concluded that the influence of fetal environment on adult total cholesterol was small compared with the influence of adult adiposity.[49] The impact of maternal hypercholesterolaemia on development of childhood arterial fatty streaks has been studied in a *post mortem* study of Italian children and it was found that children of mothers who had hypercholesterolaemia in pregnancy had a greater rise with age in area of aortic arch lesions than that of mothers with normal cholesterol.[50] Although shared 'pro-hypercholesterolaemic' genes between mother and child may provide an explanation this study has been supported by similar investigations in genetically identical animals (see below).

Studies in Animals

Unfortunately, in rodent models of developmental programming, most investigators have not determined plasma lipid concentrations in adult offspring. Rodents carry most cholesterol as HDL cholesterol and measurement of LDL cholesterol is relatively meaningless. Studies in offspring of protein restricted rodents are few. In one, plasma HDL cholesterol and triglyceride concentrations were reduced in 6month old female offspring compared to controls.[51] In contrast, male offspring of calorie restricted (85% of ad lib diet) guinea pig sows have increased plasma total cholesterol concentrations compared with controls but female litter mates have normal values.[52] These offspring also demonstrate a greater increment in plasma cholesterol upon a cholesterol dietary challenge compared to controls. We have reported an atherogenic lipid profile in adult offspring of rats fed a fat-rich diet[28,53] and Palinksi et al[54] have shown atherosclerotic plaques in offspring of rabbits fed cholesterol rich diet during pregnancy. Very few studies of developmental programming have been carried out in primates, but an investigation in baboons has shown clear evidence for altered lipid metabolism in breast *versus* formula fed animals, with HDL cholesterol concentrations being lower in breast fed animals.[55]

Vascular Compliance

Studies in Human Populations

Reduced vascular compliance is an identified risk factor for cardiovascular morbidity[56] and it has been proposed that impaired growth in early development may be linked to reduced compliance in adulthood through early and permanent alteration in structure of the aorta and other larger arteries. A study in 9 year old low birth weight children in Sweden has reported increased carotid artery stiffness compared with normal birthweight controls[13] and a study of 331 young adults (20-28 years of age) has suggested that extended breastfeeding (and therefore prolonged exposure to a fat-rich diet) is also linked to poor endothelial dilator function.[57] Martyn et al have reported an inverse association between birth size and pulse-wave velocity (PWV) of the femoral and radial arteries (an estimate of compliance) in middle aged men and women[58] whereas others have found no association.[59,60] However it should be appreciated that elevation of blood pressure per se can lead to altered vascular compliance and that studies of compliance in adults may be confounded by the presence of preexisting hypertension.

Animal Studies

Evidence for alteration in vascular compliance has to our knowledge not been reported in animal models of developmental programming. In one relevant investigation Berry and Looker[61] demonstrated that intrauterine growth restriction (induced by maternal methotrexate administration on day 15 of pregnancy followed by folinic acid administration as a specific methotrexate antagonist 16 hours later) resulted in low birthweight without catch-up growth. Collagen and elastin content were reduced in offspring of methotrexate treated animals at 26 weeks of age and it was hypothesised that failure to synthesise adequate amounts of elastin during a critical period cannot be rectified later in life. Altered smooth muscle structure, particularly in myosin heavy chain (SM_2) content may also lead to altered compliance but has not been investigated. The measurement of vascular distensibility can readily be undertaken in isolated arteries and assessment of compliance and collagen, elastin and smooth muscle content in young and adult offspring would be worthwhile in the different animal models.

Other Parameters of Cardiovascular Dysfunction

Endothelial dysfunction may play a role in elevation of the blood pressure and insulin resistance in developmental programming. However blood pressure could also rise through enhanced constrictor responses in the vasculature, thereby increasing peripheral resistance, or from elevation of the cardiac output. Altered renal function and perturbation of volume control could also play an important role, and the observation of reduced nephron number (reviewed elsewhere in this book) in several animal models could be of fundamental importance. Several groups have assessed constrictor function of isolated arteries, and other than one report of enhanced endothelin constrictor responses in fetal sheep exposed to glucocorticoids[32] there is little evidence for a fundamental defect in constrictor function amongst the different animal models. However there is emerging evidence to suggest that altered sympathetic activity, which has been associated with essential hypertension in man, may contribute in some models. The adult offspring of nutritionally deprived sheep show altered baroreceptor responses[62] and there is evidence in low birthweight man of altered sympathetic outflow.[63] Chronically hypoxic chick embryos develop altered sympathetic function[64] and vascular adrenergic responses are altered in the offspring of rats subjected to protein restriction during pregnancy.[65] Furthermore, in normotensive male offspring of rat dams fed a fat rich diet, we have observed a reduction of basal heart rate which may be indicative of an altered baroreceptor response.[28]

Conclusions

In conclusion, the demonstration of altered endothelium dependent dilatation in human and animal studies suggest that endothelial dysfunction may play a central role in developmental programming of adulthood cardiovascular disease. Further studies are required in the different animal models to elaborate similarities/differences with human disease in relation to activation of the inflammatory response and abnormalities of lipid metabolism. Very little effort has been directed towards investigation of sympathetic activity in the animal models and this is an area of potentially fruitful study.

References

1. Hunt BJ, Jurd KM. The endothelium in health and disease. In: Hunt B, Poston L, Schachter M et al, eds. An Introduction to Vascular Biology. Cambridge University Press, 2002.
2. Galley HF, Webster NR. Physiology of the Endothelium. Br J Anaesth 2004; 93:105-113.
3. Landmesser U, Hornig B, Drexler H. Endothelial function: A critical determinant in atherosclerosis? Circulation 2004; 109:27-33.
4. Hsueh WA, Lyon CJ, Quinones MJ. Insulin resistance and endothelium. Am J Med 2004; 117:109-117.
5. Chan N, Vallance P. Nitric Oxide. In: Hunt B, Poston L, Schachter M et al, eds. An Introduction to Vascular Biology. Cambridge University Press, 2002.
6. Busse R, Edwards G, Feletou M et al. EDHF: Bringing the concepts together. Trends in Pharmacol Sci 2002; 23:374-380.
7. Halcox JPJ, Schenke WH, Zalos G et al. Prognostic value of coronary vascular endothelial dysfunction. Circulation 2002; 106:653-658.
8. Libby P. Inflammation and atherosclerosis. Nature 2002; 420:868-874.
9. Anderson TJ, Uehata A, Gerhard MD et al. Close relation of endothelial function in the human coronary and peripheral circulations. J Am Coll Cardiol 1995; 26:1235-1241.
10. Leeson CP, Whincupp PH, Cook DG et al. Flow-mediated dilatation in 9-11 year old children: The influence of intrauterine and childhood factors. Circulation 1997; 96:2233-2238.
11. Leeson CP, Katternhorm M, Morley R et al. Impact of low birthweight and cardiovascular risk factors on endothelial function in early adult life. Circulation 2001; 103:1264-1268.
12. Goodfellow J, Bellamy NF, Gorman ST et al. Endothelial function is impaired in fit young adults of low birthweight. Cardiovascular Research 1998; 40:600-606.
13. Martin H, Hu J, Gennser G et al. Impaired endothelial function and increased carotid stiffness in 9-year old children with low birthweight. Circulation 2000; 102:2739-2744.
14. Singhal A, Lucas A. Early origins of cardiovascular disease: Is there a unifying hypothesis? Lancet 2004; 363:1642-1645.
15. Singhal A, Cole TJ, Fewtrell M et al. Breastmilk feeding and lipoprotein profile in adolescents born preterm: Follow-up of a prospective randomised study. Lancet 2004; 15(363):1571-1578.
16. Brawley L, Itoh S, Torrens C et al. Dietary protein restriction in pregnancy induces hypertension and vascular defects in rat male offspring. Pediatr Res 2003; 54:83-90.
17. Brawley L, Torrens C, Anthony FW et al. Glycine rectifies vascular dysfunction induced by dietary protein imbalance in pregnancy. J Physiol 2004; 554:497-504.
18. Torrens C, Brawley L, Barker AC et al. Maternal protein restriction in the rat impairs resistance artery but not conduit artery function in pregnant offspring. J Physiol 2003; 15:77-84.
19. Jackson AA, Dunn RL, Marchand MC et al. Increased systolic blood pressure in rats induced by a maternal low-protein diet is reversed by dietary supplementation with glycine. Clin Sci (Lond) 2002; 103:633-639.
20. Lamireau D, Nuyt AM, Hou X et al. Altered vascular function in fetal programming of hypertension. Stroke 2002; 33:2992-2998.
21. Ozaki T, Hawkins P, Nishina H et al. Effects of undernutrition in early pregnancy on systemic small artery function in late-gestation fetal sheep. Am J Obstet Gynecol 2000; 183:1301-1307.
22. Holemans K, Gerber R, Meurrens K et al. Maternal food restriction in the second half of pregnancy affects vascular function but not blood pressure of rat female offspring. Br J Nutr 1999; 81:73-79.

23. Franco MDC, Arruda RM, Dantas AP et al. Intrauterine undernutrition: Expression and activity of the endothelial nitric oxide synthase in male and female adult offspring. Cardiovasc Res 2002; 56:145-153.
24. Franco MDC, Akamine EH, Di Marco GS et al. NADPH oxidase and enhanced superoxide generation in intrauterine undernourished rats: Involvement of the renin-angiotensin system. Cardiovasc Res 2004; 59:767-775.
25. Hamilton CA, Brosnan MJ, Al-Benna A et al. NAD(P)H oxidase inhibition improves endothelial function in rat and human blood vessels. Hypertension 2002; 40:755-762.
26. Touyz RM. Reactive oxygen species, vascular oxidative stress and redox signalling in hypertension. What is the clinical significance? Hypertension 2004; 44:248-252.
27. Ozaki T, Nishina H, Hanson MA et al. Dietary restriction in pregnant rats causes gender-related hypertension and vascular dysfunction in offspring. J Physiol 2001; 530:141-152.
28. Khan IY, Taylor PD, Dekou V et al. Gender-linked hypertension in offspring of lard-fed pregnant rats. Hypertension 2003; 41:168-175.
29. Taylor PD, Khan IY, Hanson MA et al. Impaired EDHF mediated vasodilatation in adult offspring of rats exposed to a fat-rich diet in pregnancy J Physiol 2004; 558:943-951.
30. Khan IY, Dekou V, Douglas G et al. A high fat diet during rat pregnancy or suckling induces cardiovascular dysfunction in adult offspring. Am J Physiol Regul Integr Comp Physiol 2005; 288:R127-133.
31. Khan I, Dekou V, Hanson M et al. Predictive adaptive responses to maternal high fat diet prevent endothelial dysfunction but not hypertension in adult rat offspring. Circulation 2004; 1001:1097-1102.
32. Gluckman PD, Hanson MA. Living with the past: Evolution, development and patterns of disease. Science 2004; 305:1733-173.
33. Molnar J, Howe DC, Nijland MJ et al. Prenatal dexamethasone leads to both endothelial dysfunction and vasodilatory compensation in sheep. J Physiol 2003; 547:61-66.
34. Szmitko PE, Wang CH, Weisel RD et al. New markers of inflammation and endothelial cell activation. Circulation 2003; 108:1917-1923.
35. Szmitko PE, Wang C-H, Weisel RD et al. Biomarkers of vascular disease linking inflammation to endothelial activation. Circulation 2003; 108:2041-2048.
36. McAllister AS, Atkinson AB, Johnston GD et al. Relationship of endothelial function to birth weight in humans. Diabetes Care 1999; 22:2061-2066.
37. Singhal A, Cole TJ, Fewtrell M et al. Is slower early growth beneficial for long term cardiovascular health? Circulation 2004; 109:1108-1113.
38. Sattar N, McConnachie A, O'Reilly D et al. Inverse association between birth weight and C-reactive protein concentrations in the MIDSPAN. Family Study. Arterioscler Thromb Vasc Biol 2004; 24:583-587.
39. Martyn CN, Meade TW, Stirling Y et al. Plasma concentrations of fibrinogen and factor VII in adult life and their relation to intra-uterine growth. Br J Haematol 1995; 89:142-146.
40. Ijzerman RG, Stehouwer CD, de Geus EJ et al. The association between birth weight and plasma fibrinogen is abolished after the elimination of genetic influences. J Thromb Haemost 2003; 1:239-242.
41. Byberg L, McKeigue PM, Zethelius B et al. Birth weight and the insulin resistance syndrome: Association of low birth weight with truncal obesity and raised plasminogen activator inhibitor-1 but not with abdominal obesity or plasma lipid disturbances. Diabetologia 2000; 43:54-60.
42. Barja-Fidalgo C, Souza EP, Silva SV et al. Impairment of inflammatory response in adult rats submitted to maternal undernutrition during early lactation: Role of insulin and glucocorticoid. Inflamm Res 2003; 52:470-6.
43. Barker DJ. The intra-uterine origins of disturbed cholesterol homeostasis. Acta Paediatr 1999; 88:483-484.
44. Owen CG, Whincup PH, Odoki K et al. Birth weight and blood cholesterol level: A study in adolescents and systematic review. Pediatrics 2003; 111:1081-1089.
45. Donker GA, Labarthe DR, Harrist RB et al. Low birth weight and serum lipid concentrations at age 7-11 years in a biracial sample. Am J Epidemiol 1997; 145:398-407.
46. Tenhola S, Martikainen A, Rahiala E et al. Serum lipid concentrations and growth characteristics in 12-year-old children born small for gestational age. Pediatr Res 2000; 48:623-628.
47. Ziegler B, Johnsen SP, Thulstrup AM et al. Inverse association between birth weight, birth length and serum total cholesterol in adulthood. Scand Cardiovasc J 2000; 34:584-588.
48. Kuzawa CW, Adair LS. Lipid profiles in adolescent Filipinos: Relation to birth weight and maternal energy status during pregnancy. Am J Clin Nutr 2003; 77:960-966.

49. Davies AA, Smith GD, Ben-Shlomo Y et al. Low birth weight is associated with higher adult total cholesterol concentration in men: Findings from an occupational cohort of 25,843 employees. Circulation 2004; 110:1258-1262.
50. Napoli C, Glass CK, Witztum JL et al. Influence of maternal hypercholesterolaemia during pregnancy on progression of early atherosclerotic lesions in childhood: Fate of Early Lesions in Children (FELIC) study. Lancet 1999; 354:1234-1241.
51. Lucas A, Barker DJ, Desai DJ et al. Nutrition in pregnant or lactating rats programs lipid metabolism in the offspring. Br J Nutrition 1996; 76:605-612.
52. Kind KL, Clifton PM, Katsman AI et al. Restricted fetal growth and the response to dietary cholesterol in the guinea pig. Am J Physiol 1999; 277:R1675-1682.
53. Ghosh P, Bitsanis D, Ghebremeskel K et al. Abnormal fatty acid composition and small artery function in offspring of rats fed a high fat diet in pregnancy. J Physiol 2001; 533:815-822.
54. Palinski W, D'Armiento FP, Witztum JL et al. Maternal hypercholesterolemia and treatment during pregnancy influence the long-term progression of atherosclerosis in offspring of rabbits. Circ Res 2001; 89:991-996.
55. Mott GE, Jackson EM, DeLallo et al. Differences in cholesterol metabolism in juvenile baboons are programmed by breast-versus formula feeding. J Lipid Res 1995; 36:299-307.
56. Kingwell BA, Gatzka CD. Aterial stiffness and prediction of cardiovascular risk. J Hypertens 2002; 20:2337-2340.
57. Leeson CP, Katternhorm M, Deanfield JE et al. Duration of breast feeding and arterial distensibility in early life: population based study. BMJ 2001; 332:643-647.
58. Martyn CN, Barker DJ, Jespersen S et al. Growth in utero, adult blood pressure, and arterial compliance. Br Heart J 1995; 73:116-121.
59. Montgomery AA, Ben-SholmoY, McCarthy A et al. Birth size and arterial compliance in young adults. Lancet 2000; 355:2136-2137.
60. Kumeran K, Fall Ch, Martyn CN et al. Blood pressure, arterial compliance and left ventricular mass; no relation to small size at birth in south Indian adults. Heart 2000; 83:272-7.
61. Berry CL, Looker T. An alteration in the chemical structure of the aortic wall induced by a finite period of growth inhibition. J Anat 1973; 114:83-94.
62. Gardner DS, Pearce S, Dandrea J et al. Peri-implantation undernutrition programs blunted angiotensin II evoke baroreflex responses in young adult sheep. Hypertension 2004; 43:1290-1296.
63. Phillips DI, Barker DJ. Association between low birthweight and high resting pulse in adult life: Is the sympathetic nervous system involved in programming the insulin resistance syndrome. Diabet Med 1997; 14:673-677.
64. Ruijtenbeek K, le Noble FA, Janssen GM et al. Chronic hypoxia stimulates periarterial sympathetic nerve development in chicken embryo. Circulation 2000; 102:2892-2897.
65. Young JB, Kaufman LN, Saville ME et al. Increased sympathetic nervous system activity in rats fed a low-protein diet. Am J Physiol 1985, 248:R627-37.
66. Sanders MW, Fazzi GE, Janssen GM et al. Reduced uteroplacental blood flow alters renal arterial reactivity and glomerular properties in the rat offspring. Hypertension 2004; 43:1283-1289.

Chapter 11

Kidney Development and Fetal Programming

Karen M. Moritz* and Luise A. Cullen-McEwen

Abstract

Alteration in the normal development of the kidney is likely to be a major contributing factor to programming of adult disease. Renal disease is reaching epidemic proportions in some sectors of the community and is often found in association with the two most characterised adult onset diseases, hypertension and noninsulin dependent diabetes mellitus. Epidemiological studies in humans have identified various maternal states such as anemia, diabetes and protein/micronutrient deficiency as causing renal abnormalities, often in association with decreased birth weight. Animal models of maternal protein/nutrient deficiency as well as maternal glucocorticoid exposure generally results in a reduction in glomerular (and thus nephron) number in the offspring along with increases in blood pressure. It is important to note that the stimulus has the greatest effect when applied at the beginning of metanephric development. The decrease in nephron endowment probably results from changes in expression of genes identified to be critical for normal branching morphogenesis in the kidney. However, other compensatory changes in the kidney may involve alteration of the renal renin-angiotensin system and changes in channels involved in sodium transport. Further research into genes/proteins altered in various "programmed" models, along with use of careful stereological methodology for determining glomerular number are essential to further our understanding of renal involvement in the programming of adult disease.

Introduction

As the concept of the developmental origins of adult disease, or the "Barker hypothesis" has gained acceptance, focus has switched from identifying "programmable" diseases to determining mechanisms through which programming can occur. Many organs and systems have been extensively investigated. Whilst there are numerous models using a range of species, maternal perturbations and timing of insults, alterations in renal development appears to be present in many models. In this chapter we will address some of the current issues/controversies relating to the potential role of the kidney in the programming of adult disease and highlight some of the areas in which further research is warranted. The reader is referred to some recent reviews that have examined the evidence for a role of the kidney in the programming of adult disease, especially in models where hypertension is present[1,2] and to other chapters in this book which demonstrate renal involvement.

An Epidemic of Kidney Disease

In all Westernised countries, the rates of chronic renal disease (CRD) are increasing. The American Kidney Foundation estimates that 20 million Americans, or 1 in 9 adults in the population have some form of CRD (source American Kidney Foundation). By 2020, it is

*Corresponding Author: Karen M. Moritz—Department of Anatomy and Cell Biology, Monash University, Clayton, Victoria, Australia, 3800. Email: karen.moritz@med.monash.edu.au

Early Life Origins of Health and Disease, edited by E. Marelyn Wintour and Julie A. Owens.

estimated that 1 in 4 people will be affected. In certain communities such as the Australian Aborigines and Native Americans, the rates are even higher. Even more concerning is the trend for CRD to be associated with diabetes and hypertension causing complex patient management issues. Similar to other adult onset diseases, such as hypertension and noninsulin dependent diabetes mellitus (type 2 diabetes), so-called lifestyle factors such as a high fat/salt diet, smoking and lack of exercise can contribute significantly to the development of renal disease. However, like these diseases, the predisposition to develop renal disease may have its roots in early prenatal development.

Is Low Birth Weight a Risk for Kidney Disease?

Much of the evidence for programming of adult disease comes from human epidemiological studies and animal models where the offspring is of low birth weight due to maternal undernutrition or placental insufficiency.[3,4] Infants born of low birth weight, that is small for gestational age (SGA) are at increased risk of developing adult diseases particularly hypertension and noninsulin dependent diabetes mellitus (NIDDM). A link between low birth weight and renal disease is not as firmly established as for cardiovascular and metabolic disease, but this is probably due to more limited examination of renal function rather than no such association being present. Also, as hypertension and NIDDM are well defined risk factors for chronic renal disease, it is often difficult to ascertain whether renal disease occurs as a result of these diseases or is independently a result of low birth weight. A recent literature review of renal disease identified low birth weight as a 'progression promoter' for renal disease.[5] Low birth weight has been shown to contribute to early onset end-stage renal disease (ESRD) in children from Southern USA[6] as well as in the Australian Aboriginal population.[7] However in other studies, males of low birth weight were not found to have impaired renal function despite having elevated blood pressure and elevated natriuresis.[8] Females born SGA did not have elevated blood pressure or altered renal function[9] whilst those born preterm had increased blood pressure with normal renal function. Low birth weight was not found to be a risk factor in the progression of diabetic nephropathy in type 1 diabetic patients.[10]

What Can Be Programmed in the Kidney?

Nephron (Glomerular) Number

The kidney has been implicated in the programming of adult disease predominantly through studies in which a low/reduced nephron number resulting from maternal intervention is associated with elevated blood pressure in the adult offspring.[2] Much interest has focused on the so called "Brenner hypothesis" which states that a reduction in nephron endowment at birth contributes to the development of hypertension.[11] However, is it really that simple? The "normal range" of glomeruli in a human kidney is between 300,000 and nearly 2 million.[12] This huge range in nephron endowment coupled with the fact that all of nephrogenesis takes place in utero in the human suggests a healthy prenatal environment is critical for adequate nephron formation. What is emerging from animal models is that a reduction in nephron endowment results from maternal insults at specific times during development with the early period of renal development being extremely susceptible. The two most common experimental models of programming, namely maternal undernutrition/ low protein and maternal glucocorticoid exposure, have both illustrated this point. In the rat, maternal low protein diet throughout gestation has been shown to cause elevations in blood pressure and reductions in nephron number although it may be sex dependant with male offspring only affected in some studies.[13,14] A study using a more severe protein restriction showed low nephron number and hypertension resulted in offspring of both sexes if present during the second half of pregnancy but not the first.[15] In sheep exposed to elevated levels of glucocorticoids for 48 hours very early in pregnancy (26-28 days out of a 150 day pregnancy) a reduction in nephron number of ~30% is found in the adult.[16] These animals were of normal birth weight but develop high blood pressure from 4 months of age.[17] At the time of the treatment the first branching of the ureteric bud has just occurred.[18] Maternal dexamethasone treatment throughout pregnancy in rats causes growth retardation, hypertension and decreased nephron number.[19] Interestingly, hypertension at 6 months of age and a reduced nephron number also resulted from short exposure (2 days) early in kidney development in rats at either 15/16 or 17/18 (males only) whilst females treated at 15/16 had elevated blood pressure at 3 weeks of age.[20,21] This treatment did not cause growth retardation. This highlights the point that birth weight (and kidney weight) can be normal but a significant nephron deficit may be present and hypertension may develop in later life. This is of great clinical relevance, as is suggests that babies born of normal birth weight may still be programmed to develop adult disease.[22]

Determining Nephron Number

One of the complications in many of the studies correlating nephron number with blood pressure is the methodology utilised. The "gold standard" for determining nephron number is an estimation of glomerular number using an unbiased stereological method using a physical dissector/fractionator system which involves random, systematic sampling of the whole organ and is the only method which the American Society of Nephrology recommends for analysis. This method makes no assumptions of size and shape and thus eliminates any bias.[23] Unfortunately this is an expensive and time consuming method and is not always employed. Classical methods of nephron estimation include maceration techniques to isolate glomeruli and profile counting in random histological sections. An accurate estimation of nephron number relies of uniform density throughout the cortex and maintenance of glomeruli integrity. This is not always possible using maceration particularly in kidneys where there is a high level of sclerosis.[24] Other stereological methods are inherently biased due to problems of size and shape of glomeruli.[23]

Is It Just Nephron Number That's Important?

Arguments against the Brenner hypothesis primarily come from human and animal experiences where removal of a kidney from an adult, thereby halving the nephron number, does not generally result in hypertension. Long term follow up studies of patients who have had a kidney removed due to a tumour or donated a kidney for transplantation have not shown an increased incidence of hypertension.[25] Other studies in the rat where the glucocorticoid treatment occurred earlier in gestation also show that hypertension can result without changes in nephron number[21] whilst the offspring of the WKY rat do not develop high blood pressure after maternal exposure to a low protein diet.[26] It is very interesting however, that unilateral nephrectomy during the period of nephrogenesis in the rat (at postnatal day1) or the sheep (at 100 days out of the 150 gestation) results in offspring with elevated blood pressure and compromised renal function.[27,28] In the case of the sheep, this resulted in some compensatory nephrogenesis in the remaining kidney but overall, a significant nephron deficit was still present.[29] This suggests other changes in the kidney apart from nephron number may be programmed and contribute to alterations in renal function and the hypertension seen in the adult. Low nephron number by itself may not result in adult disease but in conjunction with exposure to a high salt after birth, a significant deficit is revealed.[15]

How May Low Nephron Number Be Programmed?

If we accept that a decrease in glomerular (thus nephron) number is indeed "programmed" in utero, then we need to examine how this could come about. What sort of genes should we be looking at and at what stage of development? To do this we need to understand the process of normal kidney development.

Kidney Development

Renal development in mammals involves the development of three sets of excretory organs, the pronephros, mesonephros (transitory organs) and the metanephros (Fig. 1). The metanephros (permanent kidney) develops from two embryonic precursor cell populations, the

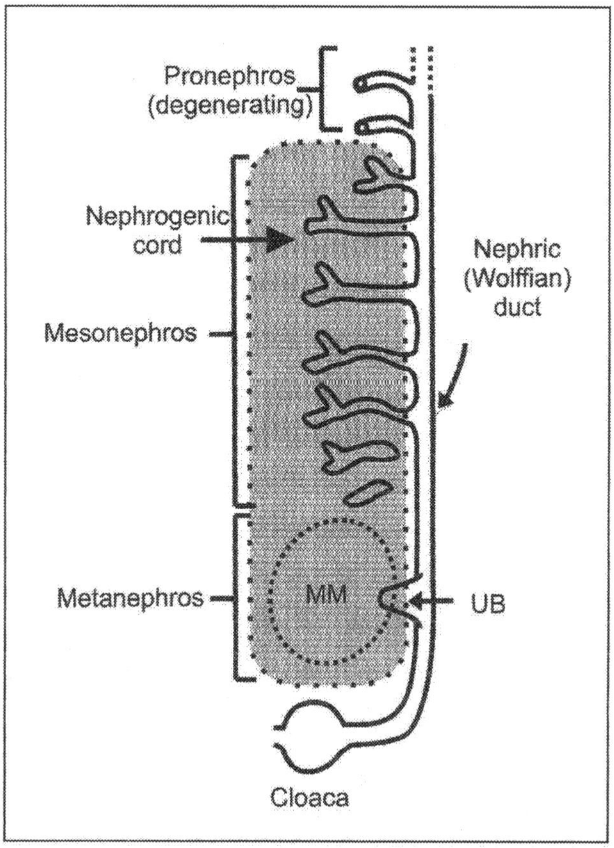

Figure 1. Development of the permanent mammalian kidney. Schematic diagram illustrating the development of the mammalian excretory system. The pronephros and mesonephros are transitory and degenerate. The metanephros forms when the ureteric bud grows (UB) laterally from the nephric (Wolffian) duct to induce the metanephric mesenchyme (MM) the caudal portion of the nephrogenic cord (grey).

ureteric bud (UB) and the metanephric mesenchyme (MM). The UB is a single outbranching of epithelium from the Wolffian duct (WD), which grows towards and penetrates the caudal portion of intermediate mesoderm known as the MM (Fig. 1). The MM consists of undifferentiated mesenchymal cells, which aggregate around the terminal portion of the advancing UB (Fig. 2A). The tip of the UB induces surrounding nephron progenitors of the MM to survive, proliferate, condense (Fig. 2B) and epithelialise forming a renal vesicle (Fig. 2C) which then develops through stages of comma shaped body (Fig. 2D), S-shaped body (Fig. 2E) and uriniferous tubule (Fig. 2F) to differentiate into nephrons (Fig. 2G,H). Simultaneously cells in direct contact with the UB stimulate the ureteric epithelium to proliferate and branch dichotomously (Fig. 2C,D). Continued iterations of branching of the ureteric epithelium ultimately gives rise to the collecting duct system, including the ureter, renal pelvis, calyces and collecting tubules with each tip capable of inducing nephrons. It is the reciprocal inductive interactions between these two tissues which generates a *finite* number of nephrons and the complete collecting duct system.

Although the number of nephrons varies extensively between species, there are two major factors that influence final nephron number: (1) the extent of UB branching; and (2) the

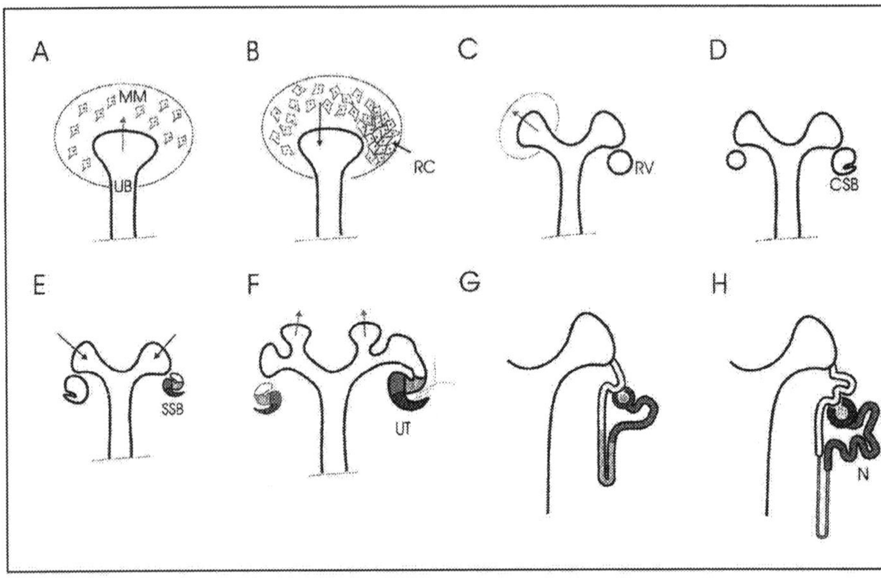

Figure 2. Ureteric branching morphogenesis and its relationship to nephron endowment. A) Ureteric bud (UB) invades the metanephric mesenchyme (MM) and induces mesenchyme to undergo nephrogenesis (arrow). B) Mesenchyme condenses at base of the tip of UB forming a renal condensate (RC). Simultaneously the non-condensing MM induces the UB to lengthen and bifurcate (arrow). C) An epithelial renal vesicle (RV) is formed when RC epithelialises. The new tip induces surrounding MM to condense (arrow). D) The RV develops into a Comma-shaped body (CSB) formed by invagination of a cleft in the proximal regionof the vesicle. E) The CSB differentiates into an S-shaped body (SSB) formed by invagination of a second cleft in the distal region. Each tip is again induced to bifurcate. F) An Uriniferous tubule (UT) is formed when the upper limb of the S-shaped body elongates and fuses to the tip of the ureteric duct. Each new tip induces a new nephron. G) The lower limb of the S-shaped body (black) ultimately forms glomerular podocytes and the Bowmans capsule. Cells lining the proximal cleft retain their cuboidal shape ultimately differentiating into glomerular podocytes. Cells on the external portion of the proximal cleft flatten forming the parietal layer of Bowman's capsule. The upper limb (white) lengthens and differentiates into the distal tubule, the middle section elongates and loops ultimately forming the loop of Henle (light grey), while the lower portion differentiates into the proximal convoluted tubul (dark grey). H) The UT undergoes growth, differentiation and elongation to form the complete nephron (N).

ability of the mesenchyme to differentiate in response to inductive signals. Thus, nephron number is likely to be highly dependent on factors regulating both ureteric bud growth and nephrogenesis during development and any change in the spatial-temporal expression of such factors may influence the final number of nephrons in kidney. Many of these signals have been identified at the molecular level (see Table 1).[30-78] These include factors regulating branching morphogenesis as well as genes controlling development and differentiation of the MM.

The exact relationship between branching events and the induction of nephrons is ill defined since development of the kidney relies on the reciprocal interaction between both these tissues. Although branching of the ureteric tree is considered directly related to the number of nephrons formed it is not considered to be a one-to-one relationship as with different species there are various types of branching in the developing kidney.[79] However, even a small decrease in branching efficiency iterated many times would result in a considerable deficit in branches forming the collecting system.[80] The efficiency of branching morphogenesis therefore has clinical consequences involving the link between nephron number and the risk of and response to renal disease. This link has stimulated investigation into the role of renal development in the

Table 1. *Key genes expressed during kidney development and their roles in regulating branching morphogenesis and nephron number as shown by in vivo and in vitro studies*

Gene	In Vivo	In Vitro	References
Activin A	Null mutants are perinatal lethal but have normal kidneys. Transgenic mice expressing truncated activin receptor contain 180% normal number of nephrons and glomeruli were 65% normal size.	Excess Activin A disrupts branch ing pattern resulting in a reduced number of secondary branch generations and enlarges the tips of UB.	30-32
AT2R	Null mutants show abnormalities in UB formation, are hypertensive and show multicystic dysplastic kidneys, hypoplastic kidneys (CAKUT).		33,34
BMP2	Null mutants die at E9.5 prior to metanephric development. Hets kidneys appear normal and contain normal nephron numbers.	Excess BMP2 inhibits branching.	35,36
BMP4	Null mutants die at E6.5 prior to metanephric development Hets display hypo/dysplastic kidneys, ectopic ureterovesical junction double collecting system. Hets with normal renal phenotype show no alteration in nephron number (unpublished data).	Excess BMP4 inhibits branching and alters branch pattern, inhibits nephrogenesis by 50%.	37-40.
BMP7	Null mutants show kidney hypoplasia, limited CD and nephron development post E12.5.	Low levels of excess BMP7 increase number of tubules formed by mIMCD-3 cells. Low levels of excess BMP7 increase & high levels inhibit branching in organ culture.	41-43
c-ret	Null mutants display bilateral/ unilateral agenesis, decreased branching and undifferentiated MM.		44
	Hets appear normal.		45
Emx2	Null mutants display bilateral agenesis, UB invades MM but fails to branch.		
Eya1	Null mutants display bilateral agenesis. Hets display renal hypoplasia/ unilateral agenesis.		46
Fgf7	Null mutants show a smaller collecting duct system (reduced UB growth), resulting in 30% fewer nephrons.	Excess FGF7 enhances branching and nephron number.	47
Foxd1 (BF2)	Null mutants display hypoplasia with large condensates, few nephrons and underdeveloped CD.		48

continued on next page

Table 1. Continued

Gene	In Vivo	In Vitro	References
GDNF	Null mutants show renal agenesis (no UB formed). HETs show dysplasia with reduced CD development, contain 30% fewer, larger glomeruli and are hypertensive.	Excess GDNF increases branch ing. GDNF neutralizing antibody inhibits branching.	49-51
Gfrα1	Null mutants show bilateral agen esis or a unilateral rudimentary kidney. Hets appear normal.		52
HGF	Null mutants show no renal phenotype.	HGF stimulates branching. Inhibiting HGF ceases UB development and nephron formation.	53-57
Hoxa11/ Hoxd11	Null mutants show variable severities from agenesis to dysplasia due to abnormal and reduced branching, elongated primary branches.		58,59
HSPG		Inhibiting synthesis of HSPG impairs UB branching. Excess HSPGs in culture reduces nephron number.	60,61
LIF		Inhibits centrifugal growth of ureteric tree, reducing number of tips in NZ and inhibiting nephron development. Triggers nephrogenesis in isolated MM.	62,63
Pax2	Null mutants display bilateral agenesis due to no WD or UB. Hets show increased apoptosis and reduced UB branching.		64,65
Pod1	Null mutants show renal hypoplasia due to decreased branching.		66
RARα/ RARβ2	Null mutants show impaired collecting duct development (decrease in UB tips and number of branch iterations).	Addition of RA increases branching and nephron formation.	67-69
TGFα		Antibodies to TGFα inhibits branching and nephrogenesis.	70
TGFβ1	Null mutants are perinatal lethal and show normal kidneys.	Excess TGFβ1 inhibits branching.	71-74
Wnt11	Null mutants show kidney hypoplasia with 36% fewer nephrons.		75
Wnt4	Null mutants have small, dysgenic kidneys with no glomeruli. UB invades the MM and undergoes some branching, MM condenses but cannot undergo MET.		76,77
WT1	Null mutants display bilateral agenesis due to the absence of UB outgrowth.		78

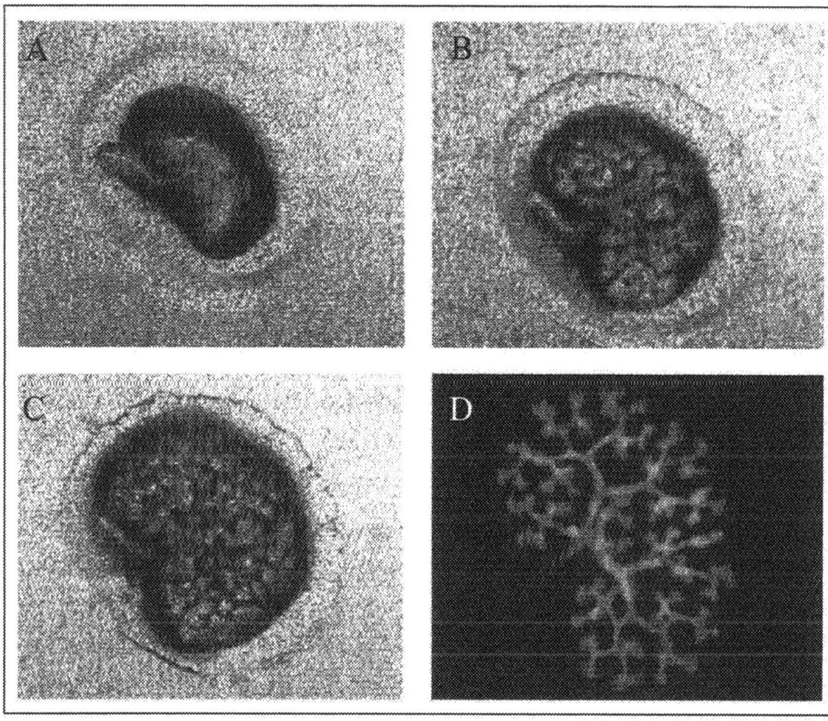

Figure 3. Whole organ culture of an E12 mouse metanephros for 48 hours. Phase contrast photomicrographs of an E12.5 metanephros at time of explanation (A); 24 hours culture (B) and 48 hours culture (C). Immunofluorescence photomicrograph of the ureteric tree at 48 hours culture (D).

determination of nephron endowment with many researchers utilizing in vitro culture systems and in vivo models (gene mutation studies) of kidney development to analyse not only the extent of branching morphogenesis but also nephron endowment (see Table 1).

Models to observe branching have been developed to allow the analysis of the process of branching at various levels as well as both the direct and indirect role branching has on nephron endowment. Both branching events and nephron induction can be followed in whole embryonic metanephric organ culture (E11-12 in the mouse and E13-14 in the rat) (Fig. 3).[81] Whole organ culture is an extremely powerful technique, in that the researcher can analyse both branching and nephron formation. However, the fact that both branching and nephron formation are highly dependent on each other makes the direct effects of branching difficult to characterize.

The direct effects on branching can also be analysed using isolated UBs where the UB is separated from the surrounding MM and maintained in culture in the presence of MM cell-derived conditioned medium.[82] These cultures are therefore independent of the inductive processes with the MM. Furthermore, branching of epithelial cells can also be observed in isolation, by observing renal epithelial cells in 3D collagen matrix gels. The renal epithelial cell lines: Madin-Darby canine kidney (MDCK) cells undergo tubulogenesis and branching morphogenesis in the presence of hepatocyte growth factor (HGF), mouse inner medullary collecting duct-3 (mIMCD-3) cells, and UB cells derived from the embryonic kidney.[83,84] These cell lines can undergo branching morphogenesis in response to certain growth factors. Growth factors, neutralizing antisera, antisense oligonucleotides etc may be added to these models systems to analyse the function and alterations of various molecular targets and their effect on branching and nephron endowment can be quantified through various means (Fig. 3).

Table 2. Maternal factors known to affect kidney development

Treatment/State	Species	Type/Experiment
Undernutrition	Human, Sheep, Rat	Global
	Rat	Protein restriction
Vitamin A deficiency	Rat	
Iron deficiency	Human, Rat	Anemia
Placental insufficency	Rat	Uterine artery ligation
	Human	Placental malfunction
Maternal stress (glucocorticoids)	Sheep	Endogenous glucocorticoids
	Sheep, Rat	Exogenous glucocorticoids
Maternal drug intake	Human	Inhibitors of renin-angiotensin system
	Human	Inhibitors of cyclo-oxygenases
	Human	Cocaine
Twin-twin transfusion syndrome	Human	
Urinary tract obstruction	Human, Sheep, Rat	
Maternal disease	Human, Rat	Diabetes

Influence of Maternal Health on Fetal Renal Development

Maternal diabetes (gestational or types 1 and 2), anemia, high blood pressure, renal insufficiency and preeclampsia are a few examples of maternal disease states that have been shown to alter fetal renal development. Table 2 documents examples where nephron number is affected by maternal health. Low protein and global undernutrition have been extensively used in experimental models but in the human it is more likely to be micronutrient deficiency or placental malfunction that results in programming effects. Hypertension was found to result in the offspring of iron-deficient pregnant rats and this was associated with a decreased nephron number in female offspring.[85,86] Half the pregnant women in the world are estimated to be anemic to some degree.[87] Preexisting maternal diabetes can have serious damaging effects on the fetal kidney including caudal regression syndrome associated with partial or total renal agenesis.[88] Another area of great interest is what happens to nephrogenesis in the very premature baby. The use of glucocorticoids and surfactants have enabled the survival of babies from as early as 22 weeks. This is some 14 weeks before the normal completion of nephrogenesis. Does nephrogenesis continue normally in these infants ex utero despite large changes in renal blood flow and renal function or is there an early cessation of nephron formation resulting in a permanent nephron deficit? Animal models for this condition are very difficult to establish due to the difficulties and costs of keeping premature animals alive. In one recent autopsy study, extreme prematurity (<1000g and 29 weeks) caused a reduction in the number of layers of glomeruli formed suggesting nephrogenesis has not continued normally outside the uterus.[89] The study was complicated by the use of nephrotoxic antibiotics and indomethacin exposure in the preterm infants. Glomerular counts using unbiased stereological techniques are required in infants that are born preterm and survive until nephrogenesis is complete to determine if a nephron deficit results.

Compensatory Changes in the Kidney

The Renal Renin-Angiotensin-System

Strong evidence is emerging in a number of animal models to suggest significant alterations in the renal renin-angiotensin system may be a common feature of programming. Interestingly, recent data suggests that in programmed animals during the period of nephrogenesis, there is a

Table 3. Evidence of compensatory changes in the renal RAS in studies of fetal programming

Species	Fetal Exposure	Age Studied	Change	Reference
Sheep	Dex (26-28d)	75 days 130 days	↓ AT1 mRNA and protein ↔ AT2 mRNA ↑ AT1 mRNA ↑ AT2 mRNA ↑ Angiotensinogen mRNA ↔ in basal renal function Altered renal response to infused ANG II in conscious fetuses	Moritz et al (unpub.) 90
Sheep	Cortisol (26-28d)	75 days 130 days	↓ AT1 mRNA and protein ↓ AT2 mRNA ↑ AT1 mRNA ↔ AT2 mRNA	Moritz et al, (unpub.) 90
Sheep	IUGR (all pregnancy)	>135 days	↑ dependence of BP on Angiotensin II	91
Sheep	Undernutrition (28-77 days)	Term	↑ AT1 mRNA ´ AT2 mRNA	92
Human	IUGR	Term	↑Cord plasma renin activity	93
Rats	Low protein	4 weeks	↑ AT1 protein ↔ in basal renal function ↑ response to angiotensin II	94
Rats	Low protein	4 weeks 10 weeks	↔ AT1a mRNA ↓ AT2 mRNA ↑ response to ANG II in anaesthetised rats	95
Rats	Low protein	Newborn	↓ tissue renin concentrations ↓ tissue angiotensin II concentrations	14
Rats	Low protein	Newborn 4 weeks	↓ AT1 protein ↑ AT2 mRNA ↓ AT2 protein ↑ AT1 protein ↔ plasma angiotensin I or II ↔ Renal angiotensin I or II	96
Rats	Low protein	12 weeks	Blood pressure normalised by ACE inhibition between weeks 2-4	97

decrease in renal expression of angiotensin receptors (AT1 and AT2) but after completion of nephron formation, there is a compensatory increase in receptor expression. This may lead to an increase in renal responsiveness to infused angiotensin II. In Table 3 the evidence for this is summarised.[14,90-97] It is speculated that a deficit in full expression of the RAS may contribute to inadequate nephrogenesis as well as a compensatory increase in components of this system in later in life and this in turn contributes to the development of hypertension. The ratio of AT1 to AT2 receptors is also likely to be important as in general these receptors mediate opposing effects in the kidney.[98]

Sodium Transporters

Long term regulation of blood pressure is dependent upon sodium and fluid homeostasis. So far, there are only two reports of 'programmed' changes in renal sodium transporters. The kidneys of offspring of rats exposed to a low protein during pregnancy, had higher levels of mRNA for both α_1 and β_1 subunits of Na/K/ATPase than those exposed to a normal protein.[99] The tubules were thus more 'adult-like' as kidneys stimulated to increase sodium retention, by aldosterone or angiotensin II, normally increase the α subunit but decrease the β subunit.[100,101] The sodium/potassium/2 chloride cotransporter (NKCC2) and the thiazide sensitive sodium/chloride (NCC) transporters were up-regulated at 4 weeks of age in offspring of rats exposed to a low protein intake (6%) from day 12 to parturition. There were no changes in the sodium/hydrogen exchanger (NHE3), or any epithelial sodium (ENaC) channels.[102] Investigation of renal sodium transporters is an area in which further research is required. These changes in postnatal gene expression may be an epigenetic modification. There is evidence for nutritional perturbation of epigenetic gene regulation in the susceptibility to chronic metabolic disease.[103] A recent observation suggests an epigenetic mechanism is involved in dexamethasone programming of metabolic disease.[104] Readers are referred to other chapters in this book dealing with epigenetic modifications.

Measuring Renal Function

Finally, in order to develop early intervention strategies, it is necessary to be able to identify individuals born with a low nephron endowment who may be predisposed to later renal disease. In an important study, birth weight in humans was found to be significantly correlated with nephron number with a predictive increase of 250,000 nephrons for each 1kg increase in birth weight.[105] Thus, should low birth weight individuals be considered a high risk for developing early renal failure? If so what can be measured or examined to detect this? Lack of agreement on the most appropriate tests to be done to test kidney function in children has led to guidelines being developed by the National Kidney Foundation's Kidney Disease Outcomes Quality Initiative. The two most critical recommendations were the measurement of protein to creatinine ratios in spot urines and estimation of glomerular filtration rate (GFR) from serum creatinine using predictive equations which take into account the patient's gender and height.[106,107] Other techniques including noninvasive measurement of GFR may also be useful in both human and experimental animal models.

Conclusions/Summary

The kidney is undoubtedly an important organ in the fetal programming of adult disease. The process of nephrogenesis in the developing fetus can be disrupted by a large number of maternal factors highlighting the susceptibility of the kidney to a sub-optimal environment. The precise mechanisms through which a reduced nephron number occurs are yet to be fully elucidated and remain a high priority for subsequent investigation. Optimal methodology in assessing nephron number needs to be employed routinely and further analysis of renal function is required to clearly define deficiencies in the adult. Clearly there are still some vital questions requiring investigation in the future. These include

1. Do extremely premature babies develop the full complement of nephrons?
2. Do animals exposed to hyperglycemia during the nephrogenic period develop hypertension as adults?
3. When a reduced nephron endowment occurs as a result of vitamin A deficiency, does hypertension result in the offspring?
4. Which genes are altered epigenetically to result in decreased nephrogenesis and/or postnatal hypertension?

References

1. Moritz KM, Dodic M, Wintour EM. Kidney development and the fetal programming of adult disease. Bioessays 25; 1-9:2003.
2. Kett MM, Bertram JF. Nephron endowment and blood pressure: What do we really know? Curr Hypertens Rep 2004; 6:133-9.
3. Huxley RR, Shiell AW, Law CM. The role of size at birth and postnatal catch up growth in determining systolic blood pressure: A systematic review of the literature. J Hypertens 2000; 18:815-831.
4. Godfrey KM. The role of the placenta in fetal programming - a review. Placenta 2002; 23:S20-S27.
5. Alebiosu CO. An update on "progression promoters" in renal disease. J Natl Med Assoc 2003; 95:30-42.
6. Lackland DT, Bendall HE, Osmond C et al. Low birth weights contribute to high rates of early onset chronic renal failure in the Southeastern United States. Arch Intern Med 2000; 160:1472-1476.
7. Hoy WE, Rees M, Kile E et al. A new dimension to the Barker hypothesis: Low birth weight and susceptibility to renal disease. Kidney Inter 1999; 56:1072-1077.
8. Vasarhelyi B, Dobos M, Reusz GS et al. Normal kidney finction and elevated natriuresis in young men born with low birth weight. Pediatr Nephrol 2000; 15:96-100.
9. Kistner A, Celsi G, Vanpee M et al. Increased blood pressure but normal renal function in adult women born preterm. Pediatr Nephrol 2000; 15:215-220.
10. Jacobsen P, Rossing P, Tarnow T et al. Birth weight- a risk factor for progression of diabetic nephropathy? J Intern Med 2003; 253:343-350.
11. Brenner BM, Garcoa DL, Anderson S. Glomeruli and blood pressure: Less of one, more of the other. Am J Hypertens 1988; 1:335-347.
12. Hoy WE, Douglas-Denton RN, Hughson MD et al. A stereological study of glomerular number and volume: Preliminary findings in a multiracial study of kidneys at autopsy. Kidney Inter Suppl 2003; 83:S31-7.
13. Langley Evans SC, Welham SJ, Jackson AA. Fetal exposure to a maternal low protein diet impairs nephrogenesis and promotes hypertension in the rat. Life Sci 1999; 64:965-974.
14. Woods LL, Ingelfinger JR, Nyengaard JR et al. Maternal protein restriction suppresses the newborn renin-angiotensin system and programs adult hypertension in rats. Pediatr Res 2001; 49:460-467.
15. Woods LL, Weeks DA, Rasch R. Programming of adult blood pressure by maternal protein restriction: role of nephrogenesis. Kidney Inter 2004; 65:1339-1348.
16. Wintour EM, Moritz KM, Johnson K et al. Reduced nephron number in adult sheep, hypertensive as a result of prenatal glucocorticoid treatment. J Physiol 2003; 3(549):929-935.
17. Dodic M, May CN, Wintour EM et al. An early prenatal exposure to glucocorticoids leads to hypertensive offspring in sheep. Clin Sci 1998; 94:149-155.
18. Moritz KM, Wintour EM. Functional development of the meso- and metanephros. Pediatr Nephrol 1999; 13:171-178.
19. Celsi G, Kistner A, Aizman R et al. Prenatal dexamethasone causes oligonephronia, sodium retention, and higher blood pressure in the offspring. Pediatr Res 1998; 44:317-22.
20. Ortiz LZ, Quan A, Weinberg A et al. Effect of prenatal dexamethasone on rat renal development. Kidney Inter 2001; 59:1663-1669.
21. Ortiz LA, Quan A, Zarzar F et al. Prenatal dexamethasone programs hypertension and renal injury in the rat. Hypertension 2003; 41:328-34.
22. Morley R, Dwyer T. Fetal origins of adult disease? Clin Exp Pharmacol Phsyiol 2000; 28:962-966.
23. Bertram J. Counting in the kidney. Kidney Inter 2001; 59:792-796.
24. Merlet-Benichou C, Gilbert T, Moreau E et al. Nephron number: Variability is the rule. Lab Invest 1999; 79:515-527.
25. NarkumBurgess DM, Nolan CR, Norman JE et al. Forty five year follow-up after uninephrectomy. Kidney Inter 1993; 43:1110-1115.
26. Zimanyi MA, Bertram JF, Black MJ. Nephron number and blood pressure in rat offspring with maternal high-protein diet. Pediatr Nephrol 2002; 17:1000-4.
27. Moritz KM, Wintour EM, Dodic M. Fetal uninephrectomy leads to postnatal hypertension and compromised renal function. Hypertens 2002; 39:1071-1076.
28. Woods LL. Neonatal uninephrectomy causes hypertension in adult rats. Am J Physiol 1999; 276:R974-R978.
29. Douglas-Denton R, Moritz KM, Bertram JF et al. Compensatory renal growth after unilateral nephrectomy of the ovine fetus. J Am Soc Nephrol 2002; 13:406-410.

30. Ritvos O, Tuuri T, Eramaa M et al. Activin disrupts epithelial branching morphogenesis in developing glandular organs of the mouse. Mech Dev 1995; 50:229-245.
31. Jhaveri S, Erzurumlu RS, Chiaia N et al. Defective whisker follicles and altered brainstem patterns in activin and follistatin knockout mice. Mol Cell Neurosci 1998; 12(4-5):206-219.
32. Kojima I, Maeshima A, Zhang YQ. Role of the activin-follistatin system in the morphogenesis and regeneration of the renal tubules. Mol Cell Endocrinol 2001; 180(1-2):179-182.
33. Brock IIIrd JW, Hunley TE, Adams MC et al. Role of the renin-angiotensin system in disorders of the urinary tract. J Urol 1998; 160:1812-1819.
34. Nishimura H, Yerkes E, Hohenfellner K et al. Role of the angiotensin type 2 receptor gene in congenital anomalies of the kidney and urinary tract, cakut, of mice and men. Mol Cell 1999; 3(1):1-10.
35. Zhang H, Bradley A. Mice deficient for BMP2 are nonviable and have defects in amnion/chorion and cardiac development. Development 1996; 122(10):2977-2986.
36. Martinez G, Mishina Y, Bertram JF. BMPs and BMP receptors in mouse metanephric development: In vivo and in vitro studies. Int J Dev Biol 2002; 46(4):525-533.
37. Winnier G, Blessing M, Labosky PA et al. Bone morphogenetic protein-4 is required for mesoderm formation and patterning in the mouse. Genes Dev 1995; 9(17):2105-2116.
38. Dunn RN, Winner G, Hargett LK et al. Haploinsufficient phenotypes in BMP-4 heterozygous null mice and modification by mutations in gli3 and alx4. Dev Biol 1997; 188:235-247.
39. Miyazaki Y, Oshima K, Fogo A et al. Bone morphogenetic protein 4 regulates the budding site and elongation of the mouse ureter. J Clin Invest 2000; 105(7):863-873.
40. Raatikainen-Ahokas A, Hytonen M, Tenhunen A et al. BMP-4 affects the differentiation of metanephric mesenchyme and reveals an early anterior-posterior axis of the embryonic kidney. Dev Dyn 2000; 217(2):146-158.
41. Jena N, Martin-Seisdedos C, M[c]Cue P et al. BMP7 null mutation in mice: Developmental defects in skeleton kidney and eye. Exp Cell Res 1997; 230:28-37.
42. Dudley AT, Lyons KM, Robertson EJ. A requirement for bone morphogenetic protein-7 during development of the mammalian kidney and eye. Genes Dev 1995; 9:2795-2807.
43. Piscione TD, Yager TD, Gupta IR et al. BMP-2 and Op-1 exert direct and opposite effects on renal branching morphogenesis. Am J Physiol Renal Physiol 1997; 42:F961-F975.
44. Schuchardt A, D'agati V, Pachnis V et al. Renal agenesis and hypoplasia in Ret-k$^-$ mutant mice result from defects in ureteric bud development. Development 1996; 122:1919-1929.
45. Miyamoto N, Yoshida M, Kuratani S et al. Defects of urogenital development in mice lacking Emx2. Development 1997; 124:1653-1664.
46. Xu PX, Adams J, Peters H et al. Eya1-deficient mice lack ears and kidneys and show abnormal apoptosis of organ primordial. Nat Gen 1999; 23:113-117.
47. Qiao J, Uzzo R, Obara-Ishihara T et al. FGF-7 modulates ureteric bud growth and nephron number in the developing kidney. Development 1999; 126(3):547-554.
48. Hatini V, Huh SO, Herzlinger D et al. Essential role of stromal mesenchyme in kidney morphogenesis revealed by targeted disruption of winged helix transcription factor BF-2. Genes Dev 1996; 10:1467-1478.
49. Pichel JG, Shen L, Sheng HZ et al. GDNF is required for kidney development and enteric innervation. Cold Spring Harb Symp Quant Biol 1996; 61:445-457.
50. Cullen-McEwen LA, Kett MM, Dowling J et al. Nephron number, renal function, and arterial pressure in aged GDNF heterozygous mice. Hypertens 2003; 41(2):335-40.
51. Cullen-McEwen LA, Drago J, Bertram JF. Nephron endowment in glial cell line-derived neurotrophic factor (GDNF) heterozygous mice. Kidney Inter 2001; 60(1):31-36.
52. Cacalano G, Farinas I, Wang L et al. GFRα-1 is an essential receptor component for GDNF in the developing nervous system and kidney. Neuron 1998; 21:53-62.
53. Montesano R, Matsumoto K, Nakamura T et al. Identification of a fibroblast-derived epithelial morphogen as hepatocyte growth factor. Cell 1991; 67(5):901-908.
54. Santos O, Barros EJ, Yang XM et al. Involvement of hepatocyte growth factor in kidney development. Dev Biol 1994; 163(2):525-529.
55. Santos OFP, Nigam SK. HGF-induced tubulogenesis and branching of epithelial cells is modulated by extracellular matrix and TGF-β. Dev Biol 1993; 160:293-302.
56. Woolf AS, Kolatsi-Joannou M, Hardman P et al. Roles of hepatocyte growth factor/scatter factor and the met receptor in the early development of the metanephros. J Cell Biol 1995; 128:171-184.
57. Van Adelsberg J, Sehgal S, Kukes A et al. Activation of hepatocyte growth factor (HGF) by endogenous HGF activator is required for metanephric kidney morphogenesis in vitro. J Biol Chem 2001; 276:15099-15106.

58. Davis AP, Witte DP, Hscih-LI HM et al. Absence of radius and ulna in mice lacking Hox-a11 and Hox-d11. Nature 1995; 375:791-195.
59. Patterson LT, Pembaur M, Potter S. Hoxa11 and Hoxd11 regulate branching morphogenesis of the ureteric bud in the developing kidney. Development 2001; 128:2153-2161.
60. Platt JL, Trescony P, Lindman B et al. Heparin and heparin sulfate delimit nephron formation in fetal metanephric kidneys. Dev Biol 1990; 139:338-348.
61. Davies JML, Gallagher J, Garrod D. Sulphated proteoglycan is required for collecting duct growth and branching but not nephron formation during kidney development. Development 1995; 12:1507-1517.
62. Bard JB, Ross AS. Lif, the ES-cell inhibition factor, reversibly blocks nephrogenesis in cultured mouse kidney rudiments. Development 1991; 113(1):193-198.
63. Barasch J, Yang J, Ware CB et al. Mesenchymal to epithelial conversion in rat metanephros is induced by Lif. Cell 1999; 99(4):377-386.
64. Torres M, Gomez-Pardo E, Dressler GR et al. Pax-2 controls multiple steps of urogenital development. Development 1995; 121(12):4057-4065.
65. Porteous S, Torban E, Cho NP et al. Primary renal hypoplasia in humans and mice with pax2 mutations: Evidence of increased apoptosis in fetal kidneys of Pax-2(1neu) +/- mutant mice. Hum Mol Genet 2000; 9(1):1-11.
66. Quaggin SE, Schwartz L, Cui S et al. The basic-helix-loop-helix protein Pod1 is critically important for kidney and lung organogenesis. Development 1999; 126(24):5771-83.
67. Vilar J, Gilbert T, Moreau E et al. Metanephros organogenesis is highly stimulated by vitamin A derivatives in organ culture. Kidney Inter 1996; 49(5):1478-1487.
68. Moreau E, Vilar J, LelievrePegorier M et al. Regulation of c-ret expression by retinoic acid in rat metanephros: Implication in nephron mass control. Am J Physiol Renal Physiol 1998; 275:F938-F945.
69. Mendelsohn C, Batourina E, Fung S et al. Stromal cells mediate retinoid-dependent functions essential for renal development. Development 1999; 126:1139-1148.
70. Rogers S, Ryan G, Hammerman MR. Metanephric transforming growth factor-alpha is required for renal organogenesis in vitro. Am J Physiol Renal Physiol 1992; 262:F533-F539.
71. Shull MM, Ormsby I, Kier AB et al. Targeted disruption of the mouse transforming growth factor-β1 gene results in multifocal inflammatory disease. Nature 1992; 359:693-699.
72. Rogers SA, Ryan G, Purchio AF et al. Metanephric transforming growth factor-β1 regulates nephrogenesis in vitro. Am J Physiol Renal Physiol 1993; 264:F996-1002.
73. Sakurai H, Nigam SK. Transforming growth factor-β selectively inhibits branching mophogenesis but not tubulogenesis. Am J Physiol Renal Physiol 1997; 272:F139-F146.
74. Clark AT, Young RJ, Bertram JF. In vitro studies on the roles of transforming growth factor-beta 1 in rat metanephric development. Kidney Inter 2001; 59(5):1641-1653.
75. Majumdar A, Vainio S, Kispert A et al. Wnt11 and RET/GDNF pathways cooperate in regulating ureteric branching during metanephric kidney development. Development 2003; 130(14):3175-3185.
76. Stark K, Vainio S, Vassileva G et al. Epithelial transformation of metanephric mesenchyme in the developing kidney regulated by Wnt-4. Nature 1994; 372:679-683.
77. Kispert A, Vainio S, Mcmahon AP. Wnt-4 is a mesenchymal signal for epithelial transformation of metanephric mesenchyme in the developing kidney. Development 1998; 125(21):4225-4234.
78. Kreidberg JA, Sariola H, Loring JM et al. Wt-1 is required for early kidney development. Cell 1993; 74(4):679-691.
79. Al-Awqati Q, Goldberg M. Architectural patterns in branching morphogenesis in the kidney. Kidney Int 1998; 54(6):1832-1842.
80. Shah MM, Sampogna RV, Sakurai H et al. Branching morphogenesis and kidney disease. Development 2004; 131(7):1449-1462.
81. Welham SJ, Woolf AS. Organ culture of intact metanephric kidneys. Methods Mol Med 2003; 86:169-177.
82. Qiao J, Sakurai H, Nigam SK. Branching morphogenesis independent of mesenchymal-epithelial contact in the developing kidney. Proc Natl Acad Sci USA 1999; 96:7330-7335.
83. Montesano R, Matsumoto K, Nakamura T et al. Identification of a fibroblast-derived epithelial morphogen as hepatocyte growth factor. Cell 1991; 67(5):901-908.
84. Santos OFP, Moura LA, Rosen EM et al. Modulation of HGF-induced tubulogenesis and branching by multiple phosphorylation mechanisms. Dev Biol 1993; 159:535-548.
85. Crowe C, Dandekar P, Fox M et al. The effects of anaemia on heart, placenta and body weight, and blood pressure in fetal and neonatal rats. J Physiol 1995; 488:515-519.
86. Lisle SJ, Lewis RM, Petry CJ et al. Effect of maternal iron restriction during pregnancy on renal morphology in the adult rat offspring. Br J Nutr 2003; 90(1):33-39.

87. Van de Broek N. Anaemia and micronutrient deficiencies. Br Med Bull 2003; 67:149-160.
88. Chugh SS, Wallner EI, Kanwar YS. Renal development in high glucose ambience and diabetic embryopathy. Seminars Nephrol 2003; 23:583-592.
89. Rodriguez MM, Gomez AH, Abitbol CL et al. Histomorphometric analysis of postnatal glomerulogenesis in extremely preterm infants. Pediatr Dev Pathol 2004:(in press).
90. Moritz KM, Johnson K, Douglas-Denton R et al. Maternal glucocorticoid treatment programs alterations in the renin-angiotensin system of the ovine fetal kidney. Endocrinol 2002; 143:4455-63.
91. Edwards LJ, McMillan IC. Periconceptual nutrition programs development of the cardiovascular system in sheep. Am J Physiol Regul Integr Comp 2002; 283:R669-R679.
92. Whorwood CB, Firth KM, Budge H et al. Maternal undernutrition during early to midgestation programs tissue-specific alterations in the expression of the glucocorticoid receptor, 11beta-hydroxysteroid dehydrogenase isoforms, and type 1 angiotensin II receptor in neonatal sheep. Endocrinol 2001; 142:2854-2864.
93. Konje JC, Bell SC, Morton JJ et al. Human fetal kidney morphometry during gestation and the relationship between weight, kidney morphometry and plasma active renin concentration at birth. Clin Sci (Lond) 1996; 91(2):169-175.
94. Sahajpal V, Ashton N. Renal function and angiotensin AT1 receptor expression in young rats following intrauterine exposure to a maternal low-protein diet. Clin Sci (Lond) 2003; 104(6):607-614.
95. McMullen S, Gardiner DS, Langley-Evans SC. Prenatal programming of angiotensin II type 2 receptor expression in the rat. Br J Nutr 2004; 91(1):133-140.
96. Vehaskari VM, Stewart T, Lafont D et al. Kidney angiotensin receptor expression in prenatally programmed hypertension. Am J Physiol 2004:(in press).
97. Sherman RC, Langley-Evans SC. Early administration of angiotensin-converting enzyme inhibitor captopril, prevents the development of hypertension programmed by intrauterine exposure to a maternal low-protein diet in the rat. Clin Sci (Lond) 1998; 94(4):373-381.
98. Siragy HM. AT1 and AT2 receptor in the kidney: Role in health and disease. Seminars Nephrol 2004; 24:93-100.
99. Bertram C, Trowern AR, Copin N et al. The maternal diet during pregnancy programs altered expression of the glucocorticoid receptor and type 2 11beta-hydroxysteroid dehyrogenase: Potential mechanisms underlying the programming of hypertension in utero. Endocrinology 2001; 142:2841-2853.
100. Knepper MA, Kim GH, Masilamani S. Renal tubule sodium transporter abundance profiling in rat kidney: Response to aldosterone and variations in NaCl intake. Ann NY Acad Sci 2003; 986:562-569.
101. Beutler KT, Masilamani S, Turban S et al. Long-term regulation of ENaC expression in kidney by angiotensin II. Hypertens 2003; 41:1143-1150.
102. Manning J, Beutler K, Knepper MA et al. Upregulation of renal BSC1 and TSC in prenatally programmed hypertension. Am J Physiol Renal Physiol 2002; 283:F202-F206.
103. Waterland RA, Jirtle RL. Early nutrition, epigenetic changes at transposons and imprinted genes, and enhanced susceptibility to adult chronic disease. Nutrition 2004; 20:63-68.
104. Drake AJ, Walker BR, Seckl JR. Intergenerational consequences of fetal programming by in utero exposure to glucocorticoids in rats. Am J Physiol Regul Integr Comp Physiol 2004:(in press).
105. Hughson M, Farris IIIrd AB, Douglas-Denton R et al. Glomerular number and size in autopsy kidneys: The relationship to birth weight. Kidney Inter 2003; 63:2113-22.
106. Hogg RJ, Furth S, Lemley KV et al. National Kidney Foundation's disease outcome quality initiative clinical practice guidelines for chronic kidney disease in children and adolescents: Evaluation, classification and stratification. Pediatr 2003; 111:1416-1421.
107. Schwartz GJ, Haycock GB, Edelman CM et al. A simple estimate of glomerular filtration rate in children derived from body length and plasma creatinine. Pediatrics 1976; 58:259-263.

CHAPTER 12

Programming of Obesity—Experimental Evidence

Bernhard H. Breier,* Stefan O. Krechowec and Mark H. Vickers

Abstract

Obesity and related metabolic disorders are prevalent health issues in modern society and are commonly attributed to lifestyle and dietary factors. However, the mechanisms by which environmental factors modulate the physiological systems that control weight regulation and the aetiology of metabolic disorders, which manifest in adult life, may have their roots before birth. The 'fetal origins' or 'fetal programming' paradigm is based on observations that environmental changes can reset the developmental path during intrauterine development leading to obesity and cardiovascular and metabolic disorders later in life. The mechanisms underlying the relationship between prenatal influences and postnatal obesity and related disorders are relatively unknown and remain speculative, as are the interactions between genetic and environmental factors. While many endocrine systems can be affected by fetal programming recent experimental studies suggest that leptin and insulin resistance are critical endocrine defects in the pathogenesis of programming-induced obesity and metabolic disorders. Recent work has also shown that offspring of undernourished mothers display reduced locomotor activity in adult life, independent of postnatal diet. However, it remains to be determined whether postnatal obesity is a consequence of programming of sedentary behaviour and appetite regulation and whether hyperphagia is the main underlying cause of the increased adiposity and the development of metabolic disorders.

Introduction

Obesity and related metabolic disorders have become a major health issue for modern society in the 21st century. It is a widely held view that dietary and lifestyle factors are the primary cause of these diseases in the general population. The mechanisms by which diet and other environmental factors influence physiological systems that control appetite, weight regulation, behaviour and the aetiology of metabolic disease are poorly understood.

Increasing evidence suggests that chronic metabolic and cardiovascular diseases which commonly manifest in adult life can have their roots before birth. The 'fetal origins of adult disease' (FOAD) or 'fetal programming' paradigm is based on the observation that adverse stimuli during the prenatal period can alter the developmental path resulting in an increased susceptibility to obesity and cardiovascular and metabolic disorders later in life. This pathogenesis is not based on genetic defects but on altered genetic expression, which occurs as a result of fetal adaptations to adverse intrauterine influences. After initial controversy when these relationships were first suggested, both prospective clinical and experimental studies have clearly shown

*Corresponding Author: Bernhard H. Breier—Liggins Institute for Medical Research, University of Auckland, Private Bag 92019, 2-6 Park Avenue, Grafton, Auckland, New Zealand.
Email: bh.breier@auckland.ac.nz

Early Life Origins of Health and Disease, edited by E. Marelyn Wintour and Julie A. Owens.

that the propensity to develop abnormalities of cardiovascular, endocrine and metabolic homeostasis in adulthood are increased when fetal development has been adversely affected.[1,2]

The mechanisms underlying the relationship between prenatal influences and postnatal outcome are relatively unknown and remain speculative, as are the interactions between genetic and environmental factors. One general thesis is that in response to an adverse intrauterine stimulus, the fetus adapts its physiological development to maximise its immediate chances for survival. These adaptations may include resetting of metabolic homeostasis and endocrine systems and the down-regulation of growth, commonly reflected in altered birth phenotype. This prenatal plasticity of the fetus may allow environmental factors to alter the physiological function of the fetus in preparation for sub-optimal environmental conditions after birth.[3] It is thought that whilst these changes in fetal physiology may be beneficial for short-term survival in utero they may be maladaptive in postnatal life, contributing to poor health outcomes when offspring are exposed to catch-up growth, diet-induced obesity and other factors.

Experimental Evidence for Programming of Obesity

Animal models have been used extensively to investigate the basic physiological principles of the FOAD hypothesis. The variety of models that have been developed is essential to the search for the mechanistic links between prenatal influences and postnatal pathophysiological outcomes. An example is the diabetic pregnant rat, in which the long-term effects on offspring following diabetic pregnancy can be investigated, a study which would not be possible in humans as treatment is ethically proscribed. Although epidemiological data suggest that fetal programming occurs within the normal range of birth size,[2] most experimental work has tended to focus on significant restriction of fetal growth in the assumption that the nature of the insults that impair fetal growth are likely to be those that trigger fetal programming. Alterations in maternal nutrition are commonly used experimentally to induce intrauterine growth retardation (IUGR), as it is an experimentally practical and reproducible way to induce nutrient limitation to the fetus and thus change its developmental trajectory. In this context IUGR is not essential to fetal programming, but is merely a surrogate for evidence that fetal development has been affected.

Several animal models of early growth restriction have been developed in an attempt to elucidate its relationship with adult onset disease and provide a framework for investigating the underlying mechanisms. Animal studies have clearly shown that prenatal undernutrition programs not only postnatal cardiovascular dysfunction but also obesity, elevated plasma leptin concentrations, glucose intolerance, and even activity levels and dietary preferences. In rats hypertension, insulin resistance and obesity have been induced in offspring by maternal undernutrition,[4-6] a low protein diet,[7] maternal uterine artery ligation,[8] maternal dexamethasone (DEX) treatment[9] or prenatal exposure to the cytokines interleukin (IL)-6 and tumour necrosis factor (TNF)-alpha.[10]

There are also increasing experimental data in other species. In guinea pigs, IUGR caused by uterine artery ligation or maternal undernutrition results in reduced glucose tolerance, increased sensitivity to cholesterol loading[11] and elevated blood pressure in offspring.[12] DEX treatment of pregnant ewes in early gestation results in elevated blood pressure[13] and altered regulation of lipolysis[14] in the adult offspring. Placental restriction by carunclectomy has also been associated with adiposity and increased insulin sensitivity during postnatal life.[15] Undernutrition for 10 days in late gestation ewes alters postnatal HPA axis function of their lambs, but not glucose metabolism. However, when the period of prenatal undernutrition is extended to 20 days, glucose metabolism is altered in adult offspring, but not HPA axis function.[16,17] Work by Bispham et al has shown that, independent of maternal nutrition in late gestation, fetuses sampled from ewes with nutrient restriction in early gestation possessed more adipose tissue, whereas when ewes were fed to appetite throughout gestation, fetal adipose tissue deposition and leptin mRNA abundance were both reduced. These changes suggested that offspring of nutrient restricted mothers were at increased risk of developing obesity in later life.[18] This study also suggested that the increased incidence of obesity in adults born to mothers exposed to the Dutch famine during early pregnancy[19] may be a direct consequence of adaptations in the endocrine sensitivity of fetal adipose tissue.

Nutritional Programming

Fetal undernutrition has been highlighted as a primary factor involved in the early life origins of adult disease. Within the laboratory, fetal undernutrition can most commonly be achieved through maternal dietary restriction during pregnancy. Manipulation of maternal nutrition during pregnancy has been known to alter fetal growth and development for some time.[20] At present, rodent models investigating the mechanistic links between maternal undernutrition and adult disease generally utilise one of two dietary protocols; global undernutrition or isocaloric low protein diets. The maternal low protein (MLP) diet during pregnancy and lactation is one of the most extensively utilised models of nutritional programming.[21-25] This model involves ad libitum feeding to pregnant rats a low protein diet containing 5-8% (w/w) protein (casein), generally a little under half the protein content but equivalent in energy of a control diet containing 18-20% (w/w) protein.[21,26] Offspring from protein restricted mothers are around 15-20% lighter at birth.[23] Maintenance of a MLP diet during lactation enhances this weight difference and permanently restricts later growth. If restricted offspring are cross-fostered to mothers fed a control diet, offspring exhibit rapid catch-up growth.[23] This catch-up growth appears to have a detrimental effect on life span, resulting in premature death which is associated with accelerated loss of kidney telomeric DNA.[27]

Restricted protein offspring exhibit significantly elevated blood pressure at an early age in comparison to controls.[22] However this finding has not been consistent and is related to differences in the composition of the low protein diet.[26] The diet used by Langley-Evans supplemented with methionine has been consistent in programming hypertension; low protein diets without methionine supplementation report either no change or a slight depression in blood pressure.[26] These results highlight the importance that the balance of micronutrients plays in determining the long-term health effects of maternal nutrition during pregnancy. Investigations into fetal programming of hypertension in offspring consistently reveal a reduction in renal mass, increased apoptosis without a balancing increase in cell proliferation and reductions in glomeruli number.[28] Changes in kidney structure and development are associated with enhanced activity of the renin-angiotensin-system (RAS).[29] Increased expression of glucocorticoid receptors and reduced expression of 11β-HSD-2 in the kidney is associated with an enhancement of glucocorticoid mediated increases in blood pressure.[30]

Carbohydrate metabolism in offspring is also altered by a MLP diet during pregnancy. Fasting plasma insulin and glucose levels are lower in MLP offspring and are associated with improved insulin sensitivity in early adulthood. However programmed offspring exhibit a greater age dependent loss of glucose tolerance.[24] By 15 months of age glucose tolerance in MLP offspring is significantly diminished compared to that of controls, and is associated with hyperinsulinemia in males and hypoinsulinemia in females.[24] The mechanisms behind these phenotypic observations include altered development of the pancreas and insulin signalling. The MLP diet programs pancreatic function in restricted offspring through a reduction in β-cell proliferation, islet size and vascularity coupled with an enhancement of β-cell apoptosis.[21] Subsequently, it has been shown that a defect in glucose-mediated insulin secretion from islets of adult MLP offspring only manifest when an additional dietary insult such as high fat feeding is introduced postnatally.[31]

Global undernutrition at various times during pregnancy is another widely used approach to induce nutritional programming. Various models have been developed with different levels of undernutrition during different periods of pregnancy. A mild nutritional restriction to 70% of normal intake in the first 18 days of pregnancy in the rat results in offspring with significant growth retardation at birth that catch up to controls at postnatal day 20.[32] Restricted offspring exhibit elevated blood pressure in adult life with an increased vasoconstriction response to

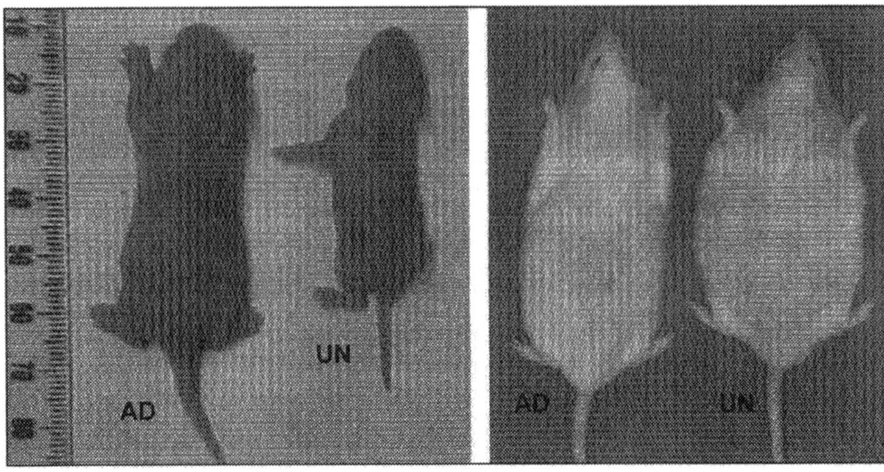

Figure 1. Postnatal phenotype at birth (left photograph) and at postnatal day 145 (right photograph) of offspring of ad libitum fed mothers (AD) and mothers undernourished throughout gestation (UN). Offspring of undernourished mothers remain shorter throughout adult life but develop obesity, independent of postnatal diet.

potassium and thromboxane A_2 mimetics. These abnormalities increase with age and are most pronounced in male offspring.[32] Another model using a nutritional restriction to 50% of standard ad libitum intake in the second half of gestation had no effect on blood pressure, with only subtle alterations in vasoconstrictive ability being observed.[33] In another study a 50% restriction of a normal diet from day 15 of pregnancy showed that 21-day-old rat fetuses had significantly decreased pancreatic insulin content.[34]

We have developed a rodent model of fetal programming using maternal undernutrition throughout pregnancy. On day one of pregnancy animals are randomly assigned to a standard rat diet ad libitum throughout pregnancy (ad libitum (AD) group) or 30% of the AD group intake of the standard diet throughout gestation (undernourished, UN group). After birth, litter size and birth weights are recorded and litter size is adjusted to 8 pups per litter. The number of pups born per litter from UN and AD mothers is identical in this experimental approach; it is not affected by maternal undernutrition. The UN offspring are cross-fostered within 24 hours of birth onto AD dams to assure adequate and standardised nutrition from birth until weaning. At birth offspring of UN mothers had fetal and placental weights that were 25–30% lower than offspring of AD mothers. A lack of catch-up growth despite a standard postnatal diet[4] was accompanied by a transient reduction in circulating IGF-I and hepatic IGF-I mRNA expression which normalised at weaning. Consistent with this observation, we also showed that UN offspring had a reduced responsiveness to growth hormone (GH) during the neonatal period, possibly reflecting delayed maturation of the somatotrophic axis, which was fully restored before puberty.[4,35] In addition, UN offspring developed elevated blood pressure in adult life.[36,37]

Postnatal Nutrition

We can distinguish two conceptually different types of interactions between prenatal influences and postnatal nutrition in the pathogenesis of metabolic disorders and obesity. Diet-induced obesity during postnatal life can amplify pathogenic mechanisms established by adverse prenatal influences.[6] Alternatively, changes caused by prenatal influences can facilitate a disease process that is induced by postnatal environmental factors such as nutrition. An example is the development of obesity and insulin resistance in individuals who are exposed to a

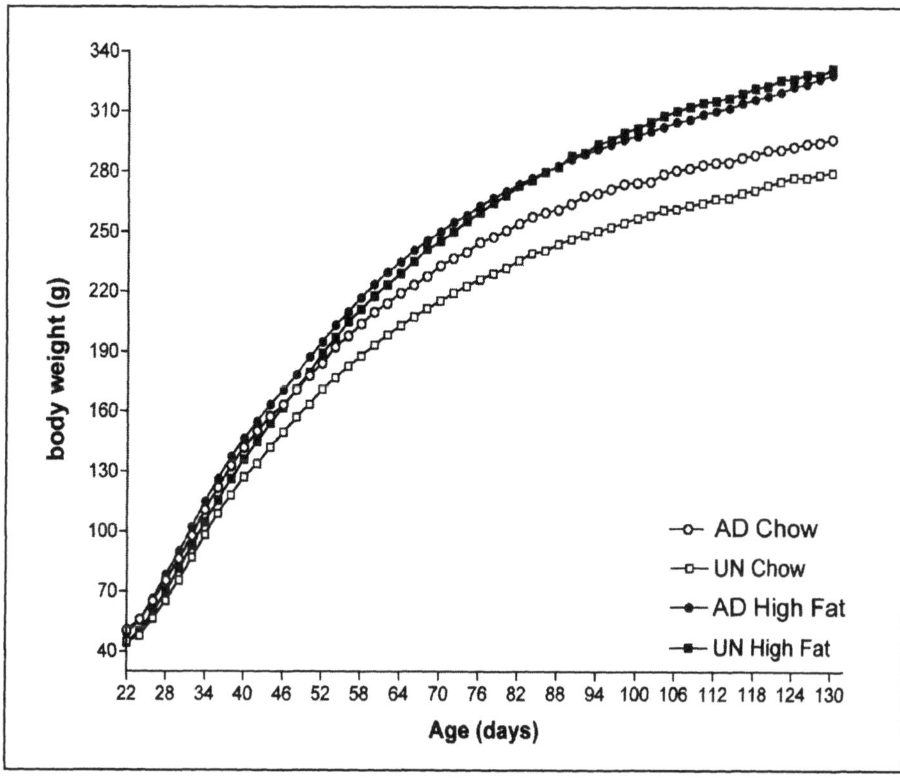

Figure 2. Postnatal growth curves of offspring from mothers fed either ad libitum (AD) or undernourished throughout gestation (UN). After weaning offspring of AD and UN mother were fed either a standard chow diet or a high fat diet. Note the rapid diet-induced catch-up growth in the UN offspring fed a high fat diet.

high fat diet during postnatal life. The direction of the interactions between prenatal and postnatal influences is most likely dependent on timing and severity of each factor.

In historically undernourished, recently urbanised populations such as India, where individuals of low birth weight are exposed to a high-fat Western diet, the incidence of obesity and type 2 diabetes is reaching epidemic proportions.[38] Work by Yajnik and colleagues has shown that although Indian babies are born of low birth weight, they exhibit relatively increased visceral adiposity and hyperinsulinemia at birth.[39] Such observations have been explained by the "thrifty-phenotype" hypothesis proposed by Hales and Barker[40] and may illustrate the long-term disadvantage of postnatal "catch-up" growth. Although there is considerable debate whether catch-up growth in early postnatal life is beneficial or not, most studies suggested that postnatal "catch-up" growth is associated with adverse outcomes in later life.[41-44]

Epidemiological studies have shown that the greatest insulin resistance is observed in people of low birth weight who develop obesity as adults.[45] In rats, the combination of prenatal undernutrition with retarded fetal growth, and good postnatal nutrition with accelerated growth, leads to a striking reduction in life span.[24,27] The well-established concept that diets high in saturated fats play a key role in the development of insulin resistance and obesity has recently been extended to the frequency of food intake. Zammit et al[46] suggested that the pathogenesis of insulin resistance may be related to a pattern of frequent snacking which results in a continuous post-prandial state for most of the day. This prevents the attainment of low basal inter-prandial insulin levels even in normal individuals. Prolonged exposure of the liver to high basal insulin, through its stimulatory effect on hepatic triglycerides and very-low-density lipoprotein secretion, may contribute to the initial induction of muscle insulin resistance.[46]

In our studies, we introduced offspring of undernourished rats to a hypercaloric (high fat/high protein) diet after weaning to investigate whether enhanced nutritional supply would facilitate postnatal catch-up growth. This led to development of obesity during adult life.[6,47] UN offspring also developed hypertension, hyperinsulinemia, hyperleptinemia and hyperphagia independent of postnatal diet. Postnatal hypercaloric nutrition amplified the existing cardiovascular, metabolic and endocrine abnormalities of UN offspring.[6] Interestingly, hyperphagia was established before puberty independent of caloric content of the diet and increased with advancing age.[6] The increased plasma insulin and leptin concentrations were paralleled by altered pancreatic histology.[48] The hyperleptinemia and hyperinsulinemia seen in UN offspring may be a mechanism induced by a nutrient-deprived fetal environment to store large quantities of triglycerides when food is plentiful, thus representing a competitive advantage ("thrifty phenotype") in preparation for a nutrient-deprived postnatal environment.[40] However, when hypercaloric nutrition persists for long periods of time, adipogenic diabetes may develop. Our work to date cannot resolve whether the primary defect in this cascade is in appetite regulation, peripheral metabolism, or altered leptin or insulin action. We have shown that therapy with insulin-like growth factor-1 or growth hormone can ameliorate obesity, hyperphagia and hypertension induced by fetal programming and high fat nutrition, but the precise mechanisms underlying these effects are yet to be resolved.[47,49] We have recently reported that maternal undernutrition can induce sedentary behaviour in offspring.[50] Hyperphagia and concomitant obesity in offspring are amplified by hypercaloric nutrition. In the course of these studies we noted that the onset of the abnormal eating behaviour occurred prior to puberty, thus preceding the development of obesity.

The "Couch Potato" Syndrome

We have recently reported experimental evidence suggesting that maternal undernutrition can induce sedentary behaviour, hyperphagia and concomitant obesity in offspring independent of postnatal dietary influences.[50] We had previously shown that maternal undernutrition throughout pregnancy in the rat results in hypertension, hyperinsulinemia and hyperleptinemia in the offspring when they reach adulthood.[6] Obesity was not present until after puberty and was associated with hyperphagia. In the course of these studies we noted that the onset of the abnormal eating behaviours occurred prior to puberty, thus preceding the development of obesity. This led us to speculate that the prenatal maternal environment might also affect other components of behaviour associated with the Metabolic Syndrome.

Voluntary locomotor activity was assessed in prenatally undernourished offspring at various ages from the peri-pubertal period to adulthood. This was done following three habituation trials using Optimax behavioral testing apparatus. The animals were fed either a standard diet or a hypercaloric diet throughout postnatal life. Offspring of undernourished mothers were significantly more sedentary in postnatal life than those born of ad-libitum fed mothers for all parameters measured and this was independent of postnatal diet. Analysis of food intake revealed hyperphagia in mature offspring that had been exposed to maternal undernutrition. This was independent of postnatal diet, although it was exacerbated by hypercaloric nutrition. Importantly, in the animals tested at a peri-pubertal age, diminished locomotor activity was already present prior to the development of maturity-onset obesity and was significantly reduced in males compared to females.

These results suggest that maternal undernutrition can lead to the development of both overeating and diminished exercise behaviour concomitant with the physiological features of the Metabolic Syndrome. The former observation raises the intriguing possibility that some behaviours and lifestyle choices that exacerbate the Metabolic Syndrome in humans may not be voluntary but may be an inherent part of the syndrome and may have a prenatal origin. Our recent studies suggest that the "couch potato" syndrome may have its origins during prenatal development. This has major implications for public health policy. Health care funding may be better spent on improving pregnancy care rather than waiting until metabolic and cardiovascular disorders manifest in offspring years or decades later.

Endocrine and Metabolic Mechanisms

The precise mechanisms underlying the programming of adult disease by maternal nutrient restriction remain a matter of debate. Maternal undernutrition or low protein diet during the last week of gestation in the pregnant rat leads to reduction of fetal pancreatic β-cell mass[51] and increased apoptosis of immature β-cells.[52] While insulin-stimulated glucose uptake in adipocytes is increased during early postnatal life due to increased insulin receptor number,[53] there is greater age-dependent loss of glucose tolerance and later insulin resistanc.[24,54] The reduced insulin action is associated with reduced phosphatidylinositol 3-kinase (PI3K) and protein kinase B (PKB) activation and altered fatty acid metabolism.[55]

Another important target of prenatal events is the liver, where glucocorticoids regulate several metabolic processes, including hepatic enzymes regulating carbohydrate and fat metabolism. Rats exposed to DEX in the last trimester of pregnancy show increased phosphoenolpyruvate carboxykinase (PEPCK) gene transcription and increased activity of this rate-limiting enzyme of gluconeogenesis in the liver.[9] These animals have adult hyperglycemia and increased hepatic glucocorticoid receptor expression. Similarly, structural changes in the liver and altered expression patterns of gluconeogenic enzymes and glucose handling have been reported after a maternal low protein diet.[56]

Neuroendocrine regulatory systems are also vulnerable to disturbances in early life which can lead to permanent structural changes, including reduced cerebral vascularity[57] and dysfunction of central nervous system regulation. Maternal low protein nutrition results in structural changes in the mediobasal hypothalamic nuclei in weanling offspring and fewer neurons immunopositive for neuropeptide Y in the arcuate hypothalamic nucleus.[58] These neuroendocrine changes are accompanied by the development of obesity and diabetogenic disturbances later in life. The HPA axis is particularly susceptible to prenatal influences. For example, neonatal disturbance of mother-pup interactions permanently alters plasma adrenocorticotropic hormone (ACTH) and corticosterone concentrations and GR levels in hippocampus, paraventricular nucleus (PVN) and pituitary.[59] Similar GR abnormalities have been described following either nutritional manipulation or glucocorticoid administration to the mother. Maternal glucocorticoids alter the utilisation of alternative exon 1 sequences coding for promotor regions on the GR gene in offspring.[60] This is strong evidence that programming can cause permanent alterations in gene expression and could explain the increase in basal plasma corticosterone levels in adulthood that may contribute directly to hypertension and hyperglycemia.[61]

It is well established that prenatal stress can influence the development of neural systems that control endocrine responses to stress and regulate behavioural traits.[59,62] Maternal glucocorticoid administration or prenatal stress in rats leads to development of decreased locomotor activity and increased defecation and avoidance behaviour in an 'open field' test.[63,64] An animal model of fetal programming by maternal tumour necrosis factor (TNF)-α administration showed reduced locomotor activity of offspring.[10] While recent studies in the rat have focused on maternal behaviour during the neonatal period which influences the HPA axis activity and anxiety behaviour[59] there is little published experimental information on the effects of maternal nutrition on offspring behaviour. One group reported that maternal low protein nutrition during pregnancy in the rat lead to changes in exploratory behaviour, social interactions and avoidance behaviour in offspring.[65,66] Since it is well established that diet-induced obesity in animals leads to a reduction in locomotor activity,[67] these studies are in general agreement with our observation of increased eating behaviour in offspring of undernourished mothers and our observation of their decreased locomotor activity, as discussed above.

Programming of the Adipoinsular Axis and Altered Adipogenesis

It is important to note that very few animal studies have addressed interactions between pre and postnatal nutrition. However, other studies that have investigated diet-induced obesity point to a link between peripheral leptin resistance and insulin resistance in the development of obesity. The physiological role of hyperleptinemia associated with caloric excess has been proposed to relate to the protection of nonadipocytes from lipid oversupply that would lead to steatosis and lipotoxicity.[68] Elevated leptin production as a result of short-term caloric excess prevents the up-regulation of lipogenesis and increases fatty acid oxidation, thus reducing lipid supply to peripheral tissue during caloric excess.[68] In diet-induced obesity, peripheral leptin function is at first normal. However, prolonged caloric excess results in dysregulation of post-receptor leptin signalling. This causes accumulation of triglycerides and lipid metabolites, providing fatty acid substrate for the damaging effects of nonoxidative metabolism leading to functional impairment of nonadipose tissue and a progression to type 2 diabetes and cardiovascular disease.[69]

A range of genetic components of obesity have been identified.[70-72] and research on alterations in biochemical pathways caused by single gene mutations in animal models has contributed significantly towards knowledge of physiological mechanisms of obesity.[73] It is well established that leptin acts at the level of the hypothalamus to regulate appetite and energy homeostasis.[74] The long-form, or signalling form, of the leptin receptor (OB-Rb) is expressed in high levels in several cell groups of the hypothalamus and in various tissues throughout the body.[48,75,76] Under normal physiological conditions, increased leptin signalling in the medial hypothalamus is associated with reduced neuropeptide Y (NPY) and agouti-related (AgRP) protein production[77] but increased cocaine- and amphetamine- regulated transcript (CART) and pro-opiomelanocortin (POMC) production.[78-80] These leptin-induced changes in neuropeptides lead to decreased food intake and increased energy expenditure.

In obese individuals, elevated plasma leptin is proposed to uncouple leptin action on its receptors in the hypothalamus, thereby disrupting signal transduction pathways that exert effects on satiety and energy expenditure.[74] Direct leptin signalling in peripheral tissues has only recently been demonstrated. For example, increased leptin signalling in muscle tissue has been shown to blunt lipogenesis and stimulate lipid oxidation.[81] There is also growing evidence for a feedback system between leptin and insulin which links the brain and the endocrine pancreas with other peripheral insulin and leptin sensitive tissues in the control of feeding behaviour, metabolic regulation and body energy balance.[82] This endocrine system has been termed the adipoinsular axis.[82] When adipose stores decrease, falling leptin concentrations permit increased insulin production resulting in the deposition of additional fat. Conversely the suppressive effects of leptin on insulin production is mediated by the autonomic nervous system and by direct actions via leptin receptor on beta-cells.[82] Our data suggest that fetal programming by maternal undernutrition throughout gestation may lead to dysregulation of the adipoinsular feedback system. Relative leptin resistance by pancreatic β–cells may contribute to hyperinsulinism which further exacerbates adipogenesis. The hyperinsulinism and hyperleptinemia may also trigger the pathogenesis of hyperphagia.

Summary and Conclusions

Numerous epidemiological studies have shown that perturbations in early life can have persistence consequences for health in later life. Both prospective clinical studies and experimental research have clearly shown that the propensity to develop abnormalities of cardiovascular, endocrine and metabolic homeostasis in adulthood is increased when fetal development has been adversely affected. The pathogenesis is not based on genetic defects but on altered genetic expression as a consequence of an adaptation to environmental changes during fetal development.

Studies of the interaction between maternal undernutrition throughout pregnancy followed by postnatal hypercaloric nutrition in the rat have shown that offspring from undernourished mothers are growth retarded at birth and develop obesity, hypertension, hyperleptinemia, hyperinsulinemia and hyperphagia during postnatal life. Close associations between a major rise in circulating insulin and leptin concentrations and a large increase in appetite and fat mass provide evidence for disturbed endocrine communication between the hypothalamus, adipose tissue and the endocrine pancreas in the pathogenesis of programming-induced obesity. Hypercaloric nutrition during postnatal life greatly amplifies prenatal effects on metabolic abnormalities, obesity, overeating and diminished exercise behaviour. However, it remains to be determined whether postnatal obesity is a consequence of programming of sedentary behaviour or whether defects in appetite regulation and hyperphagia are the main underlying cause of the increased adiposity and the development of metabolic disorders.

Fetal programming research offers a novel approach to investigate the mechanistic basis of metabolic disorders, hyperphagia and diminished exercise behaviour which in human populations predominantly arise from environmental factors and lifestyle choices. The use of animal models can establish model conditions that will reliably provide high contrasts of phenotypes. Such studies offer an exciting potential for advances in our understanding of critical determinants and mechanisms for human obesity and metabolic disorders.

Acknowledgements

The authors acknowledge support from the Health Research Council of New Zealand. Stefan O. Krechowec is the recipient of a Bright Future Scholarship from the Foundation for Research, Science and Technology.

References

1. Godfrey KM, Barker DJP. Fetal nutrition and adult disease. Am J Clin Nutr 2000; 1344s-52s.
2. Barker DJP. Mothers, Babies and Health in Later Life. Edinburgh: Churchill Livingstone, 1998:1-217.
3. Gluckman PD. Editorial: Nutrition, glucocorticoids, birth size, and adult disease. Endocrinology 2001; 142:1689-91.
4. Woodall SM, Breier BH, Johnston BM et al. A model of intrauterine growth retardation caused by chronic maternal undernutrition in the rat: Effects on the somatotropic axis and postnatal growth. J Endocrinol 1996; 150:231-42.
5. Woodall SM, Johnston BM, Breier BH et al. Chronic maternal undernutrition in the rat leads to delayed postnatal growth and elevated blood pressure of offspring. Pediatr Res 1996; 40:438-43.
6. Vickers MH, Breier BH, Cutfield WS et al. Fetal origins of hyperphagia, obesity and hypertension and its postnatal amplification by hypercaloric nutrition. Am J Physiol 2000; 279:E83-E87.
7. Langley-Evans SC, Gardner DS, Jackson AA. Association of disproportionate growth of fetal rats in late gestation with raised systolic blood pressure in later life. J Reprod Fertil 1996; 106:307-12.
8. Rajakumar PA, He J, Simmons RA et al. Effect of uteroplacental insufficiency upon brain neuropeptide Y and corticotropin-releasing factor gene expression and concentrations. Pediatr Res 1998; 44:168-74.
9. Nyirenda MJ, Lindsay RS, Kenyon CJ et al. Glucocorticoid exposure in late gestation permanently programs rat hepatic phosphoenolpyruvate carboxykinase and glucocorticoid receptor expression and causes glucose intolerance in adult offspring. J Clin Invest 1998; 101:2174-81.
10. Dahlgren J, Nilsson C, Jennische E et al. Prenatal cytokine exposure results in obesity and gender-specific programming. Am J Physiol Endocrinol Metab 2001; 281:E326-E334.
11. Kind KL, Clifton PM, Katsman AI et al. Restricted fetal growth and the response to dietary cholesterol in the guinea pig. Am J Physiol 1999; 277:R1675-R1682.
12. Persson E, Jansson T. Low birth weight is associated with elevated adult blood pressure in the chronically catheterised guinea pig. Acta Physiol Scand 1992; 145:195-96.
13. Dodic M, Wintour EM, Whitworth JA et al. Effect of steroid hormones on blood pressure. Clin Exp Pharmacol Physiol 1999; 26:550-52.
14. Gatford KL, Wintour EM, De Blasio MJ et al. Differential timing for programming of glucose homoeostasis, sensitivity to insulin and blood pressure by in utero exposure to dexamethasone in sheep. Clin Sci 2000; 98:553-60.
15. Owens JA. Catch-up growth in early life: Causes and consequences in experimental paradigms. Pediatr Res 2004; 53:8a.

16. Oliver MH, Breier BH, Gluckman PD et al. Birth weight rather than maternal nutrition influences glucose tolerance, blood pressure and IGF-1 levels in sheep. Pediatr Res 2002; 52:516-24.
17. Bloomfield FH, Oliver MH, Giannoulias D et al. Brief undernutrition in late-gestation sheep programmes the hypothalamic-pituitary adrenal axis in adult offspring. Endocrinology 2003; 144:2933-40.
18. Bispham J, Gopalakrishnan GS, Dandrea J et al. Maternal endocrine adaptation throughout pregnancy to nutritional manipulation: Consequences for maternal plasma leptin and cortisol and the programming of fetal adipose tissue development. Endocrinology 2003; 144:3575-85.
19. Roseboom TJ, van der Meulen JH, Osmond C et al. Plasma lipid profiles in adults after prenatal exposure to the Dutch famine. Am J Clin Nutr 2000; 72:1101-06.
20. Dobbing J. Maternal nutrition in pregnancy-eating for two? Early Hum Dev 1981; 5(2):113-115.
21. Snoeck A, Remacle C, Reusens B et al. Effect of a low protein diet during pregnancy on the fetal rat endocrine pancreas. Biol Neonate 1990; 57:107-18.
22. Langley SC, Jackson AA. Increased systolic blood pressure in adult rats induced by fetal exposure to maternal low protein diets. Clin Sci 1994; 86:217-22.
23. Desai M, Crowther NJ, Lucas A et al. Organ-selective growth in the offspring of protein-restricted mothers. Br J Nutr 1996; 76:591-603.
24. Hales CN, Desai M, Ozanne SE et al. Fishing in the stream of diabetes: From measuring insulin to the control of fetal organogenesis. Biochem Soc Trans 1996; 24:341-50.
25. Petry CJ, Dorling MW, Pawlak DB et al. Diabetes in old male offspring of rat dams fed a reduced protein diet. Int J Exp Diabetes Res 2001; 2(2):139-143.
26. Langley-Evans SC. Critical differences between two low protein diet protocols in the programming of hypertension in the rat. Int J Food Sci Nutr 2000; 51:11-17.
27. Jennings BJ, Ozanne SE, Dorling MW et al. Early growth determines longevity in male rats and may be related to telomere shortening in the kidney. FEBS Lett 1999; 448:4-8.
28. Langley-Evans SC, Welham SJ, Jackson AA. Fetal exposure to a maternal low protein diet impairs nephrogenesis and promotes hypertension in the rat. Life Sci 1999; 64:965-74.
29. Langley-Evans SC, Sherman RC, Welham SJ et al. Intrauterine programming of hypertension: The role of the renin-angiotensin system. Biochem Soc Trans 1999; 27:88-93.
30. Bertram C, Trowern AR, Copin N et al. The maternal diet during pregnancy programs altered expression of the glucocorticoid receptor and type 2 11beta-hydroxysteroid dehydrogenase: Potential molecular mechanisms underlying the programming of hypertension in utero. Endocrinology 2001; 142(7):2841-2853.
31. Ozanne S. Metabolic programming in animals. Br Med Bull 2001; 60(143):152.
32. Ozaki T, Nishina H, Hanson MA et al. Dietary restriction in pregnant rats causes gender-related hypertension and vascular dysfunction in offspring. J Physiol 2001; 530:141-52.
33. Holemans K, Gerber R, Meurrens K et al. Maternal food restriction in the second half of pregnancy affects vascular function but not blood pressure of rat female offspring. Br J Nutr 1999; 81(1):73-79.
34. Blondeau B, Lesage J, Czernichow P et al. Glucocorticoids impair fetal beta-cell development in rats. Am J Physiol Endocrinol Metab 2001; 281:E592-E599.
35. Woodall SM, Bassett NS, Gluckman PD et al. Consequences of maternal undernutrition for fetal and postnatal hepatic insulin-like growth factor-I, growth hormone receptor and growth hormone binding protein gene regulation in the rat. J Mol Endocrinol 1998; 20:313-26.
36. Weder AB, Schork NJ. Adaptation, allometry, and hypertension. Hypertension 1994; 24:145-56.
37. Woodall SM, Breier BH, Johnston BM et al. Administration of growth hormone or IGF-I to pregnant rats on a reduced diet throughout pregnancy does not prevent fetal intrauterine growth retardation and elevated blood pressure in adult offspring. J Endocrinol 1999; 163:69-77.
38. Yajnik C. Interactions of perturbations in intrauterine growth and growth during childhood on the risk of adult-onset disease. Proc Nutr Soc 2000; 59:257-65.
39. Yajnik CS, Lubree HG, Rege SS et al. Adiposity and hyperinsulinemia in Indians are present at birth. J Clin Endocrinol Metab 2002; 87:5575-80.
40. Hales CN, Barker DJ. Type 2 (noninsulin-dependent) diabetes mellitus: The thrifty phenotype hypothesis. Diabetologia 1992; 35:595-601.
41. Ong KK, Dunger DB. Thrifty genotypes and phenotypes in the pathogenesis of type 2 diabetes mellitus. J Pediatr Endocrinol Metab 2000; 13(Suppl 6):1419-1424.
42. Bonora M, Boule M, Gautier H. Ventilatory strategy in hypoxic or hypercapnic newborns. Biol Neonate 1994; 65:198-204.
43. Ong KKL, Ahmed ML, Emmett PM et al. Association between postnatal catch-up growth and obesity in childhood: Prospective cohort study. BMJ 2000; 320:967-71.

44. Eriksson JG, Forsen T, Tuomilehto J et al. Catch-up growth in childhood and death from coronary heart disease: Longitudinal study. BMJ 1999; 318:427-31.
45. Phillips DI. Birth weight and the future development of diabetes. A review of the evidence. Diabetes Care 1998; 21(Suppl 2):B150-B155.
46. Zammit VA, Waterman IJ, Topping D et al. Insulin stimulation of hepatic triacylglycerol secretion and the etiology of insulin resistance. J Nutr 2001; 131:2074-77.
47. Vickers MH, Ikenasio BA, Breier BH. IGF-1 treatment reduces hyperphagia, obesity, and hypertension in metabolic disorders induced by fetal programming. Endocrinology 2001; 142:3964-73.
48. Vickers MH, Reddy S, Ikenasio BA et al. Dysregulation of the adipoinsular axis - a mechanism for the pathogenesis of hyperleptinemia and adipogenic diabetes induced by fetal programming. J Endocrinol 2001; 170:323-32.
49. Vickers MH, Ikenasio BA, Breier BH. Adult growth hormone treatment reduces hypertension and obesity induced by an adverse prenatal environment. J Endocrinol 2002; 175:615-23.
50. Vickers M, Breier B, McCarthy D et al. Sedentary behavior during postnatal life is determined by the prenatal environment and exacerbated by postnatal hypercaloric nutrition. Am J Physiol 2003; 285:R271-R273.
51. Petrik J, Reusens B, Arany E et al. A low protein diet alters the balance of islet cell replication and apoptosis in the fetal and neonatal rat and is associated with a reduced pancreatic expression of insulin-like growth factor-II. Endocrinology 1999; 140:4861-73.
52. Cherif H, Reusens B, Ahn MT et al. Effects of taurine on the insulin secretion of rat fetal islets from dams fed a low-protein diet. J Endocrinol 1998; 159:341-48.
53. Ozanne SE, Smith GD, Tikerpae J et al. Altered regulation of hepatic glucose output in the male offspring of protein-malnourished rat dams. Am J Physiol 1996; 270(4 Pt 1):E559-E564.
54. Ozanne SE, Dorling MW, Wang CL et al. Impaired PI 3-kinase activation in adipocytes from early growth-restricted male rats. Am J Physiol Endocrinol Metab 2001; 280:E534-E539.
55. Ozanne SE, Martensz ND, Petry CJ et al. Maternal low protein diet in rats programmes fatty acid desaturase activities in the offspring. Diabetologia 1998; 41:1337-42.
56. Burns SP, Desai M, Cohen RD et al. Gluconeogenesis, glucose handling, and structural changes in livers of the adult offspring of rats partially deprived of protein during pregnancy and lactation. J Clin Invest 1997; 100:1768-74.
57. Bennis-Taleb N, Remacle C, Hoet JJ et al. A low-protein isocaloric diet during gestation affects brain development and alters permanently cerebral cortex blood vessels in rat offspring. J Nutr 1999; 129:1613-19.
58. Plagemann A, Harder T, Rake A et al. Hypothalamic nuclei are malformed in weanling offspring of low protein malnourished rat dams. J Nutr 2000; 130:2582-89.
59. Meaney MJ. Maternal care, gene expression, and the transmission of individual differences in stress reactivity across generations. Annu Rev Neurosci 2001; 24:1161-92.
60. McCormick JA, Lyons V, Jacobson MD et al. 5' heterogeneity of glucocorticoid receptor messenger RNA is tissue specific: Differential regulation of variant transcripts by early-life events. Mol Endocrinol 2000; 14:506-17.
61. Seckl JR, Nyirenda MJ, Walker BR et al. Glucocorticoids and fetal programming. Biochem Soc Trans 1999; 27:74-78.
62. Ladd CO, Huot RL, Thrivikraman KV et al. PM. Long-term behavioral and neuroendocrine adaptations to adverse early experience. Prog Brain Res 2000; 122:81-103.
63. Welberg LA, Seckl JR. Prenatal stress, glucocorticoids and the programming of the brain. J Neuroendocrinol 2001; 13:113-28.
64. Welberg LAM, Seckl JR, Holmes MC. Prenatal glucocorticoid programming of brain corticosteroid receptors and corticotrophin-releasing hormone: Possible implications for behaviour. Neuroscience 2001; 104:71-79.
65. Almeida SS, Tonkiss J, Galler JR. Prenatal protein malnutrition affects exploratory behavior of female rats in the elevated plus-maze test. Physiol Behav 1996; 60:675-80.
66. Almeida SS, Tonkiss J, Galler JR. Prenatal protein malnutrition affects avoidance but not escape behavior in the elevated T-maze test. Physiol Behav 1996; 60:191-95.
67. Remke H, Wilsdorf A, Muller F. Development of hypothalamic obesity in growing rats. Exp Pathol 1988; 33:223-32.
68. Unger RC, Orci L. Lipotoxic diseases of nonadipose tissues in obesity. Int J Obes Relat Metab Disord 2001; 24:S28-S32.
69. Unger RH. Lipotoxicity in the pathogenesis of obesity-dependent NIDDM. Genetic and clinical implications. Diabetes 2001; 44:863-70.

70. Akiyama T, Tachibana I, Shirohara H et al. High-fat hypercaloric diet induces obesity, glucose intolerance and hyperlipidemia in normal adult male Wistar rat. Diabetes Res Clin Pract 1996; 31:27-35.
71. Kawano K, Hirashima T, Mori S et al. Spontaneous long-term hyperglycemic rat with diabetic complications. Otsuka Long-Evans Tokushima Fatty (OLETF) strain. Diabetes 1992; 41:1422-28.
72. Flatt PR, Bailey CJ. Development of glucose intolerance and impaired plasma insulin response to glucose in obese hyperglycaemic (ob/ob) mice. Horm Metab Res 1981; 13:556-60.
73. Campfield LA, Smith FJ, Burn P. Strategies and potential molecular targets for obesity treatment. Science 1998; 280(5368):1383-1387.
74. Ahima RS, Flier JS. Leptin. Annu Rev Physiol 2000; 413-37.
75. Baskin DG, Hahn TM, Schwartz MW. Leptin sensitive neurons in the hypothalamus. Horm Metab Res 1999; 31:345-50.
76. Baskin DG, Breininger JF, Schwartz MW. Leptin receptor mRNA identifies a subpopulation of neuropeptide Y neurons activated by fasting in rat hypothalamus. Diabetes 1999; 48:828-33.
77. Hahn TM, Breininger JF, Baskin DG et al. Coexpression of Agrp and NPY in fasting-activated hypothalamic neurons. Nat Neurosci 1998; 1:271-72.
78. Schwartz MW, Seeley RJ, Woods SC et al. Leptin increases hypothalamic pro-opiomelanocortin mRNA expression in the rostral arcuate nucleus. Diabetes 1997; 46:2119-23.
79. Thornton JE, Cheung CC, Clifton DK et al. Regulation of hypothalamic proopiomelanocortin mRNA by leptin in ob/ob mice. Endocrinology 1997; 138:5063-66.
80. Wang ZW, Zhou YT, Kakuma T et al. Comparing the hypothalamic and extrahypothalamic actions of endogenous hyperleptinemia. Proc Natl Acad Sci USA 1999; 96:10373-78.
81. Muoio DM, Dohm GL, Fiedorek FTJ et al. Leptin directly alters lipid partitioning in skeletal muscle. Diabetes 1997; 46:1360-63.
82. Kieffer TJ, Habener JF. The adipoinsular axis: Effects of leptin on pancreatic beta-cells. Am J Physiol 2000; 278:E1-E14.

CHAPTER 13

Perinatal Programming of Adult Metabolic Homeostasis:

Lessons From Experimental Studies

Kathryn L. Gatford,* Miles J. De Blasio, Miodrag Dodic, Dane M. Horton and Karen L. Kind

Abstract

Poor fetal growth and associated neonatal catch-up growth are independent risk factors for metabolic disease in later life. Epidemiological studies in humans consistently show associations of small size at birth and later glucose intolerance and/or diabetes. A primary defect is thought to be insulin resistance, which is associated with both small size at birth and neonatal catch-up growth. The available evidence suggests that this resistance may result from a signalling defect downstream of the insulin receptor in peripheral tissues. Recent evidence also suggests that insulin secretion may be impaired in the individual who was small at birth. Most of the contemporary data in humans relates later outcomes to size at birth rather than to specific exposures. Experimental models that restrict fetal growth or produce variation in size at birth have therefore been used to explore these associations between small size at birth, neonatal catch-up growth and later metabolic disease. In this chapter we will review what has been learnt from human and experimental studies about the mechanistic basis for poor metabolic homeostasis following restricted fetal growth and neonatal catch-up growth, and comment on future directions in this area.

Introduction

Rates of non-insulin dependent diabetes (NIDDM) and obesity, the most common disorders of metabolic homeostasis and major contributors to cardiovascular disease risk, are rapidly increasing in Australia and globally.[1,2] Some of this increase undoubtedly reflects changes in lifestyle.[3] Nevertheless, in mature adults at least, early life experiences can account for similar proportions of the incidence of the metabolic syndrome as do lifestyle factors.[4] Small-size-for-gestational age, indicated by low weight or thinness at birth, is consistently associated with increased risk of NIDDM in populations from developed countries.[5] The association of small size at birth with later obesity is less consistent, but in children and young adults is usually positive or, in some cases, J- or U-shaped, suggesting that individuals who were heavy at birth are at highest risk of obesity.[6] Individuals of low birth weight do however have altered fat deposition, with more central or abdominal fat,[6] identified as a stronger predictor of cardiovascular disease than total fat (reviewed by ref. 7). Both genetic and environmental factors affect size at birth. Small-size-for-gestational-age, however, indicates the failure of a fetus to achieve its genetic potential for growth, and results from maternal and placental environmental

*Corresponding Author: Kathryn L. Gatford—Department of Obstetrics and Gynaecology, University of Adelaide, South Australia 5005, Australia. Email: kathy.gatford@adelaide.edu.au

Early Life Origins of Health and Disease, edited by E. Marelyn Wintour and Julie A. Owens.

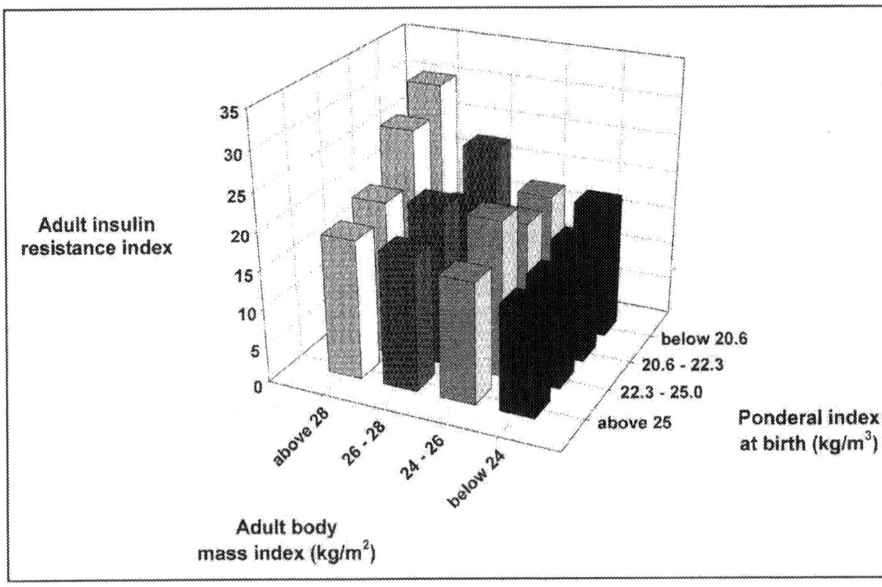

Figure 1. Thinness at birth, adult body mass index and adult insulin resistance across the population range in adult fatness (body mass index) in the Preston cohort studied by Phillips, Barker and colleagues, the increase in the insulin resistance index from the thinnest to the fattest adults was 71% in adults who were thinnest at birth but only 31% in adults who were fattest at birth (adapted from Phillips et al, 1994[114]).

factors that restrict fetal substrate supply.[8] This implicates the prenatal environment and poor substrate supply in particular, as initiating factors in development of impaired metabolic homeostasis in later life. There is also evidence that pre and post-natal factors act synergistically to increase the risk of later metabolic disease, so that the individual who grew poorly before birth is most susceptible to the impact of adverse adult lifestyle factors (Fig. 1). Changes in lifestyle are therefore likely to increase the impact of small size at birth on disease risk in contemporary cohorts. In addition to the risks of later disease directly associated with their small size at birth, the majority of IUGR infants undergo accelerated growth (catch-up growth) as neonates.[9-12] This neonatal or childhood catch-up growth is an independent risk factor for later development of insulin resistance,[13] NIDDM,[14] obesity,[13,15,16] the central metabolic syndrome[4] and cardiovascular disease.[17,18] Understanding the initiating factors and causal pathways linking poor fetal growth, neonatal catch-up growth and impaired metabolic homeostasis in later life is essential to enable the development of strategies to reduce the impact of these diseases. In this chapter, we will review what has been learnt from human and experimental studies about the mechanistic basis for poor metabolic homeostasis following restricted fetal growth and neonatal catch-up growth, and comment on future research directions in this area.

Can the Perinatal Environment Influence Adult Metabolic Homeostasis?

The concept that events in early life might be causally associated with later outcomes for health and disease has gained widespread acceptance since the seminal epidemiological studies in humans by Barker and others (reviewed in ref. 19). Lucas defined this process as "programming", where an event occurring at a critical period in development has long-lasting or permanent effects.[20] The initial evidence that the intrauterine environment could program subsequent metabolic health derived from geographical associations between high rates of neonatal death and subsequent high rates of coronary heart disease and stroke in the survivors of the same cohort.[21] These and other co-localisations of markers of maternal nutrition, fetal and childhood growth with later disease incidence caused Barker to hypothesise that undernutrition caused permanent changes to the physiology and metabolism of the fetus which led to coronary heart disease and stroke in later life.[19] Epidemiological studies tested this hypothesis initially by investigating the causes and rates of death in individuals for whom records of size at birth and in childhood were available. Subsequent studies in the UK measured blood pressure and markers of metabolic homeostasis in individuals of known birth weight. Most of these studies reported that individuals who were light or thin at birth had increased rates of death from cardiovascular disease and also an increased incidence of risk factors for cardiovascular disease, including high blood pressure and NIDDM or insulin resistance (reviewed by ref. 19). Subsequent larger epidemiological studies tested the hypothesis that size at birth was not related to later outcomes, but instead reported[22,23] strong negative associations between size at birth and risks of heart disease and of diabetes! Although these epidemiological studies clearly demonstrated that small size at birth was a risk factor for metabolic and cardiovascular disease in later life, most lacked information on gestational age at birth, and therefore could not distinguish between restricted fetal growth and prematurity as causes of small size at birth. The inclusion of gestational age data and information on fetal growth trajectories in contemporary cohorts will assist in resolving this issue, although such cohorts are presently children or young adults, and it will be several decades before their risks of metabolic and cardiovascular disease in later life can be assessed.

Maternal nutrient supply and her ability to transport these via the placenta to the fetus are the determinants of fetal substrate supply and therefore have a major impact on fetal growth.[8] The associations between small-size-at-birth and later impairment of metabolic homeostasis therefore implicate maternal and placental factors as initiating factors. There is limited data of a direct nature in humans on the long-term consequences of exposure to individual maternal or placental factors during pregnancy that might be the initiating factors for poor adult metabolic homeostasis, however. Severe acute maternal undernutrition in women who were pregnant during the Dutch famine of World War 2 increased rates of obesity in young adult men, and increased body mass index in women but not men in middle age, when fetal exposure occurred during the first half of pregnancy.[24,25] Interestingly, fetal exposure to famine during the last trimester of pregnancy reduced the incidence of obesity in young men and did not affect BMI in middle-aged men and women.[24,25] Glucose tolerance was impaired in middle-aged men and women following fetal exposure in the middle or last trimesters of pregnancy.[26] Fetal exposure to such severe and delimited maternal undernutrition has few, if any, contemporary equivalents, but does support the hypothesis drawn from the consistent association between small size at birth and later adverse outcomes, that factors able to influence fetal growth have a long-term impact on metabolic homeostasis in humans. This has been further confirmed by experimental studies described below, which have shown that interventions which restrict fetal growth, including severe restriction of maternal feed or protein intake throughout pregnancy and placental insufficiency, cause defects in metabolic homeostasis in the progeny.

How Could the Perinatal Environment Influence Adult Metabolic Homeostasis?

It is possible that systemic, tissue and/or cellular adaptations to an adverse environment in the fetus might persist to later life and contribute to later metabolic disease. The neonatal environment may also interact with fetal adaptations and result in changes that further affect later disease risk. For example, if abundant nutrients are available to the neonate that was growth restricted in utero, the adaptations driving catch-up growth may lead to accumulation of lipid in adipose and other tissues,[13,15,16] contributing to increased risk of diabetes.[3] At the systemic level, the abundance of circulating hormones[5] or activity of the sympathetic nervous system[27] might be altered, possibly reflecting changes to the sensitivity of endocrine or nerve tissues to their regulatory signals. At a tissue level, the fetal environment might alter organ size or structure or the proportions of different cell types, as shown in many experimental studies (reviewed by ref. 28). Cellular phenotypes within a tissue might also be altered by clonal selection. Within cells, epigenetic changes in chromatin structure and DNA methylation state might alter expression of particular genes (see also Whitelaw in this book), and metabolic capacity of individual cells might be changed, for example due to differing numbers of mitochondria. Cellular responses to systemic signals might be changed, for example due to changes in receptor expression or downstream signalling components.[29] Which aspects of metabolism are programmed may vary according to the timing and duration of exposure to an adverse intrauterine environment, as well as the severity and type of insult imposed (reviewed by refs. 30, 31).

Mechanistic Basis for Programming of Metabolic Homeostasis—Evidence from Human Studies

Is there evidence that particular systemic, tissue or cellular changes underlying metabolic disease are related to size at birth in humans? A recent systematic review by Newsome et al[5] reported that indicators of impaired glucose tolerance or diabetes were negatively associated with size at birth in the majority of studies in children, young and old adults (Table 1). Small size at birth was also associated with indirect measures of insulin resistance in children and adults in most studies (Table 1). Few studies have directly measured insulin resistance in relation to size at birth, and the negative relationship between birth weight and insulin resistance has been observed in half of those to date (Table 1). In contrast to the reasonably consistent associations of small size at birth and insulin resistance in adults and children, there is limited, and conflicting, data available on insulin sensitivity in neonates following restricted fetal growth, some of which suggests that insulin sensitivity might be increased in the SGA neonate (Table 1). Unlike insulin resistance, failure of insulin secretion, which must occur concomitantly with insulin resistance before progression to NIDDM occurs, was not consistently associated with size at birth (Table 1). Indeed, children and adults of low birth weight had elevated fasting or glucose-stimulated plasma insulin concentrations in most studies (Table 1), probably indicating some degree of compensatory hyperinsulinaemia. The insulin disposition index (product of insulin secretion and insulin sensitivity) provides a more accurate and unbiased assessment of insulin secretory capacity than plasma insulin concentrations, since it provides a measure of whether insulin secretion is appropriate given the individual's insulin sensitivity.[32] Use of the disposition index in biological studies has been limited because independent and direct measures of each parameter are required. Two of four recent studies suggested that the level of insulin secretion in IUGR individuals may be lower than appropriate for their level of insulin resistance, indicating impaired β-cell capacity for compensatory insulin hypersecretion. The other two studies did not find any effect of IUGR on insulin secretion relative to sensitivity, however (Table 1). Further studies of IUGR individuals with independent measures of insulin secretion and sensitivity are needed to determine whether β-cell defects play a role in their increased susceptibility to diabetes. We also do not know whether β-cell defects are programmed directly by the fetal environment or result from the long-term challenge imposed by insulin resistance and the consequent need to maintain compensatory hyperinsulinaemia. Overall, studies of low or variable birth weight in humans implicate defects in insulin action, and insulin resistance in particular, as a common mechanism leading to NIDDM and cardiovascular disease following poor intrauterine growth.

Less is known about the mechanisms underlying the increased risks of metabolic and cardiovascular disease associated with neonatal or childhood catch-up growth. In children who were small for gestational age (SGA) at birth, a rapid increase in BMI between 2 and 9 years of age was associated with lower whole-body insulin sensitivity and higher first and second phase insulin secretion at 9 years of age.[33] In children of normal birth weight, the rate of BMI gain during this period did not affect insulin sensitivity or secretion.[33] Two other studies[13,34] have reported that relative gains in weight or ponderal index during childhood are positively associated with the HOMA insulin resistance index across the range of population birth weights, and

Table 1. Metabolic disease and defects at the whole-body level are associated with small size at birth in humans

Defect	Relationship with Birth Weight
Indicators of glucose intolerance and non-insulin dependent diabetes	
Fasting plasma glucose concentration	Negative (15 of 25 studies of children and adults)[5]
Plasma glucose 2 h after glucose load	Negative (20 of 25 studies of children and adults)[5]
Prevalence of NIDDM	Negative (13 of 16 studies of children and adults)[5]
Whole-body insulin resistance	
Indirect & direct measures of insulin resistance	Negative (17 of 22 studies of children and adults)[5] Reduced[90] or increased[91] in SGA cf. AGA neonates
Direct measure of insulin resistance by hyperinsulinaemic euglycaemic clamp	Negative (2 of 4 studies), inverted U-shaped (1 of 4 studies), or not related (1 of 4 studies of children and adults)[5]
Insulin secretion	
Fasting plasma insulin concentration	Negative (20 of 26 studies of children and adults)[5] Reduced[90] or increased[91] in SGA cf. AGA neonates
Insulin secretion after glucose load	Negative (16 of 24 studies), not related (6 of 24 studies), positive (7 of 24 studies of children and adults)[5]
Insulin secretion relative to sensitivity	
Slope of secretion vs sensitivity	Negative in all children; steeper in children of normal cf. low birth weight[92]
Insulin disposition index	30% lower in 19-year-old men of low birth weight cf normal birth weight.[93] Not different between SGA and AGA pre-pubertal children.[33,92] Not different between IUGR and controls in 25 year old men and women.[94]

not only following IUGR. Thus, catch-up growth, like IUGR, is associated with insulin resistance in children. Catch-up growth in terms of length, but not in terms of weight, was positively associated with fasting insulin and insulin secretion during an IVGTT in 1-year old children,[35] but it is not clear to what extent this reflects compensatory insulin secretion in response to resistance. To date, insulin secretion in older children, and insulin resistance and secretion in adults has not been related to their neonatal and childhood growth patterns.

What is known about the tissue and cellular changes during and after human IUGR that might underlie altered insulin sensitivity and/or secretion? Decreased whole-body insulin sensitivity might reflect decreased peripheral or hepatic insulin action. Individuals who were small at birth have reduced muscle mass as adults.[36] Since muscle is a major insulin-sensitive tissue, reduced muscle mass might contribute to decreased glucose uptake in response to insulin. Some, but not all studies, have demonstrated that insulin sensitivity of peripheral tissues, specifically that of muscle, is reduced in low birth weight individuals, which would also decrease the action of insulin per unit mass of muscle (Table 2). Insulin appears to stimulate endothelial-mediated vasodilation in muscle normally in individuals of low birth weight, however (Table 2). The available evidence in humans suggests that a signalling defect downstream of the insulin receptor impairs insulin-stimulation of glucose transport and hence glucose uptake in peripheral tissues (Table 2). The limited reports of effects of birth weight on hepatic insulin sensitivity have been inconsistent, with either increased or decreased suppression of endogenous glucose production by insulin in low birth weight individuals compared to those

Table 2. Effects of IUGR and size at birth on tissue and cellular defects that might underlie postnatal metabolic disease in humans

Defects that might underlie whole-body insulin resistance

- Insulin resistance of peripheral tissue (yes):
  - Young adults who were IUGR had:
    - Decreased forearm glucose uptake in response to insulin[95]
    - Reduced up-regulation of GLUT4 expression by insulin in muscle and adipose tissue[96]
    - Normal basal and insulin-stimulated gene expression of insulin-receptor and signalling pathway components in muscle[96]
  - Variable size at birth:
    - No relationship with postprandial glycogen synthase activity in muscle in middle-aged women[97]
- Decreased insulin-stimulation of muscle blood flow (no):
  - Adults and children who were light or thin at birth had:
    - Normal forearm basal and insulin-stimulated endothelium-dependent vasodilation[95]
    - Normal skeletal muscle capillary density and resting blood flow[98]
    - Faster muscle re-oxygenation[98,99]
  - Variable size at birth:
    - No relationship with endothelium-dependent and -independent vasodilation in prepubertal children
    - Positive relationship with post-occlusive capillary recruitment in prepubertal children[100]
- Changes in muscle fibre composition (unknown):
  - Thinness at birth did not predict skeletal muscle fibre density and proportions of Type 1 and 2 fibres in middle-aged women[98]
- Hepatic insulin resistance (unknown):
  - Low birth weight INCREASED insulin-suppression of endogenous glucose production and glycolytic flux in young men[93]
  - Low birth weight DECREASED insulin-suppression of endogenous glucose production in Pima Indians as young adults[101]

Defects that might underlie impaired insulin secretion

- Decreased numbers of islets or β-cells (unknown):
  - IUGR decreased[102] or did not change[103] islet density and β-cell fraction in pancreas of late gestation fetuses and term neonates
- Decreased perfusion of β-cells (unknown):
  - 'Less pronounced vasculature', in pancreas from IUGR infants than in those of normal weight[102]

of normal birth weight (Table 2). The structural changes that might affect insulin secretion by the pancreas have been little studied, with one, but not another study reporting decreases in the proportions of islet tissue and numbers of β-cells in IUGR babies (Table 2). Further studies in humans are thus required to confirm muscle as a primary site of insulin resistance in the individual who was born SGA, and to explore possible defects in hepatic insulin sensitivity and in pancreatic development that might underlie impaired whole-body insulin sensitivity and insulin secretion respectively.

Use of Experimental Models in the Investigation of Programming of Metabolic Homeostasis in Humans

Since most of the contemporary data in humans relates size at birth to later outcomes, relevant experimental models are likely to be those that restrict fetal growth or produce variation in size at birth. Although genetic factors determine the expected size of the fetus, the intra-uterine environment regulates the extent to which potential size is achieved or exceeded.

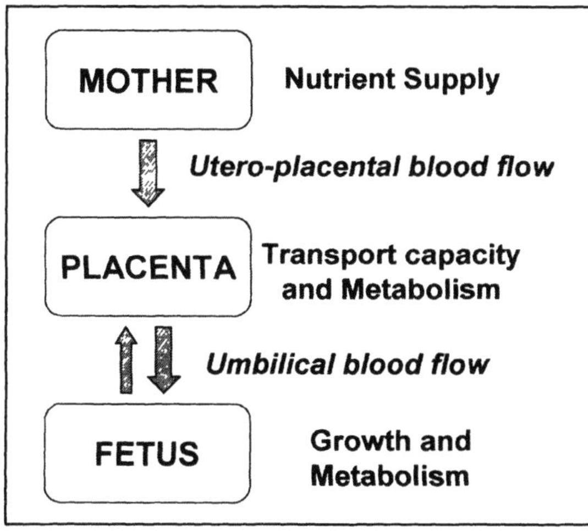

Figure 2. Control of fetal growth.

Thus, the supply of nutrients and oxygen to the fetus depends on maternal factors that affect their availability, but also on placental metabolism and capacity to transport nutrients to the fetus (Fig. 2). Factors that limit maternal capacity to supply nutrients and those that impact on placental growth or function are associated with small size at birth or IUGR in humans (Table 3). This includes total food or energy availability to the mother, but also the composition of the maternal diet, although detailed studies of their effects on fetal growth are limited. Recently, Moore et al[37] reported that birth weight and placental weight were positively related to the percentage of energy derived from protein in the maternal diet in early pregnancy in a population living in Adelaide, Australia, confirming that even in contemporary Western societies nutrient composition can affect fetal growth. The form in which nutrients are consumed may also be important. For example, iso-energetic diets containing most carbohydrates in high glycaemic index forms produce larger babies and placentae than diets where the carbohydrates were predominantly in low glycaemic index forms.[38] The placenta itself presents significant nutrient demands related to its own growth and also plays an important role in modifying maternally- and even fetally-derived factors for fetal consumption, for example in the interconversion of amino acids.[39,40] The peri-conceptional environment is also emerging as an important factor impacting subsequent fetal development, and possibly programming postnatal function. The use of Assisted Reproductive Technology (ART), and hence extra-uterine culture in the first few cycles of cell division, is associated with lower average birth weights and increased rates of clinical IUGR, as well as with increased perinatal mortality[41,42] (see also Thompson et al in this book). Experimental models will be required here to differentiate the effects of ART from those of factors that may have contributed to the decision to use ART, since a history of infertility is itself associated with decreased birth weight and an increased incidence of IUGR (for example, see ref. 43). Experimental models may also provide insight regarding the longer-term outcomes following ART, including metabolic homeostasis in these individuals. Similarly, experimental models may provide insights into the long-term consequences following other clinical scenarios or interventions that may reduce birth weight (Table 3), and the mechanisms underlying these consequences.

In addition to reducing size at birth, the long-term outcomes of experimental interventions should be broadly consistent with the effects of small size at birth in humans, in order for these models to provide specific insights into the underlying mechanisms. In particular, appropriate experimental models will have negative associations of birth weight with insulin resistance and

Table 3. Factors contributing to small size at birth and/or clinical IUGR in humans

Maternal factors	
	Small or underweight mother
	First pregnancy
	Multiple pregnancy (twins or higher multiples)
	Chronic undernutrition or severe undernutrition in late gestation
	Adolescent pregnancy
	Smoking
	Socio-economic disadvantage
	Infection
Placental factors	
	Small placental size
	Low uteroplacental blood flow
	Impaired placental transport of amino acids
	Pre-eclampsia
Clinical interventions	
	Assisted reproductive technology, including in vitro fertilisation
	Repeated corticosteroid treatment for threatened premature delivery[82]

Adapted from Robinson et al.[104]

obesity, and progression to impaired glucose tolerance and NIDDM with aging. The available experimental models for the programming of metabolic homeostasis can be considered as manipulations of (1) maternal or (2) placental factors, or models of (3) clinical scenarios and interventions.

Maternal Factors

The fetal rat[44-47] and guinea pig[48,49] that are subject to maternal food restriction from before and throughout pregnancy develop obesity and insulin resistance after birth (Table 4). In the case of the guinea pig, progeny of feed-restricted mothers are also glucose intolerant despite increased insulin secretion,[49] and have an impaired capacity to maintain cholesterol homeostasis when subjected to a dietary challenge.[48] Few studies in the rat have explored the metabolic consequences for progeny of maternal food-restriction prior to and maintained throughout pregnancy, however. One inconsistency between experimental models of food-restriction throughout pregnancy in nonhuman species and human outcomes following IUGR is the failure of neonatal progeny to undergo catch-up growth. Moderate food restriction for part of pregnancy only does not consistently reduce birthweight in the rat, so is difficult to extrapolate from this model to the currently described human phenomenon. In rats, birth weight was not altered by a 50-70% reduction in maternal food supply during early pregnancy only,[50,51] or during either the first or last 2/3 of pregnancy.[52,53] In contrast to other rat studies, a 50% restriction of maternal feed intake during only the last 1/3 of pregnancy in the rat did reduce birth weight.[54-56] Similarly, a 50% restriction of maternal feed intake during the second half of pregnancy reduced birth weight in the guinea pig,[57] as did severe acute food restriction in late pregnancy in the sheep.[58] Limited data from some of these models suggests that imposing maternal feed restriction for part of pregnancy only does not impair glucose homeostasis in the progeny (Table 4). The metabolic programming effects of restricted maternal protein intake have been explored in far greater detail than those of restricted maternal feed intake. Protein restriction (6 to 8% cf. 20%) throughout pregnancy in the rat,[59-63] reduces birth weight and may decrease insulin secretory capacity of progeny, but does not impair adult glucose tolerance, and unlike the human is not characterised by neonatal catch-up growth (Table 4). Prolonging protein restriction (8% cf. 20%) throughout lactation as well as

Table 4. Size at birth and metabolic programming in experimental models of fetal growth restriction[1]

Intervention and Species	Reported Effect	Outcomes: Size at Birth	Neonatal Growth Rates	Adult Size	Obesity	Basal Glucose	Basal Insulin	Glucose Intolerance	Insulin Secretion	Insulin Sensitivity of Glucose Metabolism	Elevated Plasma Cholesterol
Constraints to Maternal Supply											
Food restriction—whole of pregnancy											
Rat (fed 30% or 50% of ad libitum)[44-47]	Effect of nutrition	↓	↓AGR ↓FGR	↔	↑	↔	↓at birth ↔ neonates & juveniles ↑↓ adults	NR	NR	↓	NR
Guinea pig (fed 60-85% of ad libitum)[48,49,105]	Effect of nutrition	↔↓	↓AGR ↔ FGR	↔↓	↑	↔	↑	↑	↑	NR	↔↑ basal ↔↑ post-challenge
	Relationship/ effect of size at birth	...	AGR +ve FGR -ve	↔↓ in LBW	NR	none	↑ in LBW	none	none	NR	↔ basal ↔↑ post-challenge in LBW
Food restriction—part of pregnancy											
Rat (fed 50% of ad libitum in early pregnancy)[50,51]	Effect of nutrition	↔	↑ males ↔ females	↑ males ↔ females	↑ males ↔ females	NR	NR	NR	NR	NR	NR
Rat (fed 50% of ad libitum for first 2/3 of pregnancy)[52]	Effect of nutrition	↔	↔ AGR	↑	NR	↔	↔	↔	↔	↔	NR

continued on next page

Table 4. Continued

Intervention and Species	Reported Effect	Size at Birth	Neonatal Growth Rates	Adult Size	Obesity	Basal Glucose	Basal Insulin	Glucose Intolerance	Insulin Secretion	Insulin Sensitivity of Glucose Metabolism	Elevated Plasma Cholesterol
			Outcomes								
Rat (fed 30% of ad libitum for last 2/3 of pregnancy)[53]	Effect of nutrition	↔	NR	NR	NR	NR	NR	NR	NR	NR	↑ post-challenge
Rat (fed 50% of ad libitum for last 1/3 of pregnancy)[54-56]	Effect of nutrition	↓	↓ AGR	↔	NR	↔ juveniles ↑ adults	↓ juveniles ↔ adults	↔	↔↓	↑ young adults	NR
Guinea pig (fed 50% of ad libitum for second 1/2 of pregnancy)[57]	Effect of nutrition	↓	↓ AGR	NR	↑	NR	NR	NR	NR	NR	NR
Sheep (fed 2-3% of ad libitum for 10 or 20 days in late pregnancy)[58]	Effect of nutrition	↓	NR	↔	NR	↔	↔	↔	↔	↔	NR
	Relationship with size at birth	...	NR	NR	NR	none	none	-ve at 5 mo ↔ at 30 mo	none	NR	NR
Food restriction—whole of pregnancy and lactation											
Rat (fed 50% of ad libitum)[44,106,107]	Effect of nutrition	↓	↓ AGR	↓	NR	↔ juveniles ↑ adults	↔ juveniles ↓ adults	↔ juveniles ↑ adults	↓ juveniles & adults	↓	NR
Protein restriction—whole of pregnancy											
Rat (maternal dietary protein range 6 to 20%)[59-63]	Effect of low protein	↓	↓↔ AGR	↓↔	NR	↔	↓ juveniles ↔ adults	↓↑ juveniles ↔ adults	↓	NR	↓ basal

continued on next page

Table 4. Continued

Intervention and Species	Reported Effect	Size at Birth	Neonatal Growth Rates	Adult Size	Obesity	Basal Glucose	Basal Insulin	Glucose Intolerance	Insulin Secretion	Insulin Sensitivity of Glucose Metabolism	Elevated Plasma Cholesterol
			Outcomes								
Pig (maternal diet 0.5% vs 13% protein)[108]	Effect of low protein	↓	↓AGR	↓	↔	NR	NR	NR	NR	NR	NR
Protein restriction—whole of pregnancy and lactation											
Rat (maternal diet 8% vs 20% protein)[29,60,62,64-67]	Effect of low protein	↓	↓AGR	↓	↔	↔	↔↓↑ adults	↓young ↔↑ old adults	↔↓↑ ↓muscle	↓whole body & liver	↓basal
Constraints to Placental Size or Function											
Natural variation in litter size											
Guinea pig[68,69]	Effect of large litter size	↓	↑FGR	NR	NR	NR	NR	NR	NR	NR	NR
	Relationship with size at birth	...	+ve AGR -ve FGR	NR	+ve	-ve	NR	NR	NR	+ve	NR
Pig[70,71]	Effect of LBW	...	↓AGR ↑FGR	↓females	NR	↓	↓males ↔females	↑	↑	↔males ↓females	NR
	Relationship with size at birth	...	+ve AGR -ve FGR	+ve females	NR	NR	+ve	-ve	NR	-ve males	NR
Sheep[72,73]	Effect of LBW	...	↓AGR ↓↑FGR	NR	↑	↔	↔at birth ↑juveniles	NR	NR	NR	NR
Surgical restriction of placental growth or function											
Sheep (surgical removal of most placental implantation sites)[74-76]	Effect of placental restriction	↓	↔AGR ↑FGR	↓females	↑ neonates	↔	↑	↑	↑absolute ↓insulin disposition in neonates	↓juvenile males ↑adult females	NR

continued on next page

Table 4. Continued

		Outcomes									
Intervention and Species	**Reported Effect**	**Size at Birth**	**Neonatal Growth Rates**	**Adult Size**	**Obesity**	**Basal Glucose**	**Basal Insulin**	**Glucose Intolerance**	**Insulin Secretion**	**Insulin Sensitivity of Glucose Metabolism**	**Elevated Plasma Cholesterol**
Sheep (surgical removal of most placental implantation sites)[74-76]	Relationship with size at birth	...	+ve	NR	+ve	NR	NR	NR	+ve for insulin disposition in adult males	NR	NR
Rat (surgical restriction of uterine blood flow in late pregnancy)[77]	Effect of restricted blood flow	↓	↓AGR	↑	↑	↑	↑	↑	↓	↓	NR
Models of Clinical Scenarios and Interventions											
Maternal gestational diabetes											
Rat (mild maternal hyperglycaemia after streptozotocin in early pregnancy)[78,79]	Effect of maternal streptozotocin	↑	↑	↓	↑	↔	↔	↑	↑	↓	NR
Rat (moderate maternal hyperglycaemia during glucose infusion in late pregnancy)[109,110]	Effect of maternal glucose infusion	↑	NR	↓	NR	↔	↔	↑	↓	NR	NR
Maternal glucocorticoid treatment											
Sheep (single dose of dexamethasone to mother in early or mid pregnancy)[111,112]	Effect of glucocorticoid	↔	NR	↔	NR	↔	↔	↔	↔	↔	NR

continued on next page

Table 4. Continued

Intervention and Species	Reported Effect	Outcomes: Size at Birth	Neonatal Growth Rates	Adult Size	Obesity	Basal Glucose	Basal Insulin	Glucose Intolerance	Insulin Secretion	Insulin Sensitivity of Glucose Metabolism	Elevated Plasma Cholesterol
Sheep (single dose of dexamethasone to mother in late pregnancy)[81]	Effect of glucocorticoid	↔	NR	↔	NR	↔	↔	↔ in juveniles ↑ in adults	↑	NR	NR
Rat (multiple doses of dexamethasone to mother in early or mid pregnancy)[83,84]	Effect of glucocorticoid	↔	↑ AGR	↑↔	↑	↔	↔	↔	↔	↔	NR
Sheep (multiple doses of dexamethasone to mother in late pregnancy)[81]	Effect of glucocorticoid	↓	NR	↔	NR	↔ in juveniles ↑ in adults	↔ in juveniles ↑ in adults	↔	↑	NR	NR
Rat (multiple doses of dexamethasone to mother in late pregnancy)[83]	Effect of glucocorticoid	↓	NR	↔	NR	↑	↔	↑	↑	NR	NR
Guinea pig (multiple doses of dexa-methasone to mother in late pregnancy)[113]	Effect of glucocorticoid	↔	NR	↔	NR	NR	NR	NR	NR	NR	NR

1 ↓= decreased; ↑ =increased; ↔= unchanged; AGR= absolute growth rate; FGR= fractional growth rate; NR= not reported; +ve= positive; -ve= negative; LBW= low birth weight; mo= months old

pregnancy in the rat[29,60,62,64-67] does impair adult insulin sensitivity, although the effects on glucose tolerance and insulin secretion are variable (Table 4). Overall, the inconsistency between outcomes following human IUGR and in progeny of dams subject to food or protein restriction during pregnancy may limit the usefulness of specific mechanistic information obtained in these models. Nevertheless they have been critical in clearly substantiating the programming phenomenon.

Placental Factors

Small-size-at-birth occurring due to spontaneous variation in litter size in guinea pigs[68,69] and pigs[70,71] produces postnatal consequences for progeny that are similar to those seen after human IUGR (Table 4). These include catch-up growth, obesity, glucose intolerance, increased insulin secretion in response to a glucose challenge, and insulin resistance, although some of these responses differ between genders (Table 4). Similar models of LBW in sheep[72,73] produce obesity, variable catch-up growth depending on neonatal nutrition, and elevated fasting insulin in juveniles, but responses to a glucose challenge have not been investigated in this model (Table 4). Surgical interventions have been used to directly investigate the progeny outcomes of impaired placental function. Surgical restriction of placental implantation in sheep[74-76] reduces progeny birth weight, and results in neonatal catch-up growth, increased fat deposition even in neonatal life, and elevated fasting and post-glucose insulin secretion, although in neonates this is still below that expected when corrected for the level of insulin resistance (Table 4). Surgical restriction of uterine blood flow in late pregnancy in rats[77] has a more severe effect on progeny phenotype than the models described previously. The progeny of these rats show elevated fasting glucose and impaired post-glucose insulin secretion in addition to obesity, glucose intolerance and insulin resistance (Table 4). The latter model appears to more closely mimic the late stages in the development of diabetes, including failure of insulin secretion as well as insulin resistance. In the three models where insulin resistance of progeny has been characterised, only surgical restriction of uterine blood flow in late pregnancy in the rat impairs the absolute insulin secretory response to glucose. The insulin secretory response to glucose is increased in absolute terms in small pigs from large litters and in placentally-restricted sheep, although secretion relative to sensitivity is still impaired in the latter model. These results suggest that surgical restriction of uterine blood flow in late pregnancy in rats affects the development and/or function of the pancreas more severely than the chronic restriction imposed by large litter size or surgical reduction in the number of placental implantation sites.

Clinical Scenarios and Interventions

Maternal hyperglycaemia induced by maternal streptozotocin treatment and consequent mild maternal hyperglycaemia throughout pregnancy in rats[78,79] increases size at birth and results in progeny with increased adiposity, glucose intolerance and insulin resistance, but no changes in fasting glucose. These models may therefore provide useful information about the mechanisms underlying the development of diabetes and obesity in children who were exposed to gestational diabetes in fetal life.[80] Data regarding the metabolic consequences of corticosteroid exposure in experimental models is currently limited (Table 4), as is the case for human data also. Exposure to a single dose of maternally-administered dexamethasone during late gestation in sheep,[81] which does not affect birth weight, increases insulin secretion after a glucose challenge (Table 4). This suggests that a single exposure to glucocorticoid during fetal life may cause insulin resistance in sheep, although this has not yet been measured directly, and these animals are able to maintain normal glucose tolerance. Progeny of ewes given multiple maternally-administered doses of dexamethasone in late pregnancy[81] also have exaggerated insulin responses to glucose, and develop glucose intolerance with aging (Table 4). Repeated doses of glucocorticoids decrease fetal growth to a greater extent in sheep[81] than that suggested in limited human trials to date,[82] suggesting that the sheep may be more sensitive to glucocorticoid exposure. Nevertheless, the data available from sheep studies suggest that repeated fetal exposure to glucocorticoids may program metabolic homeostasis and this should therefore be investigated in human cohorts.

Mechanistic Basis for Programming of Metabolic Homeostasis—Evidence from Experimental Models

What additional information has the use of these experimental models provided about the mechanisms that underlie impaired insulin sensitivity and development of obesity following restriction of fetal growth in humans?

One interesting feature of many of the models described above is that progeny outcomes relate much more closely to size at birth than to treatment group. For example, in our studies in sheep[74-76] we have found that size at birth is a stronger predictor of neonatal growth rates and adult insulin sensitivity than whether the mother was a control ewe or had undergone surgery to restrict placental implantation. Similarly, measures of glucose homeostasis were more closely related to size at birth than to litter size in the guinea pig with spontaneous variation in size at birth.[68,69] This might suggest that multiple factors, such as placental restriction and restriction due to litter size, act via a common mechanism to induce fetal programming, and that size at birth reflects the sum of these factors to a degree. Changes in fetal glucocorticoid exposure, with consequences for accelerated differentiation but reduced proliferation of cells within a range of tissues, have been suggested as a common mechanism that might mediate the effects of multiple environmental factors on the fetus.[30] Certainly, maternal glucocorticoid administration in rats[83,84] and sheep[81] produce progeny outcomes with some similarities to those following human IUGR.

What about the systematic, tissue or cellular changes underlying the changes in postnatal growth and metabolism in experimental models? Several experimental models of constraints to placental size or function, produce progeny with reduced size at birth followed by neonatal catch-up growth (Table 4). Increased neonatal appetite appears to be a central feature of catch-up growth,[72] although the underlying mechanism is largely unknown. In rats, fetal growth restriction induced by maternal undernutrition during pregnancy increases the appetite of progeny after weaning, which may contribute to the later development of obesity and metabolic dysfunction.[47] Neonatal catch-up growth is also associated with increased whole-body sensitivity to the actions of insulin and insulin-like growth factor, important anabolic hormones for neonatal growth, in sheep and guinea pigs that were small at birth.[85-87] Despite increased insulin sensitivity as neonates, insulin resistance develops with ageing, as young adults who were small at birth are insulin-resistant in both experimental models.[85,88,89] At least in the guinea pig that was small at birth, whole-body insulin resistance is primarily peripheral[69] and we have preliminary data to show that this is in muscle (JA Owens et al, unpublished data). Impairment of glucose tolerance with aging probably also reflects a reduced capacity for compensatory insulin secretion, as in our studies insulin disposition is positively related to size at birth in young adult male sheep.[76] Further studies in experimental models are needed to delineate the mechanistic basis for catch-up growth and increased appetite, and how these contribute to later disease. The structural, cellular and molecular changes underlying decreased whole-body insulin resistance and probable impaired insulin secretory capacity also require further investigation.

Conclusions

The characterisation of experimental prenatal perturbations that restrict fetal growth and lead to similar pre and post-natal phenotypes to human IUGR has confirmed the mechanistic information available from human studies. Human and experimental studies consistently implicate whole-body and skeletal muscle insulin resistance as primary defects underlying the association of metabolic disease with small size at birth. Evidence from experimental studies also supports the more limited evidence of impaired insulin secretory capacity in humans of low birth weight. The available experimental models should allow a more detailed and hopefully more rapid exploration of the systemic, tissue and cellular changes underlying altered insulin resistance and impaired secretion, than may be possible in humans. These models additionally allow the effects of specific factors that influence fetal growth on progeny outcomes to be tested, rather than using the summative measures of birth weight or dimensions. The real power of experimental models, however, may be that they allow us to test pre- and post-natal interventions to prevent or ameliorate the deleterious consequences of exposure to an adverse fetal environment for impaired metabolic homeostasis in later life.

Acknowledgements

The authors wish to acknowledge the support of the National Health and Medical Research Council of Australia. Kathryn L. Gatford currently holds the Hilda Farmer Medical Research Associateship awarded by the Faculty of Health Sciences, University of Adelaide.

References

1. Fagot-Campagna A. Emergence of type 2 diabetes mellitus: Epidemiological evidence. J Pediatr Endocrinol Metab 2000; 13:1395-402.
2. Aye T, Levitsky LL. Type 2 diabetes: An epidemic disease in childhood. Curr Opin Pediatrics 2003; 15:411-15.
3. Schulze MB, Hu FB. Primary prevention of diabetes: What can be done and how much can be prevented? Annu Rev Public Health 2005, (Review in advance: Doi: 10.1146/annurev.publhealth.26.021304.144532).
4. Parker L, Lamont DW, Unwin N et al. A lifecourse study of risk for hyperinsulinaemia, dyslipidaemia and obesity (the central metabolic syndrome) at age 49-51 years. Diabet Med 2003; 20:406-15.
5. Newsome CA, Shiell AW, Fall CHD et al. Is birth weight related to later glucose and insulin metabolism? - a systemic review. Diabet Med 2003; 20:339-48.
6. Rogers I, Group E-BS. The influence of birthweight and intrauterine environment on adiposity and fat distribution in later life. Int J Obes Relat Metab Disord 2003; 27:755-77.
7. Pi-Sunyer FX. The epidemiology of central fat distribution in relation to disease. Nutr Rev 2004; 62:S120-S26.
8. Robinson JS, Owens JA. Pathophysiology of intrauterine growth failure. In: Gluckman PD, Heymann MA, eds. Pediatrics and Perinatology. The Scientific Basis. 2nd ed. London: Arnold, 1996:290-97.
9. Fitzhardinge PM, Steven EM. The small-for-date infant I. Later growth patterns. Pediatrics 1972; 49:671-81.
10. Tenovuo A, Kero P, Piekkala P et al. Growth of 519 small for gestational age infants during the first two years of life. Acta Paediatrica Scandinavica 1987; 76:636-46.
11. Albertsson-Wikland K, Wennergren G, Wennergren M et al. Longitudinal follow-up of growth in children born small for gestational age. Acta Paediatrica 1993; 82:438-43.
12. Hokken-Koelega ACS, De Ridder MAJ, Lemmen RJ et al. Children born small for gestational age: Do they catch up? Pediatric Res 1995; 38:267-71.
13. Crowther NJ, Cameron N, Trusler J et al. Association between poor glucose tolerance and rapid post natal weight gain in seven-year-old children. Diabetologia 1998; 41:1163-67.
14. Forsén T, Eriksson J, Tuomilehto J et al. The fetal and childhood growth of persons who develop type 2 diabetes. Ann Intern Med 2000; 133:176-82.
15. Ong KKL, Ahmed ML, Emmett PM et al. Association between postnatal catch-up growth and obesity in childhood: Prospective cohort study. BMJ 2000; 320:967-71.
16. Parsons TJ, Power C, Manor O. Fetal and early life growth and body mass index from birth to early adulthood in 1958 British cohort: Longitudinal study. BMJ 2001; 323:1331-35.
17. Eriksson JG, Forsen T, Tuomilehto J et al. Catch-up growth in childhood and death from coronary heart disease: Longitudinal study. BMJ 1999; 318:427-31.
18. Forsén T, Eriksson JG, Tuomilehto J et al. Growth in utero and during childhood among women who develop coronary heart disease: Longitudinal study. BMJ 1999; 319:1403-07.
19. Barker DJP. Mothers, babies and health in later life. 2nd ed. Edinburgh: Churchill Livingstone, 1998.
20. Lucas A. Programming by early nutrition in man. In: Bock GR, Whelan J, eds. The Childhood Environment and Adult Disease. Chicester: Wiley, 1991:38-55.
21. Barker DJP, Osmond C. Infant mortality, childhood nutrition, and ischaemic heart disease in England and Wales. Lancet 1986; 1(8489):1077-81.
22. Rich-Edwards JW, Stampfer MJ, Manson JE et al. Birth weight and risk of cardiovascular disease in a cohort of women followed up since 1976. BMJ 1997; 315:396-400.

23. Rich-Edwards JW, Colditz GA, Stampfer MJ et al. Birthweight and the risk for type 2 diabetes mellitus in adult women. Ann Intern Med 1999; 130:322-24.
24. Ravelli G-P, Stein ZA, Susser MW. Obesity in young men after famine exposure in utero and early infancy. N Eng J Med 1976; 295:349-53.
25. Ravelli ACJ, ven der Meulen JHP, Osmond C et al. Obesity at the age of 50 y in men and women exposed to famine prenatally. Am J Clin Nutr 1999; 70:811-16.
26. Ravelli ACJ, van der Meulen JHP, Michels RPJ et al. Glucose tolerance in adults after prenatal exposure to famine. Lancet 1998; 351:173-77.
27. Flanagan DE, Vaile JC, Petley GW et al. The autonomic control of heart rate and insulin resistance in young adults. J Clin Endocrinol Metab 1999; 84:1263-67.
28. Fowden AL, Hill DJ. Intra-uterine programming of the endocrine pancreas. British Medical Bulletin 2001; 60:123-42.
29. Ozanne SE, Nave BT, Wang CL et al. Poor fetal nutrition causes long-term changes in expression of insulin-signaling components in adipocytes. Am J Physiol 1997; 273:E46-E51.
30. Fowden AL, Forhead AJ. Endocrine mechanisms of intrauterine programming. Reproduction 2004; 127:515-26.
31. Gluckman PD, Hanson MA. Living with the past: Evolution, development, and patterns of disease. Science 2004; 305:1733-36.
32. Bergman RN, Ader M, Huecking K et al. Accurate assessment of β-cell function. The hyperbolic correction. Diabetes 2002; 51(Suppl 1):S212-S20.
33. Veening MA, van Weissenbruch MM, Heine RJ et al. β-cell capacity and insulin sensitivity in prepubertal children born small for gestational age. Influence of body size during childhood. Diabetes 2003; 52:1756-60.
34. Whincup PH, Cook DG, Adshead F et al. Childhood size is more closely related than size at birth to glucose and insulin levels in 10-11-year-old children. Diabetologia 1997; 40:319-26.
35. Soto N, Bazaes RA, Pena V et al. Insulin sensitivity and secretion are related to catch-up growth in small-for-gestational-age infants at age 1 year: Results from a prospective cohort. J Clin Endocrinol Metab 2003; 88:3645-50.
36. Phillips DIW. Relation of fetal growth to adult muscle mass and glucose tolerance. Diabet Med 1995; 12:686-90.
37. Moore VM, Davies MJ, Willson KJ et al. Dietary composition of pregnant women is related to size of the baby at birth. J Nutrition 2004; 134:1820-26.
38. Clapp JFR. Maternal carbohydrate intake and pregnancy outcome. Proc Nutr Soc 2002; 61:45-50.
39. Battaglia FC. Fetal liver and the placenta: An interactive system. In: Battaglia FC, ed. Placental Function and Fetal Nutrition. Philadelphia: Lippincott-Raven Publishers, 1997:47-57.
40. Meschia G. Placental delivery of amino acids. Utilization and production vs. transport. In: Battaglia FC, ed. Placental Function and Fetal Nutrition. Philadelphia: Lippincott-Raven Publishers, 1997:21 30.
41. Helmerhorst FM, Perquin DA, Donker D et al. Perinatal outcome of singletons and twins after assisted conception: A systematic review of controlled studies. BMJ 2004; 328:261.
42. Jackson RA, Gibson KA, Wu YW et al. Perinatal outcomes in singletons following in vitro fertilization: A meta-analysis. Obstet Gynecol 2004; 103:551-63.
43. Ghazi HA, CS, Kallen B. Delivery outcome after infertility - a registry study. Fertil Steril 1991; 55(4):726-32.
44. Holemans K, Verhaeghe J, Dequeker J et al. Insulin sensitivity in adult female rats subjected to malnutrition during the perinatal period. J Soc Gynecol Investig 1996; 3:71-77.
45. Woodall SM, Breier BH, Johnston BM et al. A model of intrauterine growth retardation caused by chronic maternal undernutrition in the rat: Effects on the somatotrophic axis and postnatal growth. J Endocrinol 1996; 150:231-42.
46. Woodall SM, Johnston BM, Breier BH et al. Chronic maternal undernutrition in the rat leads to delayed postnatal growth and elevated blood pressure of offspring. Pediatr Res 1996; 40:438-43.
47. Vickers MH, Breier BH, Cutfield WS et al. Fetal origins of hyperphagia, obesity, and hypertension and postnatal amplification by hypercaloric nutrition. Am J Physiol 2000; 279(1):E83-E87.
48. Kind KL, Clifton PM, Katsman AI et al. Restricted fetal growth and the response to dietary cholesterol in the guinea pig. Am J Physiol 1999; 277:R1675-R82.
49. Kind KL, Clifton PM, Grant PA et al. Effect of maternal feed restriction during pregnancy on glucose tolerance in the adult guinea pig. Am J Physiol 2003; 284:R140-R52.
50. Jones AP, Friedman MI. Obesity and adipocyte abnormalities in offspring of rats undernourished during pregnancy. Science 1982; 215:1518-19.
51. Jones AP, Simson EL, Friedman MI. Gestational undernutrition and the development of obesity in rats. J Nutrition 1984; 114:1484-92.

52. Portha B, Kergoat M, Blondel O et al. Underfeeding of rat mothers during the first two trimesters of gestation does not alter insulin action and insulin secretion in the progeny. Eur J Endocrinol 1995; 133:475-82.
53. Szitanyi P, Hanzlova J, Poledne R. Influence of intrauterine undernutrition on the development of hypercholesterolemia in an animal model. Physiological Res 2000; 49(6):721-24.
54. Garofano A, Czernichow P, Bréant B. In utero undernutrition impairs rat beta-cell development. Diabetologia 1997; 40:1231-34.
55. Garofano A, Czernichow P, Bréant B. Beta-cell mass and proliferation following late fetal and early postnatal malnutrition in the rat. Diabetologia 1998; 41:1114-20.
56. Bertin E, Gangnerau MN, Bailb D et al. Glucose metabolism and beta-cell mass in adult offspring of rats protein and/or energy restricted during the last week of pregnancy. Am J Physiol 1999; 277:E11-E17.
57. Ashwell M, Purkins L, Cowen T et al. Pre and postnatal development of adipose tissue at four sites in the guinea pig: Effect of maternal diet restriction during the second half of pregnancy. Ann Nutr Metab 1987; 31:197-210.
58. Oliver MH, Breier BH, Gluckman PD et al. Birth weight rather than maternal nutrition influences glucose tolerance, blood pressure, and IGF-I levels in sheep. Pediatr Res 2002; 52(4):516-24.
59. McLeod KI, Goldrick RB, Whyte HM. The effect of maternal malnutrition on the progeny of the rat. Aust J Exp Biol Med Sci 1972; 50:435-46.
60. Dahri S, Snoeck A, Reusens-Billen B et al. Islet function in offspring of mothers on low-protein diet during gestation. Diabetes 1991; 40(suppl 2):115-20.
61. Langley SC, Browne RF, Jackson AA. Altered glucose tolerance in rats exposed to maternal low protein diets in utero. Comp Biochem Physiol 1994; 109A:223-29.
62. Lucas A, Baker BA, Desai M et al. Nutrition in pregnant or lactating rats programs lipid metabolism in the offspring. Br J Nutrition 1996; 76:605-12.
63. Muaku SM, Beauloye V, Thissen J-P et al. Long-term effects of gestational protein malnutrition on postnatal growth, insulin-like growth factor (IGF)-I, and IGF-binding proteins in rat progeny. Pediatr Res 1996; 39:649-55.
64. Hales CN, Desai BM, Ozanne SE et al. Fishing in the stream of diabetes: From measuring insulin to the control of fetal organogenesis. Biochem Soc Trans 1996; 24:341-50.
65. Ozanne SE, Smith GD, Tikerpae J et al. Altered regulation of hepatic glucose output in the male offspring of protein-malnourished rat dams. Am J Physiol 1996; 2701:E559-E64.
66. Ozanne SE, Wang CL, Coleman N et al. Altered muscle insulin sensitivity in the male offspring of protein-malnourished rats. Am J Physiol 1996; 271:E1128-E34.
67. Sugden MC, Holness MJ. Gender-specific programming of insulin secretion and action. J Endocrinol 2002; 175:757-67.
68. Horton DM, Kind KL, Thavaneswaran P et al. Large size at birth and neonatal catch-up growth independently predict increased adiposity and reduced muscle mass in the guinea pig. Paper presented at: Perinatal Society of Australia and New Zealand 6th Annual Congress. New Zealand: Christchurch, 2002.
69. Horton DM, Kind KL, Walker MR et al. Fetal growth restriction and accelerated postnatal growth independently predict insulin resistance in the adult guinea pig. Am J Physiol 2005; in press.
70. Poore K, Fowden AL. The effect of birth weight on glucose tolerance in pigs at 3 and 12 months of age. Diabetologia 2002; 45:1247-54.
71. Poore K, Fowden AL. Insulin sensitivity in juvenile and adult Large White pigs of low and high birthweight. Diabetologia 2004; 47:340-48.
72. Greenwood PL, Hunt AS, Hermanson JW et al. Effects of birth weight and postnatal nutrition on neonatal sheep: I. Body growth and composition, and some aspects of energetic efficiency. J Anim Sci 1998; 76:2354-67.
73. Greenwood PL, Hunt AS, Slepetis RM et al. Effects of birth weight and postnatal nutrition on neonatal sheep: III. Regulation of energy metabolism. J Anim Sci 2002; 80:2850-61.
74. Gatford KL, Clarke IJ, De Blasio MJ et al. Perinatal growth and plasma GH profiles in adolescent and adult sheep. J Endocrinol 2002; 173:151-59.
75. De Blasio MJ, Gatford KL, Fielke SL et al. Placental restriction of fetal growth reduces size at birth and increases postnatal growth and adiposity in the young lamb. Am J Physiol 2005; in press.
76. Gatford KL, De Blasio MJ, Walker M et al. Restriction of placental and fetal growth impairs insulin secretory capacity in the sheep postnatally. Paper presented at: 12th International Congress of Endocrinology. Portugal: Lisbon, 2004.
77. Simmons RA, Templeton LJ, Gertz SJ. Intrauterine growth retardation leads to the development of type 2 diabetes in the rat. Diabetes 2001; 50:2279-86.

78. Gelardi NL, Cha C-JM, Oh W. Glucose metabolism in adipocytes of obese offspring of mild hyperglycemic rats. Pediatr Res 1990; 28:641-45.
79. Holemans K, van Bree R, Verhaeghe J et al. Maternal semistarvation and streptozotocin-diabetes in rats have different effects on the in vivo glucose uptake by peripheral tissues in their female adult offspring. J Nutr 1997; 127:1371-76.
80. Weintrob N, Karp M, Hod M. Short- and long-range complications in offspring of diabetic mothers. Journal of Diabetes and its Complications 1996; 10(5):294-301.
81. Moss TJM, Sloboda DM, Gurrin LC et al. Programming effects in sheep of prenatal growth restriction and glucocorticoid exposure. Am J Physiol 2001; 281:R960-R70.
82. Crowther C, Harding J. Repeat doses of prenatal corticosteroids for women at risk of preterm birth for preventing neonatal respiratory disease. The Cochrane Database of Systematic Reviews 2003, (Issue 2(2):Art. No.: CD003935. DOI: 10.1002/14651858.CD003935).
83. Nyirenda MJ, Lindsay RS, Kenyon CJ et al. Glucocorticoid exposure in late gestation permanently programs rat hepatic phosphoenolpyruvate carboxykinase and glucocorticoid receptor expression and causes glucose intolerance in adult offspring. Journal of Clinical Investigation 1998; 101:2174-81.
84. Dahlgren J, Nilsson C, Jennische E et al. Prenatal cytokine exposure results in obesity and gender-specific programming. Am J Physiol 2001; 281(2):E326-E34.
85. De Blasio MJ, Gatford KL, Fielke SL et al. Fetal growth restriction increases growth rate and insulin sensitivity in the postnatal lamb. Paper presented at: 11th International Congress of Endocrinology. Australia: Sydney, 2000.
86. De Blasio MJ, Bradbury MR, Adams DH et al. Placental restriction increases postnatal growth rate and sensitivity to IGF-I in the neonatal lamb. Paper presented at: Endocrine Society of Australia Annual Scientific Meeting. Australia: Gold Coast, 2001.
87. Lloyd NK, Thavaneswaran P, Grover S et al. IGF-1 sensitivity and catch-up growth following intrauterine growth restriction in the weanling guinea pig. Paper presented at: Second World Congress on Fetal Origins of Adult Disease. United Kingdom: Brighton, 2003.
88. Horton DM. Prenatal growth and postnatal insulin sensitivity in the guinea pig. Paper presented at: Perinatal Origins of Adult Disease Workshop. Australia: Melbourne, 1999.
89. Gatford KL, De Blasio MJ, McMillen IC et al. Placental restriction and ontogeny of insulin-regulated glucose homeostasis in sheep. Paper presented at: Endocrine Society of Australia Annual Scientific Meeting. Australia: Adelaide, 2002.
90. Bazaes RA, Salazar TE, Pittaluga E et al. Glucose and lipid metabolism in small for gestational age infants at 48 hours of age. Pediatrics 2003; 111:804-09.
91. Wang X, Cui Y, Tong X et al. Effects of the Trp64Arg polymorphism in the b3-adrenergic receptor gene on insulin sensitivity in small for gestational age neonates. J Clin Endocrinol Metab 2004; 89:4981-85.
92. Li C, Johnson MS, Goran MI. Effects of low birth weight on insulin resistance syndrome in Caucasian and African-American children. Diabetes Care 2001; 24:2035-42.
93. Jensen CB, Storgaard H, Dela F et al. Early differential defects of insulin secretion and action in 19-year-old Caucasian men who had low birth weight. Diabetes 2002; 51:1271-80.
94. Jaquet D, Gaboriau A, Czernichow P et al. Insulin resistance early in adulthood in subjects born with intrauterine growth retardation. J Clin Endocrinol Metab 2000; 85:1401-06.
95. Hermann TS, Rask-Madsen C, Ihlemann N et al. Normal insulin-stimulated endothelial function and impaired insulin-stimulated muscle glucose uptake in young adults with low birth weight. J Clin Endocrinol Metab 2003; 88:1252-57.
96. Jaquet D, Vidal H, Hankard R et al. Impaired regulation of glucose transporter 4 gene expression in insulin resistance associated with in utero undernutrition. J Clin Endocrinol Metab 2001; 86:3266-71.
97. Phillips DIW, Borthwick AC, Stein C et al. Fetal growth and insulin resistance in adult life: Relationship between glycogen synthase activity in adult skeletal muscle and birthweight. Diabet Med 1996; 13:325-29.
98. Thompson CH, Sanderson AL, Sandeman D et al. Fetal growth and insulin resistance in adult life: Role of skeletal muscle morphology. Clin Sci 1997; 92:291-96.
99. Arrowsmith F, Ward J, Ling A et al. Fetal nutrition and muscle oxygen supply in childhood. Metabolism 2002; 51:1569-72.
100. IJzerman RG, van Weissenbruch MM, Voordouw JJ et al. The association between birth weight and capillary recruitment is independent of blood pressure and insulin sensitivity: A study in prepubertal children. J Hypertens 2002; 20:1957-63.
101. Stefan N, Weyer C, Levy-Marchal C et al. Endogenous glucose production, insulin sensitivity, and insulin secretion in normal glucose-tolerant Pima Indians with low birth weight. Metabolism 2004; 53(7):904-11.

102. Van Assche FA, De Prins F, Aerts L et al. The endocrine pancreas in small-for-dates infants. Br J Obstetr Gynaecol 1977; 84:751-53.
103. Béringue F, Blondeau B, Castellotti MC et al. Endocrine pancreas development in growth-retarded human fetuses. Diabetes 2002; 51:385-91.
104. Robinson JS, Moore VM, Owens JA et al. Origins of fetal growth restriction. Eur J Obstetr Gynecol Reprod Biol 2000; 92:13-19.
105. Dwyer CM, Madgwick AJA, Ward SS et al. Effect of maternal undernutrition in early gestation on the development of fetal myofibres in the guinea-pig. Reprod Fertil Dev 1995; 7:1285-92.
106. Bedi KS, Birzgalis AR, Mahon M et al. Early life undernutrition in rats 1. Quantitative histology of skeletal muscles from underfed young and refed adult animals. Br J Nutr 1982; 47:417-31.
107. Garofano A, Czernichow P, Bréant B. Effect of aging on beta-cell mass and function in rats malnourished during the perinatal period. Diabetologia 1999; 42:711-18.
108. Schoknecht PA, Pond WG, Mersmann HJ et al. Protein restriction during pregnancy affects postnatal growth in swine progeny. J Nutr 1993; 123:1818-25.
109. Bihoreau M-T, Ktorza A, Kinebanyan MF et al. Impaired glucose homeostasis in adult rats from hyperglycemic mothers. Diabetes 1986; 35:979-84.
110. Gauguier D, Bihoreau M-T, Ktorza A et al. Inheritance of diabetes mellitus as consequence of gestational hyperglycemia in rats. Diabetes 1990; 39:734-39.
111. Dodic M, May CN, Wintour EM et al. An early prenatal exposure to excess glucocorticoid leads to hypertensive offspring in sheep. Clin Sci 1998; 94:149-55.
112. Gatford KL, Wintour EM, De Blasio MJ et al. Differential timing for programming of glucose homeostasis, metabolic sensitivity to insulin and blood pressure by in utero exposure to dexamethasone in sheep. Clin Sci 2000; 98:553-60.
113. Banjanin S, Kapoor A, Matthews SG. Prenatal glucocorticoid exposure alters hypothalamic-pituitary-adrenal function and blood pressure in mature male guinea pigs. J Physiol 2004; 558:305-18.
114. Phillips DIW, Barker DJP, Hales CN et al. Thinness at birth and insulin resistance in adult life. Diabetologia 1994; 37:150-54.

Chapter 14

Programming Effects of Excess Glucocorticoid Exposure in Late Gestation

Timothy J.M. Moss* and Deborah M. Sloboda

Abstract

Glucocorticoids are powerful hormones that play a crucial role in normal maturation of fetal organs in preparation for life outside the womb. However, exposure of the fetus to elevated levels of glucocorticoids, or exposure at inappropriate times, subtly disturbs normal fetal development. Experimental studies have demonstrated that late gestational exposure to excess glucocorticoids causes programming of a number of organ systems.

Introduction

In the late 1960s and early 1970s a series of experiments conducted by Sir Graeme 'Mont' Liggins in New Zealand led to the use of maternal glucocorticoid treatment for the prevention of respiratory distress syndrome in preterm babies.[1] He had shown that preterm lambs born after infusion of the synthetic glucocorticoid, dexamethasone, were able to breathe effectively, despite delivery at an age at which immaturity of their lungs would normally make this impossible.[2] By harnessing the potent maturational effects of glucocorticoids, the therapy devised by Mont Liggins has improved or saved the lives of thousands of preterm infants.

In recent years it has become evident that exposure of the fetus to inappropriately high levels of either endogenous or exogenous glucocorticoids can cause programming, whereby the long term health of an individual is affected by subtle developmental alterations that cause permanent changes to the body's tissues, organs and systems.[3] Furthermore, elevated fetal exposure to glucocorticoids is a common characteristic of a variety of interventions with programming effects, such as maternal under nutrition.[4,5] In this chapter we highlight the major programming effects of excess glucocorticoid exposure in late gestation, rather than exposure early in gestation or throughout its entirety. We focus on programming effects of the clinical use of glucocorticoids, as demonstrated by human studies, and on our own investigations in sheep, using glucocorticoid treatments designed to mimic those used clinically (see Table 1).

Normal Glucocorticoid Levels in Late Gestation

Normally, glucocorticoid production by the fetal adrenal gland is high in early gestation and becomes reduced during mid-late gestation.[6,7] Maintenance of these normal low levels of glucocorticoids is essential for normal fetal growth and development. The fetus is usually 'protected' from exposure to circulating maternal cortisol by the presence of the enzyme 11βhydroxysteroid dehydrogenase type 2 (11βHSD2) in the placenta, which converts active cortisol to inactive cortisone.

*Corresponding Author: Timothy J.M. Moss—School of Women's and Infants' Health, University of Western Australia, Box m094, 35 Stirling Highway, Crawley, Western Australia 6009, Australia. Email: tmoss@cyllene.uwa.edu.au

Early Life Origins of Health and Disease, edited by E. Marelyn Wintour and Julie A. Owens.

Table 1. Timing of late gestational glucocorticoid treatment used in our sheep experiment

Treatment Group	Gestational Age 104 Days	111 Days	118 Days	124 Days
Control	Saline	Saline	Saline	Saline
Single treatment	Betamethasone	Saline	Saline	Saline
Repeated treatment	Betamethasone	Betamethasone	Betamethasone	Betamethasone

Maternal or ultrasound-guided fetal intramuscular betamethasone injections (0.5 mg/kg maternal or fetal bodyweight, respectively) are given at gestational ages of 65-85% of full term. Antenatal corticosteroid treatment is recommended for women at risk of preterm delivery between 24-34 weeks' gestation (60-85% of full term).[79] In recent years, the use of repeated courses of antenatal corticosteroid treatment was common.[80]

In many mammalian species circulating fetal cortisol (the principal glucocorticoid in most mammals, including humans) levels rise exponentially in the days leading up to birth (Fig. 1)[8] as a result of increased activation of the fetal hypothalamic-pituitary-adrenal (HPA) axis. These rising glucocorticoid levels not only contribute to the initiation of parturition but also cause a switch in many cells from division to differentiation, thus slowing fetal growth and causing maturation of a variety of organs in order to prepare the late gestation fetus for extrauterine life. An extensive list of cellular functions affected by glucocorticoids in utero is provided in an excellent review of the programming effects of various endocrine factors.[9]

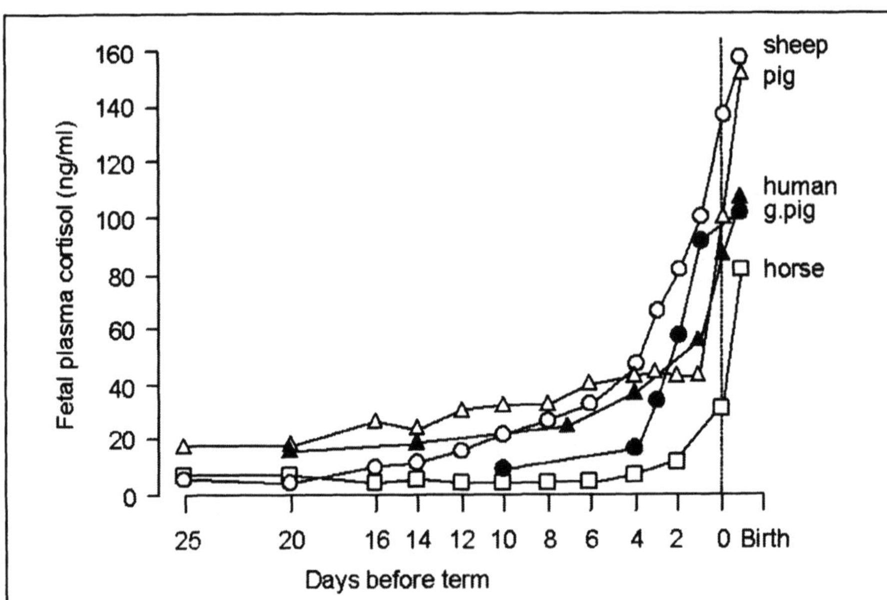

Figure 1. Endogenous glucocorticoid levels in late gestation for various mammalian species. Reprinted with permission from The Nutritional Society. Fowden AL, Li J, Forhead AJ. Glucocorticoids and the preparation for life after birth: are there long-term consequences of the life insurance? The Procedings of The Nutritional Society, 1998; 57:113-122. ©1998 The Nutritional Society.

Excess Glucocorticoids and Direct Effects on Fetal Growth

The normal late gestational slowing of fetal growth in sheep can be abolished by adrenalectomy, and premature increases in cortisol by exogenous infusion causes the fetal growth rate to slow early.[10] Thus, in many species, intrauterine growth restriction (IUGR) is a direct consequence of late gestation exposure of the fetus to excess glucocorticoids.

Administration of the synthetic glucocorticoid, betamethasone, to pregnant sheep, in doses and at times that mimics clinical use in pregnant women, causes IUGR.[11] While single courses of antenatal glucocorticoids used in obstetric practice do not cause IUGR,[12] repeated treatment given to women at continual risk of preterm birth can have adverse effects on fetal growth.[13,14] Determination of the effects of repeated glucocorticoid treatments, the use of which arose mainly from an incomplete understanding of long term efficacy of the therapy and the supposition that 'more is better,[15] are currently the subject of randomised controlled trials.

In contrast to the direct fetal growth restricting effects of maternal betamethasone we have observed, direct ultrasound-guided fetal injections of betamethasone do not cause IUGR.[11,16] We believe the growth restricting effect of maternal betamethasone is due to the prolonged duration of fetal exposure to betamethasone that occurs after maternal injection.[17] Consistent with this, previous studies in which fetuses received intravenous cortisol infusions demonstrated that fetal growth begins to slow only after approximately a day of exposure.[10]

Excess Glucocorticoids and Programming

Early studies by David Barker, which showed associations between low birthweight and subsequent 'adult-onset' diseases, suggest that the direct growth restricting effects of late gestation glucocorticoid exposure might be accompanied by later programmed physiological perturbations. This indeed seems to be the case[3,9] and it is evident that fetal exposure to glucocorticoids is a common, and critical, consequence of a number of experimental interventions that affect fetal programming.[4,5,18] However, like any other programming stimulus, the effects of exposure to excess glucocorticoids are dependent on the timing of the insult and the sex of the fetus. The programming effects of excess glucocorticoid exposure in late gestation are summarised below.

Late Gestational Glucocorticoids and Programming of Metabolism

In recent years the health of millions throughout the world has been threatened by an upsurge in the incidence of 'the metabolic syndrome' or 'Syndrome X', the constituents of which include type 2 diabetes (glucose intolerance/insulin resistance), hyperlipidemia, hypertension and obesity.[19] Since Prof. David Barker's initial investigations, demonstrating an association between birthweight and the incidence of type 2 diabetes,[20] it has become well established that an individual's intrauterine environment influences their risk of developing the metabolic syndrome and that fetal exposure to glucocorticoids is the likely mediator of this effect.[21]

Using a protocol designed to mimic clinical use of glucocorticoids for women at risk of preterm birth, we have shown that single or repeated maternal betamethasone injections, given to pregnant sheep during the final third of pregnancy, cause alterations in postnatal glucose metabolism of their offspring.[16] In this same study we showed that direct fetal injections had similar effects on postnatal glucose metabolism; prenatal betamethasone exposure resulted in elevated insulin responses to intravenous glucose administration, suggesting these sheep were insulin resistant. The different effects on fetal growth of maternal or fetal betamethasone injection (outlined above) allowed us to demonstrate that effects on postnatal glucose metabolism of late gestation glucocorticoid exposure occur independently of effects on fetal growth.[16]

Our studies in sheep are consistent with investigations conducted using rats, which have begun to illustrate the likely molecular mediators of the programming effects of late gestation glucocorticoids on postnatal metabolism. Administration of the synthetic glucocorticoid, dexamethasone, to rats late in pregnancy resulted in fetal growth restriction and adult offspring that were hyperglycaemic at rest with elevated insulin responses to glucose challenge.[22] These rats had increased hepatic expression of glucocorticoid receptor (GR) and phosphoenopyruvate carboxykinase (PEPCK), the rate limiting enzyme in gluconeogenesis, which would be expected to result in increased hepatic glucose production. These effects of late gestation glucocorticoid exposure were not observed when dexamethasone was administered earlier in pregnancy.[22]

Other mechanisms that may underlie the association between late gestation glucocorticoid exposure and altered postnatal metabolism are incompletely understood. Effects of excess glucocorticoids on postnatal insulin sensitivity of peripheral tissues are complex.[23] A critical role for glucocorticoids in fetal pancreatic development is established, and fetal pancreatic insulin content is related to fetal glucocorticoid levels,[24] but there has been no thorough investigation of the effects of excess glucocorticoids in late gestation on pancreatic development. Our initial observations indicate that normal late gestational fetal pancreatic islet remodelling is altered in fetal sheep as a result of repeated maternal betamethasone injections in late gestation (unpublished).

Intergenerational programming of dexamethasone-induced growth restriction, elevated hepatic PEPCK activity and abnormal glucose homeostasis have recently been demonstrated in rats.[25] Male and female offspring of pregnant rats treated with dexamethasone during late pregnancy themselves had male offspring that were of low birthweight, with elevated hepatic PEPCK activity and abnormal glucose homeostasis, without exposure to dexamethasone during gestation. These second generation effects were not present in third generation offspring. The fact that intergenerational programming occurred in male offspring of either males or females whose mothers were treated with dexamethasone demonstrates that this phenomenon cannot be attributed to the intrauterine environment. Rather, these findings raise the intriguing possibility of epigenetic effects of late gestational dexamethasone exposure, and the persistence of genomic 'imprinting' in subsequent generations.[25]

To date, there are no published data from studies of human subjects relating to the effects on glucose metabolism of antenatal corticosteroid treatment.

Late Gestational Glucocorticoids and Programming of the Hypothalamic-Pituitary-Adrenal Axis

Normal physiology is dependent on adequate function of the HPA axis, which is responsible for regulating synthesis and release of a variety of corticosteroid hormones. Cortisol is the principle corticosteroid (in mammals other than rodents) produced by the adrenal cortex; it regulates metabolic, immune and behavioural processes, and the body's response to stressful stimuli. Cortisol acts through glucocorticoid and mineralocorticoid receptors (GR and MR, respectively), which are present in many organs. Permanent alterations in GR and/or MR gene expression have been observed in a variety of organs after manipulation of fetal glucocorticoid exposure[26] and such changes are likely to underlie some of the programming effects on HPA axis function of late gestational glucocorticoid exposure.

Administration of dexamethasone to pregnant rhesus monkeys, late in pregnancy, results in elevated basal and stress-induced circulating cortisol concentrations in juvenile offspring.[27] Rats born after maternal dexamethasone treatment throughout the last third of gestation have elevated circulating corticosteroid concentrations in adulthood, accompanied by hypertension.[28] These same rat offspring had lower GR mRNA expression levels in discrete hippocampal regions responsible for HPA axis feedback regulation, but not in brain nuclei associated with central cardiovascular control,[28] therefore the physiological programming effects of late gestation glucocorticoids in these animals appear to result from GR-mediated alterations in HPA axis regulation. In contrast, HPA axis programming does not appear to underlie postnatal hypertension induced by dexamethasone treatment in early pregnancy in sheep.[29] A short period of maternal dexamethasone treatment in late gestation did not alter early postnatal corticosteroid levels in young rats but hypothalamic CRH/AVP content was altered, which may increase HPA axis responsiveness.[30] Evidence from experiments using guinea pigs indicates that programming effects of late gestational glucocorticoid exposure on the HPA axis are age-[31]

and sex-dependent.[32,33] HPA axis function in young adult female guinea pigs, born after repeated maternal dexamethasone treatments in late pregnancy, depended on the stage of their reproductive cycle, reflecting the influence of female sex hormones.[33]

Our own longitudinal studies using sheep, and others' experiments using guinea pigs, have demonstrated that the capacity of late gestational glucocorticoid exposure to alter HPA responsiveness is dependent on postnatal age. We have demonstrated that responsiveness of the HPA axis to stimulation by corticotrophin releasing hormone (CRH) and arginine vasopressin (AVP) is elevated in one-year-old lambs after gestational exposure to maternal or fetal injection of betamethasone.[34] Later in postnatal life, adrenal responsiveness is reduced in sheep exposed in late gestation to single or repeated maternal betamethasone injections.[35] Male guinea pigs born after repeated maternal dexamethasone injections, studied as young adults, had reduced HPA axis function compared to control.[33] Older adult males, exposed to identical prenatal treatments, displayed more normal HPA axis function, despite abnormalities in hippocampal MR expression. These older male guinea pigs, born after repeated maternal dexamethasone treatments, had higher blood pressure than controls,[31] an effect not observed in younger animals.[33] These longitudinal experiments demonstrate that the effects of late gestational glucocorticoid exposure on postnatal HPA axis responsiveness are dynamic.

Published accounts of the effects of antenatal corticosteroid treatment on postnatal HPA axis function are limited to studies of neonates, which indicate that stress-induced activation of the HPA axis may be impaired, despite infants' ability to maintain normal cortisol concentrations under basal conditions.[36] Whether or not abnormalities persist throughout postnatal life remains to be determined.

Late Gestational Glucocorticoids and Programming of Blood Pressure

Administration of dexamethasone to rats throughout the final third of gestation results in offspring that are growth restricted in utero, hypertensive as adults, and have abnormal HPA function.[28] Betamethasone treatment of pregnant rats throughout this same period reduced birth weight but did not alter postnatal blood pressure.[37] There are numerous potential explanations for the discrepant findings from these two studies.[37] Other experiments using rats, in which dexamethasone treatment was restricted to shorter periods during the final third of pregnancy, have demonstrated an association between postnatal hypertension and reduced nephron endowment; but only when dexamethasone treatment occurred between 70-85% of gestation, and not at earlier or later times.[38] As mentioned above, repeated maternal dexamethasone treatments during late pregnancy (which do not significantly alter birthweight) result in hypertension in mature, but not younger adult, male offspring[31,33] but the underlying cause is unknown.

In sheep, single or repeated, maternal or fetal, betamethasone injections in late gestation do not cause hypertension in adult offspring,[39] in contrast to the established and well-characterised effects of glucocorticoid administration early in prenatal life, which causes hypertension in sheep.[40,41] Postnatal hypertension resulting from early gestational dexamethasone exposure is associated with a reduction in nephron number,[42] similar to observations from studies of rats, outlined above.[38] The gestational timing of corticosteroid treatment resulting in postnatal hypertension in rats and sheep is quite different; however, in terms of renal development, dexamethasone treatment is given at similar times, prior to commencement of nephrogenesis.[42,43] Our investigations in sheep, a species with a similar gestational nephrogenic profile to humans,[44] indicate that late gestational betamethasone exposure in sheep does not reduce nephron number (unpublished observations); this suggests that antenatal corticosteroid treatment in humans is unlikely to alter nephron number.

The entire spectrum of possible effects of antenatal corticosteroid treatment on postnatal blood pressure in humans has been observed. Systolic blood pressure was lower in a group of 20 year olds born after 1 course of antenatal corticosteroids than in controls;[45] higher systolic and diastolic pressures were observed in 14 year old children born after 1 course of antenatal corticosteroid treatment;[46] no effect of antenatal corticosteroid treatment was observed on follow up of 6-year-olds whose mothers were enrolled in Liggins's original randomised controlled trial.[47] Such differences between studies could theoretically be due to changes associated with age but this seems unlikely. Certainly, any reported effects of antenatal corticosteroid treatment on postnatal blood pressure are small.[45,46]

Late Gestation Glucocorticoids and Programming of Immune Function

Function of the immune system is influenced by basal glucocorticoid levels and by HPA axis responsiveness,[48] raising the possibility that alterations in postnatal HPA axis function induced by exposure to excess glucocorticoids in late gestation might alter susceptibility to postnatal inflammatory/immune disease.

Investigations in rats and pigs have demonstrated that prenatal stress results in postnatal immunosuppression.[49-51] These effects are likely mediated by prenatal exposure to glucocorticoids but alterations in postnatal HPA axis function do not necessarily account for altered postnatal immune function in these studies,[50,51] suggesting that prenatal stress has direct programming effects on development of the immune system.

Only a few experimental studies have examined effects on postnatal immune function, beyond the immediate neonatal period, of glucocorticoid exposure in late gestation. Mice aged 5 months, born after prolonged maternal dexamethasone treatment during the final half of gestation, had impaired immunological function, associated with low thyroxine levels and anatomically abnormal thymus, adrenal and thyroid glands.[52] Immunosuppression was also observed in juvenile monkeys born at term, 1 month after 2 days of maternal dexamethasone administration.[53]

A few studies of human neonates, born after antenatal corticosteroid therapy, indicate that immediate postnatal immune function is impaired. Total numbers of circulating white cells are decreased by antenatal corticosteroids, with particular effects on T cells.[54] T cell proliferation is impaired in infants born after antenatal corticosteroids but natural killer cell activity is increased.[55] These effects do not result in an increased (or decreased) incidence of infection in neonates born after a single course of antenatal corticosteroids[12] but data suggest that the incidence of infection in childhood may be increased by the therapy.[56] Multiple courses of antenatal corticosteroids increase the risk of early onset sepsis and sepsis-related neonatal death.[57] The long term effects in humans of single or repeated courses of antenatal corticosteroids remain to be determined.

Late Gestational Glucocorticoids and Programming of the Brain and Behaviour

Late gestation excess glucocorticoid exposure reduces fetal brain growth (Fig. 2)[58-60] and we have shown recently that such glucocorticoid-induced reductions in brain weight persist until adulthood.[39] The brain regions and cell types that are affected by late gestational glucocorticoids are unknown but reductions in fetal brain myelination in sheep[61-63] and reductions in cytoskeletal microtubule associated proteins and the synapse associated protein, synaptophysin, in fetal baboons[64] have been observed. Neuronal number in the brains of fetal and juvenile primates is reduced after late gestational dexamethasone treatment(s), apparently due to neuronal degeneration.[27,65] The hippocampus is likely the most affected brain region because it has a high density of glucocorticoid receptors.[66]

Reductions in brain weight of experimental animals exposed to late gestational glucocorticoids are consistent with observations of human infants, showing dose-dependent reductions in neonatal head circumference after antenatal corticosteroids.[14,13] Magnetic resonance imaging of a small group of infants born after repeated antenatal corticosteroid treatments showed that cortical folding and surface area are reduced, suggesting delayed brain maturation.[67]

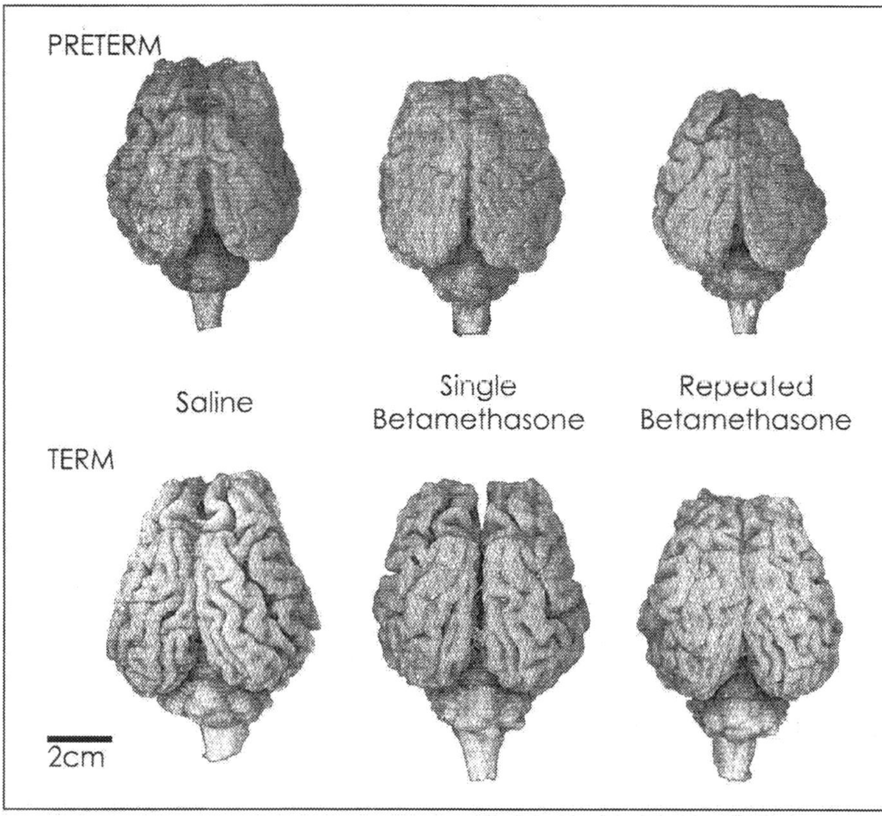

Figure 2. Effects of single or repeated injections of betamethasone in late gestation on fetal brain growth. Reprinted with permission from Huang WL, Beazley LD, Quinlivan JA et al. Effect of corticosteroids on brain growth in fetal sheep. Obstetrics and Gynecology 1999; 94:213-218. ©1999 The American College of Obstetricians and Gynecologists.

The effects on postnatal behaviour of late gestational glucocorticoid exposure have been the subject of only a few studies. Subtle behavioural effects of prenatal betamethasone or dexamethasone exposure have been observed in mice but there were no major adverse consequences.[68] This is consistent with data from humans showing that there are no adverse long-term neurological or cognitive effects of a single course of antenatal corticosteroids.[69-71] However, evidence of psychomotor delay[72] and hyperactivity[73] in children have been associated with repeated courses of antenatal corticosteroids.

There are a number of adverse cognitive or behavioural outcomes associated with reduced head size, including low cognitive ability,[74] and low developmental and intelligence quotients[75] in childhood. Small head circumference at birth is associated with an increased risk of developing schizophrenia;[76] characteristics of which include reductions in brain weight and hippocampal volume,[77] and altered function of the hypothalamic-pituitary-adrenal axis.[78] Such changes are consequences of antenatal corticosteroid treatments in sheep or primates[11,34,65] and together these data raise the possibility that antenatal corticosteroid treatments could contribute to the development of schizophrenia and related disorders; to our knowledge, this possibility has never been investigated.

Conclusion

The ubiquitous effects of late gestational glucocorticoid exposure on fetal development have programming consequences for many physiological functions. The extent to which maternal stress during late pregnancy and antenatal corticosteroid therapy contribute to postnatal health and wellbeing will be elucidated by ongoing and future investigations. Future experimental studies will be necessary to determine the mechanisms responsible for the programming effects of late gestational glucocorticoid exposure, and to investigate potential strategies for preventing or ameliorating adverse effects.

References

1. Liggins GC, Howie RN. A controlled trial of antepartum glucocorticoid treatment for prevention of the respiratory distress syndrome in premature infants. Pediatrics 1972; 50:515-525.
2. Liggins GC. Premature delivery of foetal lambs infused with glucocorticoids. J Endocrinol 1969; 45:515-523.
3. Seckl JR. Glucocorticoid programming of the fetus; adult phenotypes and molecular mechanisms. Mol Cell Endocrinol 2001; 185:61-71.
4. Clark PM. Programming of the hypothalamo-pituitary-adrenal axis and the fetal origins of adult disease hypothesis. Eur J Pediatr 1998; 157(Suppl 1):S7-10.
5. Langley-Evans SC. Fetal programming of cardiovascular function through exposure to maternal undernutrition. Proc Nutr Soc 2001; 60:505-513.
6. Glickman JA, Challis JR. The changing response pattern of sheep fetal adrenal cells throughout the course of gestation. Endocrinology 1980; 106:1371-1379.
7. Wintour EM, Brown EH, Denton DA et al. The ontogeny and regulation of corticosteroid secretion by the ovine foetal adrenal. Acta Endocrinologica 1975; 79:301-316.
8. Fowden AL, Li J, Forhead AJ. Glucocorticoids and the preparation for life after birth: Are there long-term consequences of the life insurance? Proc Nutr Soc 1998; 57:113-122.
9. Fowden AL, Forhead AJ. Endocrine mechanisms of intrauterine programming. Reproduction 2004; 127:515-526.
10. Fowden AL, Szemere J, Hughes P et al. The effects of cortisol on the growth rate of the sheep fetus during late gestation. J Endocrinol 1996; 151:97-105.
11. Newnham JP, Evans SF, Godfrey M et al. Maternal, but not fetal, administration of corticosteroids restricts fetal growth. J Matern Fetal Med 1999; 8:81-87.
12. Crowley P. Prophylactic corticosteriods for preterm birth. Cochrane Library 2004; 4.
13. French NP, Hagan R, Evans SF et al. Repeated antenatal corticosteroids: Size at birth and subsequent development. Am J Obstet Gynecol 1999; 180:114-121.
14. Thorp JA, Jones PG, Knox E et al. Does antenatal corticosteroid therapy affect birth weight and head circumference? Obstet Gynecol 2002; 99:101-108.
15. Antenatal corticosteroids revisited: Repeat courses - National Institutes of Health Consensus Development Conference Statement, August 17-18, 2000. (Obstet Gynecol 2001; 98:144-150).
16. Moss TJ, Sloboda DM, Gurrin LC et al. Programming effects in sheep of prenatal growth restriction and glucocorticoid exposure. Am J Physiol Regul Integr Comp Physiol 2001; 281:R960-970.
17. Moss TJ, Doherty DA, Nitsos I et al. Pharmacokinetics of betamethasone after maternal or fetal intramuscular administration. Am J Obstet Gynecol 2003; 189:1751-1757.
18. Seckl JR. Glucocorticoids, feto-placental 11 beta-hydroxysteroid dehydrogenase type 2, and the early life origins of adult disease. Steroids 1997; 62:89-94.
19. Zimmet P, Alberti KGMM, Shaw J. Global and societal implications of the diabetes epidemic. Nature 2001; 414:782-787.
20. Hales CN, Barker DJ, Clark PM et al. Fetal and infant growth and impaired glucose tolerance at age 64. BMJ 1991; 303:1019-1022.
21. Bertram CE, Hanson MA. Prenatal programming of postnatal endocrine responses by glucocorticoids. Reproduction 2002; 124:459-467.
22. Nyirenda MJ, Lindsay RS, Kenyon CJ et al. Glucocorticoid exposure in late gestation permanently programs rat hepatic phosphoenolpyruvate carboxykinase and glucocorticoid receptor expression and causes glucose intolerance in adult offspring. J Clin Invest 1998; 101:2174-2181.
23. Cleasby ME, Kelly PAT, Walker BR et al. Programming of rat muscle and fat metabolism by in utero overexposure to glucocorticoids. Endocrinology 2003; 144:999-1007.
24. Blondeau B, Lesage J, Czernichow P et al. Glucocorticoids impair fetal beta-cell development in rats. Am J Physiol Endocrinol Metab 2001; 281:E592-599.
25. Drake AJ, Walker BR, Seckl JR. Intergenerational consequences of fetal programming by in utero exposure to glucocorticoids in rats. Am J Physiol Regul Integr Comp Physiol 2004:00106-02004.

26. Cleasby ME, Livingstone DE, Nyirenda MJ et al. Is programming of glucocorticoid receptor expression by prenatal dexamethasone in the rat secondary to metabolic derangement in adulthood? Eur J Endocrinol 2003; 148:129-138.
27. Uno H, Eisele S, Sakai A et al. Neurotoxicity of glucocorticoids in the primate brain. Horm Behav 1994; 28:336-348.
28. Levitt NS, Lindsay RS, Holmes MC et al. Dexamethasone in the last week of pregnancy attenuates hippocampal glucocorticoid receptor gene expression and elevates blood pressure in the adult offspring in the rat. Neuroendocrinology 1996; 64:412-418.
29. Dodic M, Peers A, Moritz K et al. No evidence for HPA reset in adult sheep with high blood pressure due to short prenatal exposure to dexamethasone. Am J Physiol Regul Integr Comp Physiol 2002; 282:R343-350.
30. Bakker JM, Schmidt ED, Kroes H et al. Effects of short-term dexamethasone treatment during pregnancy on the development of the immune system and the hypothalamo-pituitary adrenal axis in the rat. J Neuroimmunol 1995; 63:183-191.
31. Banjanin S, Kapoor A, Matthews SG. Prenatal glucocorticoid exposure alters hypothalamic-pituitary-adrenal function and blood pressure in mature male guinea pigs. J Physiol (Lond) 2004; 558:305-318.
32. Dean F, Matthews SG. Maternal dexamethasone treatment in late gestation alters glucocorticoid and mineralocorticoid receptor mRNA in the fetal guinea pig brain. Brain Research 1999; 846:253-259.
33. Liu L, Li A, Matthews SG. Maternal glucocorticoid treatment programs HPA regulation in adult offspring: Sex-specific effects. Am J Physiol Endocrinol Metab 2001; 280:E729-739.
34. Sloboda DM, Moss TJ, Gurrin LC et al. The effect of prenatal betamethasone administration on postnatal ovine hypothalamic-pituitary-adrenal function. J Endocrinol 2002; 172:71-81.
35. Sloboda DM, Moss TJM, Nitsos I et al. Antenatal glucocorticoid treatment in sheep results in adrenal supression in adulthood. J Soc Gynecol Investig 2003; 10:233A.
36. Davis EP, Townsend EL, Gunnar MR et al. Effects of prenatal betamethasone exposure on regulation of stress physiology in healthy premature infants. Psychoneuroendocrinology 2004; 29:1028-1036.
37. McDonald T, Franko K, Brown J et al. Betamethasone in the last week of pregnancy causes fetal growth retardation but not adult hypertension in rats. J Soc Gynecol Investig 2003; 10:469-473.
38. Ortiz LA, Quan A, Weinberg A et al. Effect of prenatal dexamethasone on rat renal development. Kidney Int 2001; 59:1663-1669.
39. Moss TJM, Doherty DA, Nitsos I et al. Effects into adulthood of single or repeated antenatal corticosteroids in sheep. Am J Obstet Gynecol 2205; 192:146-152.
40. Dodic M, May CN, Wintour EM et al. An early prenatal exposure to excess glucocorticoid leads to hypertensive offspring in sheep. Clin Sci (Lond) 1998; 94:149-155.
41. Dodic M, Peers A, Coghlan JP et al. Altered cardiovascular haemodynamics and baroreceptor-heart rate reflex in adult sheep after prenatal exposure to dexamethasone. Clin Sci (Lond) 1999; 97:103-109.
42. Wintour EM, Moritz KM, Johnson K et al. Reduced nephron number in adult sheep, hypertensive as a result of prenatal glucocorticoid treatment. J Physiol 2003; 549:929-935.
43. Bertram JF, Young RJ, Spencer K et al. Quantitative analysis of the developing rat kidney: Absolute and relative volumes and growth curves. Anat Rec 2000; 258:128-135.
44. Moritz KM, Wintour EM. Functional development of the meso- and metanephros. Pediatr Nephrol 1999; 13:171-178.
45. Dessens AB, Haas HS, Koppe JG. Twenty-year follow-up of antenatal corticosteroid treatment. Pediatrics 2000; 105:E77.
46. Doyle LW, Ford GW, Davis NM et al. Antenatal corticosteroid therapy and blood pressure at 14 years of age in preterm children. Clin Sci (Lond) 2000; 98:137-142.
47. Dalziel SR, Liang A, Parag V et al. Blood pressure at 6 years of age after prenatal exposure to betamethasone: Follow-up results of a randomized, controlled trial. Pediatrics 2004; 114:e373-377.
48. Wilckens T, De Rijk R. Glucocorticoids and immune function: Unknown dimensions and new frontiers. Immunology Today 1997; 18:418-424.
49. Sobrian SK, Vaughn VT, Bloch EF et al. Influence of prenatal maternal stress on the immunocompetence of the offspring. Pharmacol Biochem Behav 1992; 43:537-547.
50. Kay G, Tarcic N, Poltyrev T et al. Prenatal stress depresses immune function in rats. Physiol Behav 1998; 63:397-402.
51. Tuchscherer M, Kanitz E, Otten W et al. Effects of prenatal stress on cellular and humoral immune responses in neonatal pigs. Vet Immunol Immunopathol 2002; 86:195-203.
52. Eishi Y, Hirokawa K, Hatakeyama S. Long-lasting impairment of immune and endocrine systems of offspring induced by injection of dexamethasone into pregnant mice. Clin Immunol Immunopathol 1983; 26:335-349.
53. Coe CL, Lubach GR. Prenatal influences on neuroimmune set points in infancy. Ann NY Acad Sci 2000; 917:468-477.

54. Chabra S, Cottrill C, Rayens MK et al. Lymphocyte subsets in cord blood of preterm infants: Effect of antenatal steroids. Biol Neonate 1998; 74:200-207.
55. Kavelaars A, van der Pompe G, Bakker JM et al. Altered immune function in human newborns after prenatal administration of betamethasone: Enhanced natural killer cell activity and decreased T cell proliferation in cord blood. Pediatr Res 1999; 45:306-312.
56. Smolders-de Haas H, Neuvel J, Schmand B et al. Physical development and medica history of children who were treated antenatally with corticosteroid to prevent respiratory distress syndrome: A 10- to 12- year follow- up. Pediatrics 1990; 86:65-70.
57. Vermillion ST, Soper DE, Newman RB. Neonatal sepsis and death after multiple courses of antenatal betamethasone therapy. Am J Obstet Gynecol 2000; 183:810-814.
58. Johnson JWC, Mitzner W, Beck JC et al. Long-term effects of betamethasone on fetal development. Am J Obstet Gynecol 1981; 141:1053-1064.
59. Huang WL, Beazley LD, Quinlivan JA et al. Effect of corticosteroids on brain growth in fetal sheep. Obstet Gynecol 1999; 94:213-218.
60. Scheepens A, Van De Waarenburg M, Van Den Hove D et al. A single course of prenatal betamethasone in the rat alters postnatal brain cell proliferation but not apoptosis. J Physiol 2003; 552:163-175.
61. Dunlop SA, Archer MA, Quinlivan JA et al. Repeated prenatal corticosteroids delay myelination in the ovine central nervous system. J Matern Fetal Med 1997; 6:309-313.
62. Huang WL, Harper CG, Evans SF et al. Repeated prenatal corticosteroid administration delays myelination of the corpus callosum in fetal sheep. Int J Dev Neurosci 2001; 19:415-425.
63. Quinlivan JA, Dunlop SA, Newnham JP et al. Repeated, but not single, maternal administration of corticosteriods delays myelination in the brain of fetal sheep. Prenat Neonat Med 1999; 4:47-55.
64. Antonow-Schlorke I, Schwab M, Li C et al. Glucocorticoid exposure at the dose used clinically alters cytoskeletal proteins and presynaptic terminals in the fetal baboon brain. J Physiol 2003; 547:117-123.
65. Uno H, Lohmiller L, Thieme C et al. Brain damage induced by prenatal exposure to dexamethasone in fetal rhesus macaques. I Hippocampus Brain Res Dev Brain Res 1990; 53:157-167.
66. Matthews SG. Antenatal glucocorticoids and the developing brain: Mechanisms of action. Semin Neonatol 2001; 6:309-317.
67. Modi N, Lewis H, Al-Naqeeb N et al. The effects of repeated antenatal glucocorticoid therapy on the developing brain. Pediatr Res 2001; 50:581-585.
68. Rayburn W, Christensen H, Gonzalez C. A placebo-controlled comparison between betamethasone and dexamethasone for fetal maturation: Differences in neurobehavioral development in mice offspring. Am J Obstet Gynecol 1997; 176:842-851.
69. Doyle L, Ford G, Rickards A et al. Antenatal corticosteroids and outcome at 14 years of age in children with birth weight less than 1501 grams. Pediatrics 2000; 106:e2.
70. Dessens A, Smolders-de Has H, Koppe J. Twenty year follow-up of antenatal corticosteroid treatment. Pediatrics 2000; 105:e77.
71. Matthews SG. Antenatal glucocorticoids and programming of the developing CNS. Pediatr Res 2000; 47:291-300.
72. Esplin M, Fausett M, Smith S et al. Multiple courses of antenatal steroids are asociated with a delay in long-term psychomotor development in children with birth weights < 1500 grams. Am J Obstet Gynecol 2000; 182:S24.
73. French N, Hagan R, Evans S et al. Repeated antenatal corticosteroids:Effects on cerebral palsy and childhood behaviour. Am J Obstet Gynecol 2004; 190:588-595.
74. Stathis S, O'Callaghan M, Harvey J et al. Head circumference in ELBW babies is associated with learning difficulties and cognition but not ADHD in the school-aged child. Dev Med Child Neurol 1999; 41:375-380.
75. Brandt I, Sticker E, Lentze M. Catch-up growth of head circumference of very low birthweight, samll for gestational age preterm infants and mental development to adulthood. J Pediatr 2003; 142:463-468.
76. MacNeil T, Cantor-Graae E, Ismail B. Obstetric complications and congential malformation in schizophrenia. Brain Res Rev 2000; 31:166-178.
77. Harrison P. The neuropathology of schizophrenia. A critical review of the data and their interpretation. Brain 1999; 122:593-624.
78. Cotter D, Pariante C. Stress and the progression of the developmental hypothesis of schizophrenia. Br J Psychiarty 2002; 181:363-365.
79. Effect of corticosteroids for fetal maturation on perinatal outcomes. Am J Obstet Gynecol 1995; 173:246-252.
80. Goldenberg RL, Wright LL. Repeated courses of antenatal corticosteroids. Obstet Gynecol 2001; 97:316-317.

CHAPTER 15

Programming Effects of Moderate and Binge Alcohol Consumption

Jeff Schwartz* and Luke C. Carey

Alcohol is a well known teratogen. Heavy, sustained consumption of alcohol by pregnant women is associated with the constellation of birth defects and symptoms known as fetal alcohol syndrome (FAS). Similarly, exposure to high concentrations of alcohol for extended periods in animal models of FAS reproduces the teratogenic effects. In contrast, far less is known regarding the more subtle effects on offspring of lesser maternal ethanol ingestion. The most widely studied permanent consequence of prenatal alcohol exposure is impaired development of the nervous system, leading to changes in brain chemistry, neurobiology and behaviour. This is reflected in the mental retardation and neurological deficits associated with FAS. Because this is more accurately described as mimicking a teratogenic outcome present at birth and is considered to be the result of high and sustained exposure to alcohol in humans, it will not be specifically covered in this essay. There are numerous excellent recent reviews on this subject.[1] Rather, this chapter will concentrate on reviewing the body of literature concerning the more subtle effects of prenatal alcohol exposure, often not apparent at birth, and impacting on the performance and regulation of various organ systems. These findings have been obtained in animal, rather than human, experiments and observations. In many ways the phenotypic changes resemble those associated with certain types of fetal programming as caused by nutrient or oxygen restriction (see other chapters in this volume), including small weight at birth and permanent alterations to the hypothalamo-pituitary-adrenal (HPA) axis, glucose metabolism, and possibly cardiovascular regulation.[2] In addition to expanding our knowledge of early physiological programming from the perspective of a novel perturbation in nonhuman species, these observations on the effects of prenatal alcohol identify possible pathophysiological consequences of human alcohol consumption during pregnancy.

The physiological effects of prenatal alcohol exposure are often not readily apparent in humans. Aside from the most obvious neurological and craniofacial manifestations of FAS, more subtle consequences of prenatal alcohol exposure in humans have rarely been looked for or identified. As Moushmoush and Abi-Mansour have written with respect to heart defects, if the high level of alcohol (responsible for the developmental alteration) occurred only during a period critical for heart (and not brain) development, then the child might only have a congenital heart defect and no other signs of FAS, and the association with alcohol would most likely be missed.[3] Hence, there are likely many human postnatal health problems that are the result of prenatal exposure to alcohol, but never so linked, because the nonFAS problems occur at lower levels of alcohol consumption and/or at times when alcohol does not exert gross effects on neural and craniofacial development.

*Corresponding Author: Jeff Schwartz—Discipline of Physiology, School of Molecular and Biomedical Science, The University of Adelaide, Adelaide, South Australia 5005
Email: jeff.schwartz@adelaide.edu.au

Early Life Origins of Health and Disease, edited by E. Marelyn Wintour and Julie A. Owens.

A note of caution: The aim of this chapter is to provide an overview of published animal studies on the long lasting effects of prenatal exposure to ethanol at levels that do not produce recognizable birth defects. Thus, terms such as high, low and relative are not used in relation to any defined standard, although efforts have been made to be consistent. The applicability of many of the studies cited to the issue of fetal programming is also complex. Many of the cited studies predate the recent articulation of the concept of programming and were therefore not specifically designed to test hypotheses related to programming. Nevertheless, they involved a prenatal treatment and examination of postnatal consequences. While they may be useful for understanding the role or actions of alcohol in eliciting programming, it must be remembered that they are not studies of programming per se. In numerous instances, the authors have included critical aspects of methodology of cited articles, which may influence interpretation of outcomes. This is necessary for studies that involve the use of a complex drug such as ethanol, which has behavioural, as well as physiological, impact, making it difficult to design adequate controls. For example, if the model of alcohol ingestion diminishes nutrient intake, the latter must be controlled for as well.

What Is Moderate Alcohol Consumption and What Are Common Pregnancy Exposures in Humans?

In humans it is generally accepted that 1-2 standard drinks (10-20 grams of ethanol) per day is not harmful and may even confer health benefits. This amount of alcohol, when consumed within a relatively short time, produces a blood alcohol concentration (BAC) in the vicinity of 0.05 % in a 50 kg woman. In countries such as Australia, the UK and US, alcohol consumption rates in women of childbearing age often appear to be somewhat higher than recommended levels.[4-6] Once women learn they have become pregnant, however, they largely decrease or cease consumption of alcohol. Nevertheless, a significant proportion of women continue to consume alcohol during pregnancy, and there are many women who, despite intentions to the contrary, consume alcohol in dangerous amounts. This is evident in a recently published Danish study, where it was found that up to 40 % of pregnant women engage in at least one episode of binge drinking (a minimum of 5 standard drinks within a short period) during the second trimester.[7]

Animal Models of Alcohol Exposure

Binge exposure also forms the basis of one model of fetal alcohol exposure in animals. One of the most widely used doses is 5 g of alcohol per kg body weight in rats or mice, delivered by oral gavage. Because of the increased metabolic rate in rodents, the maximum BACs achieved are far less than would occur if the same dose were administered to humans. A regression based on maximum BACs in pregnant rats fed alcohol at various doses suggests that maximum BAC = 0.05 X dose (g/kg). Thus 5 g alcohol per kg body weight produces a maximum BAC in the order of 0.25 %.[8] West has suggested in other work that, by virtue of the concentrations reached and time of exposure, binge exposure may be the most relevant model for human alcohol consumption and effects on the fetus.[9]

Perhaps the most widely used protocol of exposure to alcohol is one in which alcohol is included as a component of a complete liquid diet, thereby mimicking moderate alcohol exposure. In these diets, alcohol constitutes approximately 5-6 % of the total volume and provides around one third of the total energy requirement. Control animals are fed an identical liquid diet, with corn starch or another energy equivalent substituted for alcohol, where the volume of the diet available to each control is equivalent to the volume of the alcohol-containing diet consumed by a paired experimental animal on the previous day. Often an ad libitum fed control is also included. The maximum BACs achieved with the 5-6 % alcohol liquid diet are in the order of 0.08-0.1 %.[10] This diet is often provided over a certain period of development, but is sometimes used throughout pregnancy.

In this chapter, the phrase *liquid diet* will be used to refer to a treatment where animals have access to a single source of alimentation in which alcohol provides approximately one third of total energy and where the diet is provided throughout pregnancy, with perhaps a 3-4 day period at the beginning of ramping up to full alcoholic strength. Any variations from this approach will be indicated. A number of other exposure models, such as inhalation of alcohol vapours, and intraperitoneal injection of alcohol, have also been used.

Endocrine System

Alcohol ingestion alters the maternal endocrine system, which is likely to induce secondary effects on fetal development. In addition, alcohol itself may act directly on the developing fetal glands and target organs for hormones. Of particular importance with respect to fetal programming in general, are alterations to the maternal, fetal and postnatal offspring HPA axes. Altered HPA axis activity is a feature in many other animal models of early programming by perinatal perturbation (reviewed elsewhere in this volume in Chapter 14), and in humans permanently altered HPA axis activity has been observed to occur as part of the small size-at-birth for gestational age phenotype in children and adults.[11,12]

Alcohol stimulates the HPA axis and, not surprisingly, this has been repeatedly demonstrated to occur in the pregnant rat on the alcohol liquid diet between gestational day (e) 8 and e20 (gestation is approximately 22 days in rats), and in those exposed to alcohol vapours between e7 and e18.[13,14] Interestingly, in contrast to the mother, fetal blood corticosterone levels are not increased relative to those in controls when the diet is fed to dams between e8 and e20.[15] Some of the effects of alcohol exposure on the development of other fetal organs, such as the heart, are attenuated if the increase in maternal corticosterone levels are prevented by adrenalectomy on e7 and replacement to maintain corticosterone at a control or normal level.[16]

Maternal adrenalectomy is associated with an increase in fetal corticosterone. This is a consequence of decreased negative feedback inhibition of fetal ACTH release, which is normally modulated by maternal corticosterone crossing the placenta. Less intuitive is the observation that the increase in fetal corticosterone, caused by maternal adrenalectomy, is attenuated by alcohol, which is normally considered to be a stimulator of the HPA axis.[15] Alcohol also attenuates the normal surge in blood testosterone observed in male fetuses on e18 and e19.[15,17,18]

The consensus of studies over the past two decades is that adult* offspring of rats that consumed alcohol during pregnancy have exaggerated responses of the HPA axis to psychological, social or physiological stress.[14,19-21] These include offspring of animals fed the liquid alcohol diet or exposed to alcohol vapours and in adulthood subjected to restraint, immune, footshock, noise, novel environment and temperature stress. There is also a sex difference associated with the effect of alcohol, with female offspring at 45-60 days of age exhibiting higher HPA responses than males to treatment with direct immune mediators, such as interleukin 1β, and activation of inflammatory responses by exposure to substances such as lipopolysaccharide and turpentine.[20] Furthermore, a significantly enhanced ACTH-response to increased brain nitric oxide was seen in adult male rats exposed to alcohol as fetuses.[22] Interestingly, less mature rats (22-24 days old) that had been prenatally exposed to alcohol via the liquid diet exhibited blunted ACTH responses to interleukin 1β compared to controls, which further emphasises the complexity of the response pattern, which may vary with age and type and timing of prenatal challenge.[23]

Prenatal alcohol also alters other endocrine functions postnatally, including the hypothalamo-pituitary-gonadal and hypothalamo-pituitary-thyroid axes. Adult offspring of ethanol-fed rat dams on the liquid diet from e8 to e20 have decreased concentrations of triiodothyronine, and increased thyroid stimulating hormone levels, suggesting permanently

*The term adult in this setting is defined as any time after the animal in question has attained sexual maturity. In rodents, this is typically at around 7–8 weeks of age.

diminished activity at the thyroid.[24] Ethanol-exposed offspring also have decreased plasma concentrations of luteinizing hormone (LH).[25] It is not yet known whether either this is the result of a brain defect in which there is diminished LH releasing hormone as well. As previously mentioned, there is a decreased/delayed testosterone surge in male fetuses during ethanol exposure, making it is tempting to speculate that ethanol exposure in utero permanently alters hypothalamic development leading to decreased activity in the gonadotropin axis.

Another interesting and apparently permanent effect of prenatal alcohol exposure has been reported with respect to the osmoregulatory system in rats. In the rat, maternal exposure to an alcohol-containing diet from e7 to e21 results in offspring at 11 weekswith a phenotype resembling central diabetes insipidus.[26] Compared to control animals, they have elevated water consumption, urine output and plasma osmolality. The hypothalamus and pituitary have decreased vasopressin mRNA and protein, respectively, while the relationship between vasopressin and plasma osmolality is shifted, reflecting a higher osmolality threshold required to elicit a vasopressin response.

Immune System

The long-lasting effects of prenatal alcohol exposure on immune function have also been studied in animal models using moderate levels of ethanol ingestion during pregnancy. As noted previously, the HPA responses to immune challenge may be elevated in fetal alcohol-exposed adult animals. Numerous studies have also demonstrated that aspects of immune system function themselves are diminished in offspring exposed to ethanol as fetuses.

The origins of this diminished immune function may be evident in alcohol-exposed fetuses. Analysis of e18 mouse fetus thymocytes from dams fed a liquid diet deriving 25 % total energy from alcohol indicate the presence of fewer L3T4-positive and Lyt2-positive cells.[27] Other indices show that alcohol retards the development of the thymus.[27] Diminished numbers of immune cells have been reported in the gastrointestinal associated lymphoid tissue of fetuses and nursing pups (examined on postnatal (p) day 14 and p18) from rats exposed to alcohol via the liquid diet.[28] The mitogenic responses of thymic and splenic lymphocytes in adult animals exposed to alcohol as fetuses also demonstrate a permanent effect on immune function. The mitogenic response of splenic lymphocytes to concanavalin-A is decreased in adult offspring rats of dams fed alcohol via the liquid diet during pregnancy.[29] In the same study, alcohol exposed males were also found to have decreased numbers of thymocytes.[29] Similarly, the response to interleukin-2 in concanavalin-A-stimulated T-lymphoblast cells was also diminished in cells from 3 month old rats exposed to alcohol as fetuses via the liquid diet over the final two weeks of gestation.[30]

In contrast, other mitogenic responses are apparently increased in cells obtained from animals exposed to alcohol as fetuses. Cells from adult prenatal alcohol exposed female rats, subjected to cold stress for one day, exhibit an increased splenic lymphocytic proliferative response to concanavalin-A or pokeweed mitogen, compared to that of pair-fed control animals.[31] Thymocytes from young, prenatal alcohol exposed male rats also have an increased mitogenic response to concanavalin-A.[32]

Metabolism

Along with altered activation of the HPA axis, an apparent resistance to insulin and glucose intolerance are arguably the postnatal consequences of prenatal exposure to moderate levels of alcohol that most resemble the general fetal programming phenotype in humans.[2] Exposure to relatively high levels of alcohol in utero (4 g per kg body weight per day throughout pregnancy) in rats results in offspring with elevated resting plasma insulin, elevated resistin mRNA in adipose tissue and protein in plasma, and decreased Glut4 transporter protein expression in skeletal muscle after a glucose challengeat 13 weeks of age.[33] These animals also have elevated plasma glucose and insulin levels in response to a glucose load. Interestingly, rats exposed prenatally to ethanol are reported to have comparable insulin resistance following an intraperitoneal glucose tolerance test to animals raised on a high fat diet.[34] The mechanisms of the insulin resistance in the two cases are likely to be distinct however, since prenatal exposure to alcohol and a high fat diet are additive.

Exposure to lesser amounts of prenatal alcohol has also been reported to lead to the development of insulin resistance. Animals born to dams who consumed alcohol in their drinking water during pregnancy had exaggerated insulin responses to an oral glucose load at 90 days of age.[35]

Aside from elevated plasma glucose and insulin, adult male rats from dams fed the liquid alcohol containing diet during pregnancy, are also reported to have elevated fasting plasma concentrations of very low density lipoproteins and triglycerides.[36] This effect may be dependent on testosterone, since it does not occur in castrate males, but does occur in females treated with exogenous testosterone.

Cardiovascular System

FAS is associated with congenital heart defects, and it is perhaps for this reason that much research on the cardiovascular system of animals in models of fetal alcohol exposure deals with the development of the heart. For example, the hearts of late gestation fetal mice from dams exposed to alcohol by gavage or intraperitoneal injection (25% alcohol at 0.015 ml/g body weight) had ventricular septal defects when the exposure was on days e8-10.[37] At the cellular level, it has been reported that the hearts of mice born to dams fed a liquid diet containing alcohol providing 20% of energy from day e8 onward have abnormal myocytes at birth, with an increase in the volume ratio of mitochondria to cytoplasm, and a decrease in mitochondrial number compared to control cells.[38] These changes, however, may reflect acute responses to alcohol, rather than programming effects. Some of the morphological changes in cardiomyocytes observed at postnatal day p7 in the hearts of rats exposed prenatally to alcohol via the maternal liquid diet are not evident at p21.[39]

In contrast, other effects of prenatal alcohol may be more subtle, but permanent, and therefore more consistent with programming. For instance, adult rats exposed to alcohol as fetuses have shorter cardiomyocytes and impaired responsiveness to extracellular calcium or stimulating frequency.[40] Impaired contractile responses of adult cardiomyocytes have also been observed in cells from rats born to dams that had been treated with 6 g alcohol per kg per day from e8 to e20 (maximum blood alcohol ~0.26%). Cells from alcohol-exposed rats in this study had decreased peak tension development and maximum contraction and relaxation velocities compared to controls. Furthermore, it has also been noted that the cardiomyocytes of adult rats exposed prenatally to alcohol have increased resting and peak intracellular calcium concentrations , and decreased calcium responses to caffeine.[41]

Similarly, prenatal exposure to alcohol may alter vascular structure and/or function. In vitro, rings cut from the aortae of rats exposed to alcohol via the maternal liquid diet were observed to be more sensitive to KCl-induced contraction (endothelium removed), and had attenuated relaxation responses to carbachol (endothelium intact).[42] The responses to norepinephrine, interestingly, were also attenuated. Significantly, the ethanol-exposed rats in this study had elevated arterial pressures as adults, compared to controls, suggesting that the observed cardiovascular changes are a manifestation of fetal programming.

We have found that female rats exposed to alcohol (liquid diet, weeks 2 and 3 of pregnancy) have left ventricular hypertrophy as adults.[16] This effect appears to be dependent on maternal glucocorticoids, since it can be prevented by maternal adrenalectomy. More recently, we have shown that a single binge exposure to alcohol (5 g per kg body weight) on e8, but not e16, results in offspring with smaller hearts and kidneys at birth (corrected for body weight), but normalized kidneys and larger hearts in adulthood.[43] The finding with respect to heart size was most evident in females, and was exacerbated with age. In addition, the smaller hearts of alcohol-exposed offspring at birth had lower levels of atrial natriuretic peptide (ANP) mRNA than those of controls, while the larger ventricles in adulthood had elevated levels of ANP mRNA. Taken together, these data suggest that alcohol triggers subtle changes in the fetus causing the heart to grow over time in the postnatal environment. These outcomes may represent myocardial hypertrophy. FAS in humans is associated with vascular malformation, in particular increased tortuosity of the retinal vessels.[44,45] This is such a prominent feature that it may be the preeminent characteristic of the syndrome.[45] It is not unreasonable, therefore, to link altered vascular development with fetal alcohol exposure, even at lesser levels than those that cause fetal alcohol syndrome. It is tempting to speculate that the cardiac hypertrophy and other changes observed in animal models may be secondary to vascular changes caused by the initial exposure to alcohol.

Gastrointestinal and Liver

Changes in liver structure and function have been reported to be associated with fetal exposure to alcohol. These are largely teratogenic or temporary acute responses, and therefore are not programming. For example, the decreases in fetal hepatic glutathione and ATP present after treatment of pregnant rats with alcohol (2.4 g per kg per day between e14 and e19) are no longer evident by p7.[46] The results of another study suggest that the livers of rats prenatally exposed to alcohol via the maternal liquid diet may be permanently altered in terms of being smaller and having decreased rates of DNA synthesis up to p14.[47] In this study, however, there was no control for decreased nutrition in alcohol-consuming dams (i.e., no pair fed controls; the alcohol-exposed offspring were compared to offspring of ad libitum fed dams). Interestingly, the changes were observed only in rats exposed to alcohol throughout pregnancy, but not in those exposed in either the first or second half of pregnancy. The pups exposed to alcohol throughout pregnancy were also smaller, had smaller brains and craniofacial features that are more consistent with teratogenesis rather than programming, where the brain is spared.

Renal

Although this area has not been the focus of extensive study, there is evidence to suggest that moderate prenatal alcohol exposure may exert effects on renal development consistent with programming. Nine day old pups born to dams that had been fed a liquid diet with ethanol during pregnancy had decreased kidney weight, protein and DNA content.[48] We have also observed that rats born to dams binge exposed to alcohol on e8 had smaller kidneys at birth.[43] The difference in renal weight between control and ethanol-exposed animals disappeared by adulthood. In a specific study of renal function, ninety day old rats prenatally exposed to alcohol via the maternal liquid diet had elevated urinary flow rates and sodium excretion rates compared to controls when placed on a low sodium diet.[49] Significantly, they also became hyperkalemic on a high potassium diet, suggesting either impaired secretion of mineralocorticoids or a more complex deficiency that impairs the rats' ability to handle altered loads of electrolytes in general.

Summary

Until recently, studies pertaining to the effect(s) of prenatal alcohol on fetal and postnatal development have largely focussed on teratogenic outcomes. These studies have generally utilized excessive alcohol exposure models, and examined resultant outcomes that are gross and can be measured in the offspring as a fetus or at birth. Very few studies have addressed the possible long term, or programming, effects of prenatal alcohol exposure. Even observational studies in humans with fetal alcohol syndrome have been limited, as noted by Day and Richardson, consisting of "too few studies to assess accurately the effect of drinking on development beyond the neonatal period".[50] These authors go on to speculate that the effects observed in humans represent only, "the more severe end of a continuum of effects". As we learn more about the origins and various mechanisms of human fetal programming, animal studies of fetal exposure to socially relevant amounts of alcohol are likely to grow in importance. Alcohol exposure is an important contributing factor to small birth weight in human populations and may also contribute to the production of the small birth weight phenotype. In addition, because the effects of moderate fetal alcohol exposure resemble the phenotype of other animal models of fetal programming, there is an experimental advantage to using alcohol to generate the phenotype and investigate the underlying initiating and mediating mechanisms. Alcohol exposure can be targeted to a single time point or spread over a longer period, and the dose can be easily titrated. For these and other reasons that will become apparent only as we learn more, the studies of developmental programming of adult disease and those of the subtle pathophysiological effects of moderate maternal alcohol consumption are bound to become increasingly linked.

References

1. Kalter H. Teratology in the 20th century: Environmental causes of congenital malformations in humans and how they were established. Neurotoxicol Teratol 2003; 25(2):131-282.
2. Barker DJP. Mothers, Babies and Health in Later Life. 2nd ed. Edinburgh: Churchill Livingstone, 1998.
3. Moushmoush B, Abi-Mansour P. Alcohol and the heart. The long-term effects of alcohol on the cardiovascular system. Arch Intern Med 1991; 151(1):36-42.
4. Australian Social Trends 1995 Health - Risk factors: Alcohol use. Canberra: Australian Bureau of Statistics, 1995.
5. Drinking to excess rising among women. Office of national statistics. Available at: www.statistics.gov.uk, 2004.
6. Morse J. Women on a binge. Time 2002; 159:56-61.
7. Kesmodel U, Kesmodel PS, Larsen A et al. Use of alcohol and illicit drugs among pregnant Danish women, 1998. Scand J Public Health 2003; 31(1):5-11.
8. Maier SE, Miller JA, West JR. Prenatal binge-like alcohol exposure in the rat results in region-specific deficits in brain growth. Neurotoxicol Teratol 1999; 21(3):285-291.
9. Maier SE, West JR. Drinking patterns and alcohol-related birth defects. Alcohol Res Health 2001; 25(3):168-174.
10. Mihalick SM, Crandall JE, Langlois JC et al. Prenatal ethanol exposure, generalized learning impairment, and medial prefrontal cortical deficits in rats. Neurotoxicol Teratol 2001; 23(5):453-462.
11. Phillips DI. Programming of adrenocortical function and the fetal origins of adult disease. J Endocrinol Invest 2001; 24(9):742-746.
12. Ward AM, Syddall HE, Wood PJ et al. Fetal programming of the hypothalamic-pituitary-adrenal (HPA) axis: Low birth weight and central HPA regulation. J Clin Endocrinol Metab 2004; 89(3):1227-1233.
13. Aird F, Halasz I, Redei E. Ontogeny of hypothalamic corticotropin-releasing factor and anterior pituitary pro-opiomelanocortin expression in male and female offspring of alcohol-exposed and adrenalectomized dams. Alcohol Clin Exp Res 1997; 21(9):1560-1566.
14. Ogilvie KM, Rivier C. Prenatal alcohol exposure results in hyperactivity of the hypothalamic-pituitary-adrenal axis of the offspring: modulation by fostering at birth and postnatal handling. Alcohol Clin Exp Res 1997; 21(3):424-429.
15. Sinha P, Halasz I, Choi JF et al. Maternal adrenalectomy eliminates a surge of plasma dehydroepiandrosterone in the mother and attenuates the prenatal testosterone surge in the male fetus. Endocrinology 1997; 138(11):4792-4797.
16. Wilcoxon JS, Schwartz J, Aird F et al. Sexually dimorphic effects of maternal alcohol intake and adrenalectomy on left ventricular hypertrophy in rat offspring. Am J Physiol Endocrinol Metab 2003; 285(1):E31-39.
17. Revskoy S, Halasz I, Redei E. Corticotropin-releasing hormone and proopiomelanocortin gene expression is altered selectively in the male rat fetal thymus by maternal alcohol consumption. Endocrinology 1997; 138(1):389-396.
18. Ward IL, Ward OB, Affuso JD et al. Fetal testosterone surge: Specific modulations induced in male rats by maternal stress and/or alcohol consumption. Horm Behav 2003; 43(5):531-539.
19. Lee S, Schmidt D, Tilders F et al. Increased activity of the hypothalamic-pituitary-adrenal axis of rats exposed to alcohol in utero: Role of altered pituitary and hypothalamic function. Mol Cell Neurosci 2000; 16(4):515-528.
20. Kim CK, Turnbull AV, Lee SY et al. Effects of prenatal exposure to alcohol on the release of adenocorticotropic hormone, corticosterone, and proinflammatory cytokines. Alcohol Clin Exp Res 1999; 23(1):52-59.
21. Yirmiya R, Chiappelli F, Tio DL et al. Effects of prenatal alcohol and pair feeding on lipopolysaccharide-induced secretion of TNF-alpha and corticosterone. Alcohol 1998; 15(4):327-335.

22. Lee S, Blanton CA, Rivier C. Prenatal ethanol exposure alters the responsiveness of the rat hypothalamic-pituitary-adrenal axis to nitric oxide. Alcohol Clin Exp Res 2003; 27(6):962-969.
23. Lee S, Rivier C. Prenatal alcohol exposure blunts interleukin-1-induced ACTH and beta-endorphin secretion by immature rats. Alcohol Clin Exp Res 1993; 17(5):940-945.
24. Wilcoxon JS, Redei EE. Prenatal programming of adult thyroid function by alcohol and thyroid hormones. Am J Physiol Endocrinol Metab 2004; 287(2):E318-326.
25. Morris DL, Harms PG, Petersen HD et al. LHRH and LH in peripubertal female rats following prenatal and/or postnatal ethanol exposure. Life Sci 1989; 44(17):1165-1171.
26. Knee DS, Sato AK, Uyehara CF et al. Prenatal exposure to ethanol causes partial diabetes insipidus in adult rats. Am J Physiol Regul Integr Comp Physiol 2004; 287(2):R277-283.
27. Ewald SJ, Walden SM. Flow cytometric and histological analysis of mouse thymus in fetal alcohol syndrome. J Leukoc Biol 1988; 44(5):434-440.
28. Zhu X, Seelig Jr LL. Developmental aspects of intestinal intraepithelial and lamina propria lymphocytes in the rat following placental and lactational exposure to ethanol. Alcohol Alcohol 2000; 35(1):25-30.
29. Weinberg J, Jerrells TR. Suppression of immune responsiveness: Sex differences in prenatal ethanol effects. Alcohol Clin Exp Res 1991; 15(3):525-531.
30. Norman DC, Chang MP, Castle SC et al. Diminished proliferative response of con A-blast cells to interleukin 2 in adult rats exposed to ethanol in utero. Alcohol Clin Exp Res 1989; 13(1):69-72.
31. Giberson PK, Kim CK, Hutchison S et al. The effect of cold stress on lymphocyte proliferation in fetal ethanol-exposed rats. Alcohol Clin Exp Res 1997; 21(8):1440-1447.
32. Wong CM, Chiappelli F, Chang MP et al. Prenatal exposure to alcohol enhances thymocyte mitogenic responses postnatally. Int J Immunopharmacol 1992; 14(2):303-309.
33. Chen L, Nyomba BL. Effect of prenatal alcohol exposure on glucose tolerance in the rat offspring. Metabolism 2003; 52(4):1454-462.
34. Chen L, Nyomba BL. Glucose intolerance and resistin expression in rat offspring exposed to ethanol in utero: Modulation by postnatal high-fat diet. Endocrinology 2003; 144(2):500-508.
35. Lopez-Tejero D, Llobera M, Herrera E. Permanent abnormal response to a glucose load after prenatal ethanol exposure in rats. Alcohol 1989; 6(6):469-473.
36. Pennington JS, Shuvaeva TI, Pennington SN. Maternal dietary ethanol consumption is associated with hypertriglyceridemia in adult rat offspring. Alcohol Clin Exp Res 2002; 26(6):848-855.
37. Webster WS, Germain MA, Lipson A et al. Alcohol and congenital heart defects: An experimental study in mice. Cardiovasc Res 1984; 18(6):335-338.
38. Uphoff C, Nyquist-Battie C, Toth R. Cardiac muscle development in mice exposed to ethanol in utero. Teratology 1984; 30(1):119-129.
39. Syslak PH, Nathaniel EJ, Novak C et al. Fetal alcohol effects on the postnatal development of the rat myocardium: An ultrastructural and morphometric analysis. Exp Mol Pathol 1994; 60(3):158-172.
40. Wold LE, Norby FL, Hintz KK et al. Prenatal ethanol exposure alters ventricular myocyte contractile function in the offspring of rats: Influence of maternal Mg2+ supplementation. Cardiovasc Toxicol 2001; 1(3):215-224.
41. Ren J, Wold LE, Natavio M et al. Influence of prenatal alcohol exposure on myocardial contractile function in adult rat hearts: Role of intracellular calcium and apoptosis. Alcohol Alcohol 2002; 37(1):30-37.
42. Turcotte LA, Aberle NS, Norby FL et al. Influence of prenatal ethanol exposure on vascular contractile response in rat thoracic aorta. Alcohol 2002; 26(2):75-81.
43. Schwartz J, Cameron V, Jarvis M et al. Lifetime cardiovascular effects of a single exposure to ethanol in utero. Paper presented at: Experimental Biology, 2004.
44. Stromland K, Pinazo-Duran MD. Ophthalmic involvement in the fetal alcohol syndrome: Clinical and animal model studies. Alcohol Alcohol 2002; 37(1):2-8.
45. Gonzalez ER. New ophthalmic findings in fetal alcohol syndrome. Jama 1981; 245(2):108.
46. Addolorato G, Gasbarrini A, Marcoccia S et al. Prenatal exposure to ethanol in rats: Effects on liver energy level and antioxidant status in mothers, fetuses, and newborns. Alcohol 1997; 14(6):569-573.
47. Meyers AF, Gong Y, Zhang M et al. Liver development in a rat model of fetal alcohol syndrome. Dig Dis Sci 2002; 47(4):767-772.
48. Gallo PV, Weinberg J. Organ growth and cellular development in ethanol-exposed rats. Alcohol 1986; 3(4):261-267.
49. Assadi FK, Manaligod JR, Fleischmann LE et al. Effects of prenatal ethanol exposure on postnatal renal function and structure in the rat. Alcohol 1991; 8(4):259-263.
50. Day NL, Richardson GA. Prenatal alcohol exposure: A continuum of effects. Semin Perinatol 1991; 15(4):271-279.

CHAPTER 16

Vitamin D in Pregnancy and Offspring Health

Marianne Tare,* Helena C. Parkington and Ruth Morley

Abstract

The prevalence of vitamin D insufficiency is increasing in western societies. The major source of vitamin D in healthy individuals of normal mobility is through the action of sunlight on the skin, but increased skin pigmentation or behaviours that reduce sun exposure, such as increased time spent indoors or extensive skin covering while outdoors, predispose to vitamin D insufficiency in the absence of dietary supplementation. Although vitamin D has been classically associated with bone mineralization, the wide distribution of vitamin D receptors provides the basis for a more extensive role for vitamin D. Thus, there is accumulating evidence for an involvement of vitamin D in the regulation of cell proliferation and differentiation, brain development, immune responses, the renin-angiotensin system and cardiovascular function. A recent disturbing recognition of startlingly low vitamin D levels amongst women of reproductive age, and indeed, in pregnant women, places in sharp focus our scant understanding of the ramifications of this on offspring health, not only in the immediate neonatal period but, as a result of the recent spotlight on the early origins of adult disease and syndrome X, on the long term outcomes of maternal vitamin D insufficiency.

Sources of Vitamin D

Vitamin D is a potent steroid hormone with a wide distribution of receptors suggesting diverse physiological roles.[1,2] Despite its name, relatively little vitamin D comes from dietary sources (apart from fortified foods or supplements). In healthy people of normal mobility under most climatic conditions, the majority of their vitamin D requirements are produced through the sunlight-mediated (via ultraviolet, UV B) photochemical conversion of 7-dehydrocholesterol in the skin to cholecalciferol (vitamin D_3).[3] The rate of this conversion is reduced by increased skin pigmentation, by covering up when outdoors or by staying indoors.[3-5] Cholecalciferol is hydroxylated in the liver to 25-hydroxycholecalciferol (25(OH)D_3) by the enzyme 25-hydroxylase. In turn, 25(OH)D_3 is hydroxylated (principally though not exclusively in the kidney) to many metabolites (Fig. 1). The main ones are the biologically active metabolite 1,25-dihydroxyvitamin D (1,25$(OH)_2D_3$), via the enzyme 1-α-hydroxylase, and 24,25-dihydroxyvitamin D (24,25$(OH)_2D_3$). Hepatic synthesis of 25(OH)D_3 is only loosely regulated, so blood levels reflect the amount of vitamin D produced in the skin or ingested. Its half-life is several weeks, whereas that of 1,25$(OH)_2D_3$ is a few hours, and its synthesis is tightly regulated.[6]

1-α-hydroxylase is found in decidual and trophoblastic cells in the placenta, with substantially higher levels in early versus late gestation, suggesting the potential importance of 1,25$(OH)_2D_3$ during early pregnancy.[7] 1,25$(OH)_2D_3$ regulates genes associated with implantation, such as HOXA10, whereas its immunosuppressive effects may have a role in

*Corresponding Author: Marianne Tare—Department of Physiology, Monash University, Clayton, Victoria, 3800, Australia. Email: marianne.tare@med.monash.edu.au

Early Life Origins of Health and Disease, edited by E. Marelyn Wintour and Julie A. Owens.

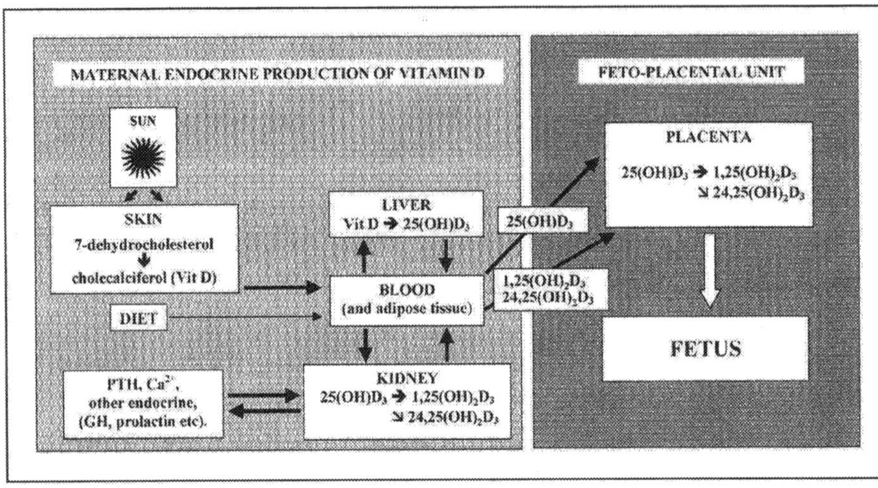

Figure 1. Irradiation of the mother's skin with UV B converts 7-dehydrocholesterol to previtamin D, which undergoes isomerization to vitamin D. Vitamin D is metabolized to $25(OH)D_3$ in the liver. Plasma $25(OH)D_3$ is converted to $1,25(OH)_2D_3$ and $24,25(OH)_2D_3$ by the kidney and also by the placenta. Parathyroid hormone (PTH) and growth hormone (GH) regulate vitamin D metabolism in the kidney. Maternal $1,25(OH)_2D_3$ and $24,25(OH)_2D_3$ also cross the placenta to supply the fetus.

implantation tolerance.[8] Both $25(OH)D_3$ and $1,25(OH)_2D_3$ cross the human placenta.[9] Placental vein $1,25(OH)_2D_3$ and $25(OH)D_3$ concentrations correlate significantly with those in the maternal circulation, implying that they diffuse easily across the placental barrier and that the vitamin D pool of the fetus depends entirely on that of the mother.[10,11] The fetal supply line for vitamin D is shown in Figure 1.

Vitamin D Insufficiency

Vitamin D deficiency has been defined in terms of bone health, and there is still debate about the levels below which there is deficiency and insufficiency.[12-14] At levels below 50 nmol/L preosteomalacic changes are seen on bone biopsy.[15] Recent studies have attempted to define sub-clinical deficiency in adults and children in terms of the basal level of $25(OH)D_3$ above which $1,25(OH)_2D_3$ did not rise and/or PTH did not fall with supplementation.[12,13] Surprisingly, this level was in the range 50-75 nmol/L in adults and 50 nmol/L in children. No study to date has determined optimal levels of vitamin D in pregnant women in terms of health and development of their offspring. However, there is no doubt that many pregnant women have low vitamin D levels and some groups are deficient. Dark-skinned women who migrate to higher latitudes or women who cover up are at high risk of vitamin D deficiency.[5,16-19]

Maternal serum concentrations of $1,25(OH)_2D_3$ are elevated during pregnancy[11,20-21] probably because of an increased production rate.[22] This is perhaps necessary, as studies in rats suggest that the fetus may be capable of actively accumulating and storing maternal vitamin D for the period of dependence on maternal milk, which has low levels of vitamin D.[23]

There is evidence that the placenta synthesizes $24,25(OH)_2D_3$ and other vitamin D metabolites as well as $1,25(OH)_2D_3$,[24] but little is known about their role. There is some evidence that the responsiveness of rat kidney to vitamin D metabolites changes during development. One study found that in embryonic and early postnatal stages, the kidney responds to $24,25(OH)_2D_3$, later to both $24,25(OH)_2D_3$ and $1,25(OH)_2D_3$, and the mature kidney only to $1,25(OH)_2D_3$.[25] Whether a similar change applies to other species or tissues is not known.

The role for vitamin D as a regulator of mineral acquisition and metabolism and bone mineralisation is well recognised, but it has recently become evident that it has far wider functions.[26-28] As with other steroid hormones, the biological effects of $1,25(OH)_2D_3$, are mediated by membrane, cytoplasmic, and nuclear receptors, with an interacting network of responses beginning immediately after exposure of cells to vitamin D and culminating in changes in gene expression affecting function and phenotype.[1] In humans vitamin D receptor (VDR) polymorphisms have been studied for association with growth and health but results to date have been somewhat inconsistent.[29-36] This may relate to differences in vitamin D status between populations that have been studied.[36]

In view of the importance of vitamin D in the regulation of cell differentiation and proliferation, the effect of maternal vitamin D deficiency in gestation on offspring health has received scant study. This is very surprising since there is increasing evidence of vitamin D deficiency in pregnant women, particularly among those who are dark skinned, cover their skin or avoid the sun.[5,16-19,37]

The Brain

There is recent evidence that low maternal vitamin D can adversely affect brain development in the fetus.[38-40] In newborn offspring of vitamin D deplete rat mothers (versus controls), the cortex had a higher length/width ratio and was proportionally thinner, and lateral ventricle volume was greater.[40] Throughout the brain there was more cell proliferation. The proportion of mitotic cells was increased while nerve growth factor content was reduced.[39] Some of these brain features altered in vitamin D deficiency are similar to those reported in patients with schizophrenia. It has been hypothesized that low maternal vitamin D may contribute to increased susceptibility to neurological disorders, including schizophrenia and multiple sclerosis in the offspring.[41-42] In a recent study using data from the Northern Finland 1966 Birth Cohort, males given at least 2000 IU of vitamin D per day in the first year of life had reduced risk of schizophrenia (Risk ratio = 0.23, 95% CI 0.06-0.95) compared with those on lower doses.[43] In a small US case control study using banked third trimester maternal serum there was weak evidence of lower maternal vitamin D among black individuals with schizophrenia than black controls.[44]

Diabetes

Vitamin D is known to have immunomodulatory actions.[45] Type I (insulin-dependent) diabetes mellitus is thought to result from the autoimmune destruction of the insulin producing β-cells of the pancreas. Although genetic factors are important in the pathogenesis of the disease, interaction with environmental factors has also been implicated. There appears to be a gradient in the prevalence of certain autoimmune diseases including type I diabetes, with increasing prevalence with increasing latitude, and ultraviolet radiation has been suggested to provide immunosuppression possibly through the actions of UV B and/or increases in serum vitamin D.[46] Several epidemiological studies have indicated a beneficial effect of vitamin D supplementation during infancy in reducing the risk of developing type I diabetes.[47,48] A small case-control study of women who took cod liver oil (a rich source of vitamin D) during pregnancy also showed that their offspring had a lower risk of developing diabetes, though it is unclear which component of cod liver oil was involved; vitamin D or long chain polyunsaturated fatty acids.[49] Interestingly, lifelong administration of the active form of vitamin D, $1,25(OH)_2D_3$ to nonobese diabetic (NOD) mice, genetically predisposed to developing type 1 diabetes, substantially reduced their risk of developing the disease by approximately 80%.[50] This was achieved using pharmacological doses of $1,25(OH)_2D_3$ with consequent effects on calcium and bone metabolism. Subsequently, using nonhypercalcaemia-inducing doses of a structural analog of $1,25(OH)_2D_3$, KH1060, the same group showed that the incidence of diabetes in these mice was still reduced by about 60%.[51] When NOD mice were exposed to vitamin D deficiency in utero and in early life, the onset of diabetes occurred earlier and the

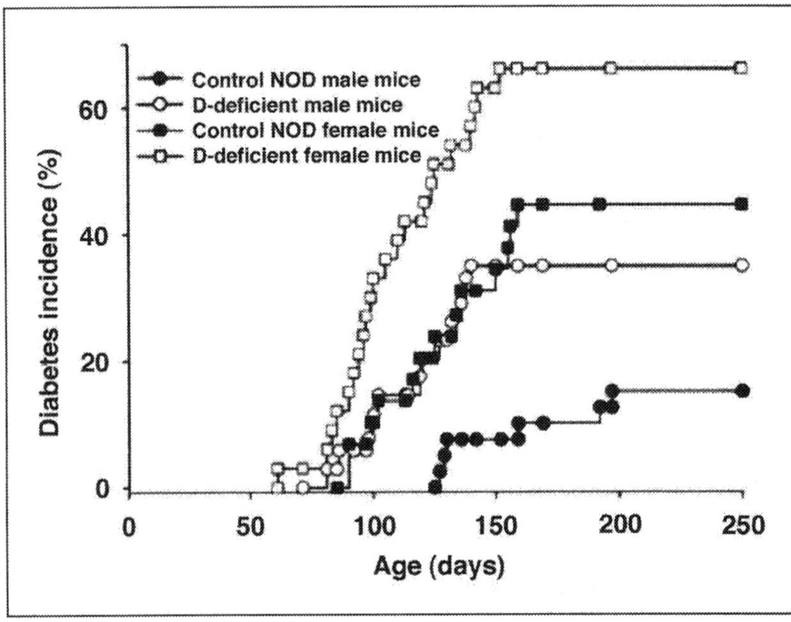

Figure 2. Incidence of diabetes in NOD mice exposed either to vitamin D deficiency or repletion in utero and during early life. Onset of diabetes was earlier and disease incidence was greater in the vitamin D deficient mice. Reproduced with permission from Springer-Verlag (Diabetologia 2004, 47:451-462; Fig. 2).[52]

final incidence was significantly greater than in control NOD mice (Fig. 2).[52] When the non-diabetic NOD mice were subjected to intraperitoneal glucose tolerance testing it was found that the vitamin D deplete mice exhibited an altered blood glucose response and glucose intolerance. In vitro testing of the function of isolated pancreatic islets from these mice revealed no differences in insulin content or secretion between vitamin D replete and deplete groups.[52] Thus, these nondiabetic vitamin D deficient NOD mice may be insulin resistant.

Evidence is beginning to emerge of a possible association between reduced vitamin D levels and glucose intolerance and type 2 diabetes. A study on elderly Dutch men revealed an inverse association between the area under the glucose curve during oral glucose tolerance test (OGTT) and serum 25(OH) vitamin D concentrations.[53] Similarly, there was an inverse association between total insulin concentrations during the OGTT and serum 25(OH) vitamin D_3. A small study of Bangladeshi Asians living in London showed a correlation between reduced glucose tolerance and low serum 25(OH) vitamin D_3 levels.[54] This population has a high incidence of type II diabetes, and in this study the prevalence of vitamin D deficiency was significantly greater in subjects classed as 'at risk' of glucose intolerance compared with those 'not at risk'.[54] Given the role of 1,25 $(OH)_2D_3$ in the regulation of insulin secretion and combined with its immunosuppressive properties, ensuring vitamin D sufficiency may have beneficial effects in reducing the prevalence or severity of diabetes in at risk populations.

Cardiovascular Function

A link between pertubations in utero and early life on the later development of disorders, including those contributing to syndrome X, has emerged from both epidemiological and animal studies. The notion that vitamin D deficiency during pregnancy may be a contributory factor to the development of insulin resistance and/or some of the other risk factors for syndrome X has been raised.[55] Cardiovascular dysfunction and/or hypertension are other risk

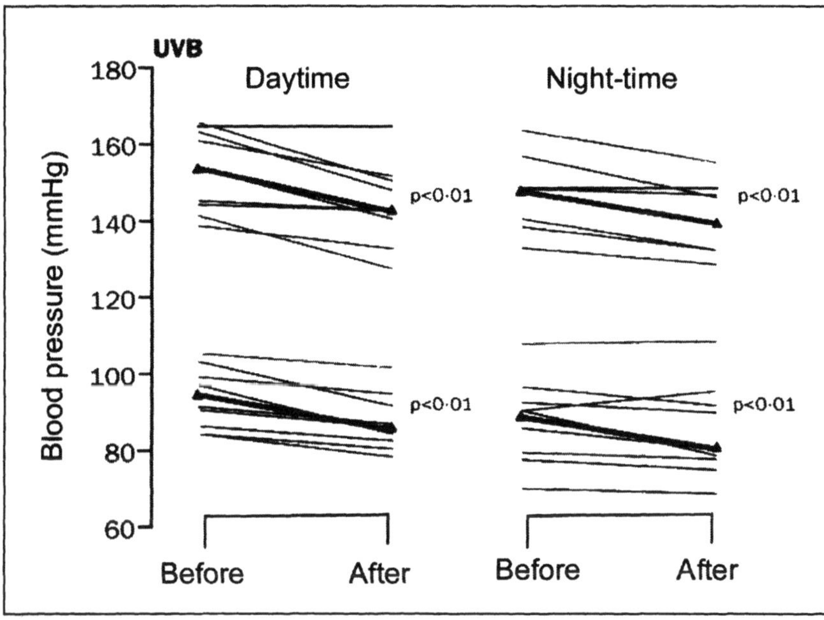

Figure 3. Effect of full body UV B irradiation on ambulatory daytime and night-time systolic and diastolic blood pressure in adults with untreated mild essential hypertension. Blood pressures were recorded before the commencement of treatment and again after 6 weeks of irradiation (3 times weekly) from a tanning bed. Reproduced with permission from Elsevier (Lancet, 1998, 352:709-710).[58]

factors linked with syndrome X and are also features linked with early origins of adult disease. Whether vitamin D may be a critical environmental factor with an aetiological/pathogenetic role in the early origins of cardiovascular disease awaits elucidation.

There is compelling evidence from epidemiological, clinical and scientific studies of a link between vitamin D and blood pressure in adults. There is an inverse relationship between blood pressure and UV B exposure. Data from the INTERSALT study indicated a positive association between the prevalence of hypertension and distance north or south of the equator.[56] There is also seasonal variation in blood pressure, with higher blood pressures reported in winter and lower blood pressures in summer, corresponding with changes in UV B levels.[27,56,57] Skin pigmentation influences the efficiency of producing vitamin D, as melanin is capable of absorbing UV B photons and thus reducing the amount available for converting 7-dehydrocholesterol to previtamin D_3. Dark skinned individuals require between 10 to 50 times the exposure to sunlight than white skinned individuals to produce similar amounts of $25(OH)_2D_3$.[4] Thus, it is not surprising that dark skinned individuals living at greater latitudes have a higher prevalence of vitamin D deficiency and higher mean blood pressure.[56] There is evidence that UV B exposure and vitamin D treatment can have beneficial effects on blood pressure. In a group of individuals with mild essential hypertension, UV B exposure three times weekly for two months reduced blood pressure by 6 mmHg (Fig. 3).[58] This treatment was also associated with increased serum vitamin D levels. There is an inverse correlation between serum vitamin D levels and blood pressure[59,60] and treatment with vitamin D has been reported to lower blood pressure in some studies.[61,62] Treatment of spontaneously hypertensive rats with vitamin D results in a significant lowering of blood pressure. Testing of vascular function in vitro shows that this effect of vitamin D treatment is accompanied by recovery of potassium channel function, as reflected in improved vascular smooth muscle relaxation and endothelial function.[63,64]

A possible mechanism linking vitamin D and blood pressure has been provided by Li and colleagues in studies on vitamin D receptor null mice (VDR-/-).[65] Systolic and diastolic blood pressures of the VDR-/- mice were 20 mmHg higher than their wild type counterparts and VDR-/- mice exhibited cardiac hypertrophy. There was evidence of an upregulation of the renin-angiotensin system in the VDR-/- mice, namely, elevated plasma angiotensin II concentrations and greater renin mRNA expression and protein levels in the kidney. All of these effects were independent of calcium metabolism. Captopril, an angiotensin-converting enzyme inhibitor, was effective in normalizing the blood pressure in the VDR-/- mice. In wild type mice, treatment with $1{,}25(OH)_2D_3$ resulted in decreased renin expression in the kidney. Based on these findings the authors proposed that vitamin D acts as a negative regulator of the renin-angiotensin system.[65,66]

Vitamin D Deficiency in Early Life

In view of the prevalence of vitamin D deficiency in women of reproductive age, the consequences of insufficiency in utero for cardiovascular function later in life warrant investigation. Male offspring of rats born to dams that were fed a low vitamin D diet during pregnancy were allocated to one of two groups: either maintained on a vitamin D deficient diet or normal chow for 9 weeks postnatally. Offspring maintained vitamin D deficient displayed alterations in cardiac and aortic function.[67] Although initially, systolic blood pressure was some 20 mmHg higher in vitamin D deficient rats compared with littermates fed a diet containing vitamin D, by the end of the 9 weeks there was no difference.[68] However, plasma renin activity was high in animals of both groups (around 30 ng compared with the more usual <10 ng angiotensin I / ml/hr).[68] Offspring of dams that were maintained on a vitamin D replete diet during pregnancy were not included in these studies.[69] At the end of the treatment period (week 9), isolated segments of aorta exhibited enhanced sensitivity to vasoconstrictors and overall constriction was greater in vessels obtained from the vitamin D deficient rats. Heart function, assessed using a Langendorff perfused preparation that included a pressure transducer in the left ventricle, revealed that the rate of pressure development during contraction and decline during relaxation was greater in the postnatally vitamin D deficient rats.[67] Surprisingly, myocardial collagen content was increased in vitamin D deficiency.[69] Subsequent studies by this group showed that the enhanced sensitivity of the aorta to constrictors could be reversed by restoring serum calcium or vitamin D, but enhanced cardiac contractility could not be rescued by these treatments.[67]

We have recently reexamined the effect of vitamin D deficiency in utero and early life on vascular function in rats (unpublished data). Female and male offspring of dams maintained on either vitamin D deplete chow during pregnancy and lactation or control chow, were studied at 6 weeks of age. The offspring were maintained on the same diet as their mothers until experimentation. Serum calcium was unchanged in all animals. Blood pressure and heart rate were significantly higher in vitamin D deplete rats compared with vitamin D replete controls. There were dramatic pro-constrictor changes in vascular function in isolated mesenteric arteries studied in vitro. The development of spontaneous tone (the myogenic response) was significantly elevated in arteries of deficient rats, especially in females. Importantly, the vasodilator function of the endothelium was halved in the vitamin D deficient animals. This involved deficits in both nitric oxide and endothelium-derived hyperpolarizing factor. Whether these vascular effects were a cause or a consequence of the higher blood pressure remains to be elucidated.

Conclusion

In modern western societies maternal undernutrition is unlikely to be a leading concern in terms of early origins of adult disease. There is however, a common misconception that young, healthy, mobile people of reproductive age are vitamin D sufficient. Yet, there is an increase in the prevalence of vitamin D insufficiency, particularly in veiled and migrant dark-skinned populations.[70,71] Alarmingly, this is being detected in pregnant women and their children, even in "sunny" countries such as Australia.[18,72] Individuals with increased skin pigmentation, those who strictly adhere to cultural/religious dress codes, white collar workers who spend little time out of doors, those who practice extensive sun protection/covering up when out of doors, or hospitalised people (as some pregnant women may be), are all at risk of vitamin D deficiency if their diet is not supplemented. The recommended daily intake for individuals is still a matter of contention[73] and likewise, that for pregnant and lactating women awaits further investigation.[74] Vitamin D is a critical environmental factor with potentially wide reaching implications for offspring health, even into adulthood. Ensuring vitamin D sufficiency during pregnancy and early life can be guaranteed easily, cheaply, safely and effectively and is likely to have positive implications for population health.

References

1. Farach-Carson MC, Davis PJ. Steroid hormone interactions with target cells: Cross talk between membrane and nuclear pathways. J Pharmacol Exp Ther 2003; 307:839-845.
2. Christakos S, Dhawan P, Liu Y et al. New insights into the mechanisms of vitamin D action. J Cell Biochem 2003; 88:695-705.
3. Norman AW. Sunlight, season, skin pigmentation, vitamin D, and 25-hydroxyvitamin D: Integral components of the vitamin D endocrine system. Am J Clin Nutr 1998; 67:1108-1110.
4. Clemens TL, Adams JS, Henderson SL et al. Increased skin pigmentation reduces the capacity of skin to synthesize vitamin D3. Lancet 1982; 1:74-76.
5. el-Sonbaty MR, Abdul-Ghaffar NU. Vitamin D deficiency in veiled Kuwaiti women. Eur J Clin Nutr 1996; 50:315-318.
6. Miller WL, Portale AA. Vitamin D 1 alpha-hydroxylase. Trends Endocrinol Metab 2000; 11:315-319.
7. Zehnder D, Evans KN, Kilby MD et al. The ontogeny of 25-hydroxyvitamin D(3) 1alpha-hydroxylase expression in human placenta and decidua. Am J Pathol 2002; 161:105-114.
8. Evans KN, Bulmer JN, Kilby MD et al. Vitamin D and placental-decidual function. J Soc Gynecol Invest 2004; 11:263-271.
9. Ron M, Levitz M, Chuba J et al. Transfer of 25-hydroxyvitamin D3 and 1,25-dihydroxyvitamin D3 across the perfused human placenta. Am J Obstet Gynecol 1984; 148:370-374.
10. Gertner JM, Glassman MS, Coustan DR et al. Feto-maternal vitamin D relationship at term. J Pediatr 1980; 97:637–640.
11. Delvin EE, Glorieux FH, Salle BL et al. Control of vitamin D metabolism in preterm infants: Feto-maternal relationships. Arch Dis Child 1982; 57:754–757.
12. Docio S, Riancho JA, Perez A et al. Seasonal deficiency of vitamin D in children: A potential target for osteoporosis-preventing strategies? J Bone Miner Res 1998; 13:544-548.
13. Malabanan A, Veronikis IE, Holick MF. Redefining vitamin D insufficiency. Lancet 1998; 351:805-806.
14. Shaw NJ, Pal BR. Vitamin D deficiency in UK Asian families: Activating a new concern. Arch Dis Child 2002; 86:147-149.
15. Parfitt A. Osteomalacia and related disorders. In: Aviolo L, Krane S, eds. Metabolic Bone Diseases and Clinically Related Disorders. Philadelphia: WB Saunders, 1990:329-339.
16. Anatoliotaki M, Tsilimigaki A, Tsekoura T et al. Congenital rickets due to maternal vitamin D deficiency in a sunny island of Greece. Acta Paediatr 2003; 92:389-391.
17. Datta S, Alfaham M, Davies DP et al. Vitamin D deficiency in pregnant women from a nonEuropean ethnic minority population—An interventional study. BJOG 2002; 109:905-908.
18. Grover SR, Morley R. Vitamin D deficiency in veiled or dark-skinned pregnant women. Med J Aust 2001; 175:251-252.
19. Mukamel MN, Weisman Y, Somech R et al. Vitamin D deficiency and insufficiency in Orthodox and nonOrthodox Jewish mothers in Israel. Isr Med Assoc J 2001; 3:419-421.
20. Fleischman AR, Rosen JF, Cole J et al. Maternal and fetal serum 1,25-dihydroxyvitamin D levels at term. J Pediatr 1980; 97:640–642.
21. Care AD. Vitamin D in pregnancy, the fetoplacental unit, and lactation. In: Feldman D, Glorieux FH, Wesley Pike J, eds. Vitamin D. New York: Academic Press, 1997:437–443.
22. Salle BL, Delvin EE, Lapillonne A et al. Perinatal metabolism of vitamin D. Am J Clin Nutr 2000; 71:1317S-1324S.
23. Clements MR, Fraser DR. Vitamin D supply to the rat fetus and neonate. J Clin Invest 1988; 81:1768-1773.
24. Rubin LP, Yeung B, Vouros P et al. Evidence for human placental synthesis of 24,25-dihydroxyvitamin D3 and 23,25-dihydroxyvitamin D3. Pediatr Res 1993; 34:98-104.

25. Somjen D, Earon Y, Harell S et al. Developmental changes in responsiveness to vitamin D metabolites. J Steroid Biochem 1987; 27:807-813.
26. Lin R, White JH. The pleiotropic actions of vitamin D. Bioessays 2004; 26:21-28.
27. Holick MF. Vitamin D: Importance in the prevention of cancers, type 1 diabetes, heart disease, and osteoporosis. Am J Clin Nutr 2004; 79:362-371.
28. Chatterjee M. Vitamin D and genomic stability. Mutat Res 2001; 475:69-88.
29. Ogunkolade BW, Boucher BJ, Prahl JM et al. Vitamin D receptor (VDR) mRNA and VDR protein levels in relation to vitamin D status, insulin secretory capacity, and VDR genotype in Bangladeshi Asians. Diabetes 2002; 51:2294-2300.
30. Malecki MT, Frey J, Moczulski D et al. Vitamin D receptor gene polymorphisms and association with type 2 diabetes mellitus in a Polish population. Exp Clin Endocrinol Diabetes 2003; 111:505-509.
31. Chang TJ, Lei HH, Yeh JI et al. Vitamin D receptor gene polymorphisms influence susceptibility to type 1 diabetes mellitus in the Taiwanese population. Clin Endocrinol 2000; 52:575-580.
32. Pani MA, Knapp M, Donner H et al. Vitamin D receptor allele combinations influence genetic susceptibility to type 1 diabetes in Germans. Diabetes 2000; 49:504-507.
33. Ban Y, Taniyama M. Vitamin D receptor gene polymorphism is associated with Graves' disease in the Japanese population. J Clin Endocrinol Metab 2000; 85:4639-4643.
34. Collins JE, Heward JM, Nithiyananthan R et al. Lack of association of the vitamin D receptor gene with Graves' disease in UK Caucasians. Clin Endocrinol 2004; 60:618-624.
35. Ntais C, Polycarpou A, Ioannidis JP. Vitamin D receptor gene polymorphisms and risk of prostate cancer: A meta-analysis. Cancer Epidemiol Biomarkers Prev 2003; 12:1395-1402.
36. Ma J, Stampfer MJ, Gann PH et al. Vitamin D receptor polymorphisms, circulating vitamin D metabolites, and risk of prostate cancer in United States physicians. Cancer Epidemiol Biomarkers Prev 1998; 7:385-390.
37. Pehlivan I, Hatun S, Aydogan M et al. Maternal vitamin D deficiency and vitamin D supplementation in healthy infants. Turk J Pediatr 2003; 45:315-320.
38. McGrath JJ, Feron FP, Burne TH et al. Vitamin D(3)-implications for brain development. J Steroid Biochem Mol Biol 2004; 89-90:557-560.
39. Brown J, Bianco JI, McGrath JJ et al. 1,25-dihydroxyvitamin D_3 induces nerve growth factor, promotes neurite outgrowth and inhibits mitosis in embryonic rat hippocampal neurons. Neurosci Lett 2003; 343:139-143.
40. Eyles D, Brown J, Mackay-Sim A et al. Vitamin D_3 and brain development. Neuroscience 2003; 118:641-653.
41. McGrath J. Hypothesis: Is low prenatal vitamin D a risk-modifying factor for schizophrenia? Schi Res 1999; 40:173-177.
42. McGrath J. Does 'imprinting' with low prenatal vitamin D contribute to the risk of various adult disorders? Med Hypotheses 2001; 56:367-371.
43. McGrath J, Saari K, Hakko H et al. Vitamin D supplementation during the first year of life and risk of schizophrenia: A finnish birth cohort study. Schizophr Res 2004; 67:237-245.
44. McGrath J, Eyles D, Mowry B et al. Low maternal vitamin D as a risk factor for schizophrenia: A pilot study using banked sera. Schizophr Res 2003; 63:73-78.
45. Briffa NK, Keogh AM, Sambrook PN et al. Reduction of immunosuppressant therapy requirement in heart transplantation by calcitriol. Transplantation 2003; 75:2133-2134.
46. Ponsonby A-L, McMichael A, van der Mei I. Ultraviolet radiation and autoimmune disease: Insights from epidemiological research. Toxicol 2002; 181-182:71-78.
47. The EURODIAB Substudy 2 Study Group. Vitamin D supplement in early childhood and risk for Type I (insulin-dependent) diabetes mellitus. Diabetologia 1999; 42:51-54.
48. Hyppönen E, Läärä E, Reunanen A et al. Intake of vitamin D and risk of type 1 diabetes: A birth-cohort study. Lancet 2001; 358:1500-1503.
49. Stene LC, Ulriksen J, Magnus P et al. Use of cod liver oil during pregnancy associated with lower risk of type 1 diabetes in the offspring. Diabetologia 2000; 43:1093-1098.
50. Mathieu C, Waer M, Laureys J et al. Prevention of autoimmune diabetes in NOD mice by 1,25 dihydroxyvitamin D_3. Diabetologia 1994; 37:552-558.
51. Mathieu C, Waer M, Casteels K et al. Prevention of Type I diabetes in NOD mice by nonhypercalcemic doses of a new structural analog of 1,25-dihydroxyvitamin D_3, KH1060. Endocrinology 1995; 136:866-872.
52. Giulietti A, Gysemans C, Stoffels K et al. Vitamin D deficiency in early life accelerates Type 1 diabetes in nonobese diabetic mice. Diabetologia 2004; 47:451-462.
53. Baynes KCR, Boucher BJ, Feskens EJM et al. Vitamin D, glucose tolerance and insulinaemia in elderly men. Diabetologia 1997; 40:344-347.

54. Boucher BJ, Mannan N, Noonan K et al. Glucose intolerance and impairment of insulin secretion in relation to vitamin D deficiency in East London Asians. Diabetologia 1995; 38:1239-1245.
55. Boucher BJ. Inadequate vitamin D status: Does it contribute to the disorders comprising syndrome 'X'? Br J Nutr 1998; 79:315-327.
56. Rostand SG. Ultraviolet light may contribute to geographic and racial blood pressure differences. Hypertension 1997; 30:150-156.
57. Sherman SS, Hollis BW, Tobin JD. Vitamin D status and related parameters in a healthy population: The effect of age, sex and season. Clin Endocrinol Metab 1990; 71:405-413.
58. Krause R, Buhring M, Hopfenmuller W et al. Ultraviolet B and blood pressure. Lancet 1998; 352:709-710.
59. Lind L, Hänni A, Lithell H et al. Vitamin D is related to blood pressure and other cardiovascular risk factors in middle-aged men. Am J Hypertens 1995; 8:894-901.
60. Kristal-Boneh E, Froom P, Harari G et al. Association of calcitriol and blood pressure in normotensive men. Hypertension 1997; 30:1289-1294.
61. Lind L, Wengle B, Wide L et al. Reduction of blood pressure during long-term treatment with active vitamin D (alphacalcidol) is dependent on plasma renin activity and calcium status. A double-blind, placebo-controlled study. Am J Hypertens 1989; 2:20-25.
62. Pfeifer M, Begerow B, Minne HW et al. Effects of a short-term vitamin D and calcium supplementation on blood pressure and parathyroid hormone levels in the elderly. J Clin Endocrinol Metab 2001; 86:1633-1637.
63. Borges ACR, Feres T, Vianna LM et al. Recovery of impaired K^+ channels in mesenteric arteries from spontaneously hypertensive rats by prolonged treatment with cholecalciferol. Br J Pharmacol 1999; 127:772-778.
64. Borges ACR, Feres T, Vianna LM et al. Effect of cholecalciferol treatment on the relaxant responses of spontaneously hypertensive rat arteries to acetylcholine. Hypertension. 1999; 34:897-901.
65. Li YC, Kong J, Wei M et al. 1,25-Dihydroxyvitamin D_3 is a negative endocrine regulator of the renin-angiotensin system. J Clin Invest 2002; 110:229-238.
66. Li YC. Vitamin D regulation of the renin-angiotensin system. J Cell Biochem 2003; 88:327-331.
67. Weishaar RE, Simpson RU. Involvement of vitamin D_3 with cardiovascular function. II. Direct and indirect effects. Am J Physiol 1987; 253:E675-E683.
68. Weishaar RE. Vitamin D_3 and cardiovascular function in rats. J Clin Invest 1987; 79:1706-1712.
69. Weishaar RE, Kim S-N, Saunders DE et al. Involvement of vitamin D_3 with cardiovascular function. III. Effects of physical and morphological properties. Am J Physiol 1990; 258:E134-E142.
70. Nesby-O'Dell S, Scanlon KS, Cogswell ME et al. Hypovitaminosis D prevalence and determinants among African American and white women of reproductive age: Third National Health and Nutrition Examination Survey, 1988-1994. Am J Clin Nutr 2002; 76:187-192.
71. Nowson CA, Diamond TH, Pasco JA et al. Vitamin D in Australia. Issues and recommendations. Aust Fam Physician 2004; 33:133-138.
72. Nozza JM, Rodda CP. Vitamin D deficiency in mothers of infants with rickets. Med J Aust 2001; 175:253-5.
73. Vieth R. Vitamin D supplementation, 25-hydroxyvitamin D concentrations, and safety. Am J Clin Nutr 1999; 69:842-856.
74. Hollis BW, Wagner CL. Assessment of dietary vitamin D requirements during pregnancy and lactation. Am J Clin Nutr 2004; 79:717-726.

CHAPTER 17

The Fetal Origins of Adult Mental Illness

Laura Bennet* and Alistair J. Gunn

"I am a crooked, twisted piece of humanity. The sooner I die the better. God will relieve me from my sufferings, as I really cannot stand it."
—Voices of the mad: Patients letters from the Royal Edinburgh Asylum 1873-1908. Allan Beveridge.[1]

Abstract

This chapter critically examines the hypothesis that the origins of some adult mental illnesses such as schizophrenia, which is the focus of this review, derive from adverse events in utero, such as maternal nutrition deficiency, infection and hypoxia. The hypothesis was originally derived from neuropathological changes in patients with established schizophrenia that are highly suggestive of impaired neural development occurring around mid-gestation. Increasingly it appears that gestational timing and the severity of the insult, rather than type of insult, plays a critical role in subsequent behavioural outcome. Supporting the neurodevelopmental hypothesis, recent studies have demonstrated that serious mental illnesses such as schizophrenia and afferent disorders are associated firstly with behavioural abnormalities that are present from early childhood, and secondly with ongoing neural injury on serial magnetic resonance imaging through late childhood and adolescence. These data suggest that alterations in brain development during fetal life lead to an evolving damage over the course of childhood before finally being overtly expressed in early adulthood. Current data suggest that the initial loss of cells in utero leads to a long-term remodelling of the brain that is mediated by upregulation of physiological apoptosis. That such adult illnesses present with early behavioural and physiological clues, are progressive and not static in nature, and that the process is potentially governed by common mechanisms regardless of cause, offers significant new opportunities for intervention and treatment.

Introduction

Schizophrenia is a surprisingly common disorder, with a lifetime incidence of around 1 in 100 people worldwide. It usually manifests its full form, with deterioration in personality, hallucinations and delusions, and cognitive impairment, in late adolescence and early adulthood.[1] It represents a major personal, social and medical burden, with costs in the billions of dollars per year. However, despite more than a hundred years of dedicated research, the aetiology of schizophrenia remains elusive. Certainly few subjects in neurobiology have generated as much fascination, controversy, and utter frustration as the hunt for the "cause" of schizophrenia—the Holy Grail of biological psychiatry.[2] Despite promising anatomical findings in the late nineteenth and early twentieth century, which suggested a neuropathological origin to the illness, subsequent research led to inconclusive and conflicting results. By the 1970s research

*Corresponding Author: Laura Bennet—Department of Physiology, The University of Auckland, Auckland, New Zealand. Email: l.bennet@auckland.ac.nz

Early Life Origins of Health and Disease, edited by E. Marelyn Wintour and Julie A. Owens.

on the neuropathology of the illness had come to a near standstill, with the general consensus on the subject succinctly summarised by Plum's now somewhat infamous dictum that schizophrenia is the *"graveyard of neuropathologists"*.[3]

This impasse reflected a number of factors, such as the relative crudeness of the methodology available, the belief that the neuropathology of schizophrenia was likely related to a chronic neurodegenerative process, and the inappropriate expectation of finding large abnormalities rather than smaller discrete ones.[4] Recent advances in imaging techniques, such as Magnetic Resonance Imaging (MRI) and Computerised Tomography scanning (CT), have allowed earlier detection and more precise investigations, which support a close association between schizophrenia and neuroanatomical abnormalities. This neuropathology literature has been extensively reviewed by others, and will only be discussed briefly in this chapter. Current data show that at the onset of schizophrenia, and thus independent of treatment effects, schizophrenic patients have enlarged cerebral ventricles with decreased volume of cortex (particularly in the prefrontal and temporal lobes) and of subcortical structures (particularly the hippocampus, amygdala, and dorsal thalamus). Further there is evidence of loss of neuropil (dendrites, spines and axons) and of extensive white matter changes which typically involves diffuse loss rather than active gliosis, as exemplified by the reduced size of the corpus callosum and prefrontal cortical white matter. There are alterations in normal cerebral asymmetries, and alterations in neuronal size, number, placement, orientation and clustering, with excessive cortical pruning, and consequent altered neurotransmitter function and aberrant functional connectivity of specific cerebral circuits.[2,4-10]

Consistent with the well-known variability in the clinical presentation of the illness, the anatomical changes are also quite variable. However, the overall nature of these neuropathological changes is consistent with a significant prenatal impairment of development. As will be discussed below, it is now evident that while there appears to be a genetic component to schizophrenia and other mental illnesses, genetics does not fully account for their development. It remains controversial whether adverse environmental events act upon a preexisting genetic predisposition (similarly to e.g., insulin dependent diabetes mellitus),[11] or modify the epigenetic status of genes, or are simply coincidental.[12,13]

The History of the Neurodevelopmental Hypothesis

The neurodevelopmental hypothesis proposes that adverse environmental events during fetal life impairs and subsequently alter neural development, leading to mental illness in adulthood.[10,14-19] Like all good theories, it has a long history. As early as 1891, the founding father of adolescent psychiatry, Scottish psychiatrist Thomas Storer Clouston, proposed that there was a developmental component to *"adolescent or developmental insanity"*.[20] He considered it a disorder of cortical development; *"the last cortical disease"*, and that the onset of psychotic symptoms was due to maturation during adolescence *"of certain parts of the brain which had lain dormant before"*.[20-23] This concept was subsequently superseded by the hypothesis proposed by Emil Kraeplin, much influenced by Alzheimer and his study of adult dementia, that the illness was a neurodegenerative organic brain disease;[24] a view Clouston *"strenuously"* objected to,[22] but one which held sway for a considerable number of years.[17,19,25] Even Kraeplin acknowledged, however, that there might be a developmental origin, at least in some cases where evidence of the illness existed in childhood,[23] as did Eugene Bleuler, who in 1911 coined the term schizophrenia; a term chosen to express the presence of schisms between thought, emotion and behaviour which characterises the "schizophrenias".[26] Bleuler reported that behavioural difficulties could be observed in childhood in more than half the patients who eventually developed schizophrenia.

This observation is fundamental, since it demonstrates that the underlying disorder that leads to schizophrenia evolves in some cases at least from early childhood if not before birth. In subsequent years these childhood clues about the potential developmental origins of schizophrenia were forgotten or dismissed, but subsequently rediscovered in the 1980's.[27,28] Prospective follow-up studies of birth cohorts have confirmed that significant impairments in neuromotor, receptive language, and cognitive development are present among children later diagnosed as having schizophreniform disorder.[29-33] These premorbid behavioural changes can be seen as early as three years of age.[29] There is also an early childhood onset version of schizophrenia.[34] Thus, while schizophrenia typically manifests in its full form in adulthood, there is now good evidence to show that the illness is already in progress much earlier and is progressive in nature.

One key conceptual difficulty is reconciling the timing between origin and onset of the disorder: how can it be that an illness which manifests in adolescent or adult life could have its origins so long ago, in fetal life?[5,25] This apparent paradox may be resolved by understanding that neurodevelopment is a continuum started at conception, but not completed until adulthood. Disturbances in particular critical windows of maturation can thus have very long lasting effects.

Neuropathological Evidence for Neural Injury before Birth in Schizophrenia

The key neuropathological data for an in utero origin to schizophrenia centre around neuronal migration, and, increasingly, glial proliferation. The presence of neuronal disarray, heterotopias and malpositioning are very suggestive since cytoarchitecture is largely determined during early fetal life, well before the last trimester.[5,35,36] Among the cellular findings are abnormal cytoarchitecture of the entorhinal cortex characterized by poorly formed layer II neuron clusters and laminar disorganization, a reduction and displacement of hippocampal and cortical pyramidal cells, and abnormal development of the subplate.[35,37-41] Such studies suggest disturbances of neuronal migration during the late first or early second trimester. An earlier time is excluded since gross abnormalities in the structure and cellular content of the cerebral cortex would be expected if neurogenesis were affected.[5]

However, these data are not conclusive, since some studies have not found evidence for abnormal migration in schizophrenia,[42-44] and other, more consistent findings such as alterations to neuronal size and synaptic and dendritic organisation may occur later in life, well after birth.[5,35] The differences between studies may reflect the methodological difficulties and subtle nature of the cytoarchitectural changes.[35] Alternatively, it could mean that in many cases the putative in utero insult may occur after mid-gestation, when migration is largely complete.[45,46] At this stage there is a marked increase in glial proliferation and if correct this would suggest that we should expect to see a consistent reduction in the amount of white matter.[45,46]

Imaging data suggests that this is indeed the case, but it has not been fully appreciated until recently because of technical difficulties,[47] although the consistent presence of ventriculomegaly in patients strongly suggests diffuse white matter atrophy.[48] Instead the focus has been on whether "lesions" exist. Traditionally, the absence of "gliosis" (i.e., astrocytic activation or scarring) in histopathological and imaging studies of patients with schizophrenia has been taken to mean two things: (1) that this must be a neurodevelopmental process and not a neurodegenerative one (which would leave tell-tale scars), and/or (2) that any changes must have taken place before the third trimester, based on the study by Friede, which supposedly showed that gliosis cannot occur until after the end of the second trimester.[5] In fact both conclusions are highly likely to be erroneous. There is evidence that that astrocytic activation can occur as early as 20 weeks of gestation,[49] and in any case a few studies have found periventricular white matter lesions in region of patients.[50,51]

Critically, modern imaging data has confirmed that the most common pathological feature of both schizophrenia and affective disorders is diffuse loss of white matter.[10,48,52-58] This loss appears to be region specific. There is, for example, loss of oligodendrocytes (the myelinating cells of the central nervous system) and astrocytes and altered oligodendrocyte ultrastructure in specific layers of the prefrontal cortex.[52,58] Consistent with these findings, there is evidence of impaired and reduced myelination in schizophrenia,[59,60] and altered expression of myelination related genes.[61] Thus there is impairment of the normal age-related development of the frontal and temporal lobes in adulthood.[62]

Loss of the supporting glia likely contributes to the atrophy of neurons that has been described in the prefrontal cortex.[52] Layers III and V of the dorsolateral prefrontal cortex, which give rise to glutamatergic projections to neostriatum, demonstrate the most structural pathology. The fundamental pathophysiology of schizophrenia remains unclear, but evidence suggests that there is excessive stimulation of striatal dopamine D2 receptors, deficient stimulation of prefrontal dopamine D1 receptors, and alterations in prefrontal connectivity involving glutamate transmission at the NMDA subtype of receptor.[63]

How Good Is the Evidence for Underlying in Utero Events?

A variety of prenatal events can adversely affect neuronal development including hypoxia, maternal undernutrition, exposure to viruses and infection, maternal stress and maternal lifestyle and other health problems (key factors are discussed below).[16,64-67] Meta-analysis suggests that schizophrenics are twice as likely to have been exposed to obstetric complications as controls.[67,68] However, most babies with obstetric complications do not develop schizophrenia and most patients with schizophrenia do not have an apparent history of these complications.[69]

The significance of these findings is highly debated. They might reflect a genetic predisposition,[6,18,70,71] which might be to obstetric complications rather than directly with mental illness.[72] For example, poor pregnancy outcomes occur more frequently among women with schizophrenia and they are at greater risk for increased interventions.[73-77] Obstetric complications do not seem to be particularly specific to schizophrenia since there now appears to be an association with affective disorders as well.[78,79] These epidemiological data are, of course, limited by lack of detail particularly with respect to gestational timing, and by an inappropriate focus on peripartum events.[80,81] It is likely significant that schizophrenia, for example, appears to be mainly related to events which occur in the first or second trimester.[5]

Since numerous adverse events are apparently equally associated with different types of mental illness, it maybe speculated that it is not the type of insult (i.e., infection versus hypoxia) which is important to outcome, but rather the gestational timing of the initial insult. Insult severity, duration, and the additive effects of interactions between insults are also likely to be key factors. The similarity in neuropathology between many illnesses is consistent with the shared symptomology of these illnesses, which often makes diagnosis difficult. It is also consistent with the increasingly accepted concept that affective disorders and schizophrenia, at least, are not distinct illnesses per se, as Kraeplin first proposed, but rather represent a psychiatric continuum ranging from unipolar to bipolar disorder to schizoaffective psychosis all the way to schizophrenia.[82] Neural impairment and injury, like psychiatric disorders, may be viewed as a continuum, with timing and severity of an insult critical factors in outcome.[83]

Hypoxia

It is increasingly clear that hypoxia can occur in the preterm fetus.[83,84] Experimentally, we now understand that despite its immaturity the preterm fetus is physiologically resilient and has a mature response to severe hypoxia.[85-89] Paradoxically, however, the capacity to survive prolonged asphyxia can place the preterm fetus at greater risk of surviving with injury than is the case later in gestation.[90] Prenatal injury, as shown by severe placental pathology such as infarction, can occur without detectable clinical signs in infants who go on to develop cerebral palsy later in childhood.[91] Studies in rodents and fetal sheep show that chronic sub-lethal hypoxia started in mid-gestation is associated with smaller brains, reduced white and grey matter volumes, ventriculomegaly and disordered neuronal migration and dendritic development.[92-96] Further, acute white matter loss after perinatal hypoxia-ischaemia leads to long-term reductions in myelination post-natally.[97] In preliminary work from our laboratory in the fetal sheep, we have observed that a sufficiently severe, but acute period of asphyxia in mid-gestation that causes subcortical injury leads to chronically evolving diffuse white matter loss, (but no cystic lesions), ventriculomegaly, and long-term, impairment of cortical development.[98] These findings were related to reduced glial proliferation and upregulation of programmed cell death.

Importantly, white matter injury, while a dominant cause of neural injury in the preterm infant, also occurs later in gestation after severe hypoxia-ischaemia.[99-101]

Clinical data, particularly studies of multiple births, are consistent with these data and suggest that exposure to hypoxia-ischaemia is a common cause of neurodevelopmental injury.[67,81,102-104] The high concordance reported in monozygotic (MZ) twins is often cited as irrefutable evidence for the etiological influence of genetics: MZ twins share 100% of their genes, and mental illness still develops if the twins are adopted. However, it is striking that far from showing 100% concordance for schizophrenia, MZ twins have reported rates between 26 and 47%, and rates as low as 6% for dizygotic (DZ) twins.[105] For schizophrenia, MZ concordance rates are significantly lower when samples are selected from a twin register as opposed to a psychiatric facility,[106] but twin registries typically record zygosity by sex rather than by explicit genotyping, and thus the real association may be even lower than reported.[107,108] The emphasis on genetics has obscured the fact that twins share a lot more than their genes, they share an in utero environment where impaired nutrition and oxygenation are common. Thus, the increased concordance of twins for mental illness might relate to their increased rate of neural injury as discussed below. Seminal imaging work by Suddath and colleagues supports this hypothesis. They demonstrated that there were significant neuroanatomical differences, including smaller anterior hippocampi and enlarged lateral and third ventricles, between twins discordant for schizophrenia and concluded that the cause of schizophrenia in those cases was at least in part not genetic.[109]

Infants born as part of multiple births have very high rates of brain injury and neurodevelopmental handicap compared to singletons.[110-112] Cerebral palsy (CP) is, for example, 5-10% more frequent in twins than singletons (1-2%), while triplets have a 47 fold higher risk.[113] The loss of a co-twin in utero is associated with a 13-15 fold higher risk for CP compared to live-born twins,[110,111,114] with an absolute risk of later neurodevelopmental impairment reaching 60%.[112] The higher relative risk is not solely due to higher rates of low-birthweight and prematurity (both of which are predictive for schizophrenia),[115] as normal birthweight twins also show increased risk of neural injury compared to singletons.[116,117]

It is likely that it is not zygosity which underlies this risk, but chorionicity; that is whether the fetuses shared a placenta.[118,119] MZ twins share the same chorion in most cases (monochorionic, MC), whereas DZ and around a third of MZ twins are of the dichorionic type (DC).[108] Fetal mortality is significantly higher and neurologic morbidity is up to 7-fold higher in MC than DC twins.[112,120] Monochorionic multiple gestations are frequently complicated by antenatal necrosis of the cerebral white matter,[121] and by abnormal cortical plate development shown by polymicrogyria or microgyric-like pattern, and heterotopias.[122] Discordant growth, the death of a twin in utero, and twin-twin transfusion are key associations.[112,121,122]

The apparent damage is typically present by 22 and 32 weeks gestation, in a pattern which is consistent with that reported in schizophrenia.[122] MZ twin pairs concordant for schizophrenia are more likely to have been monochorionic. Pairwise concordances for MZ twins without monochorionic markers averaged 10.7%, whereas concordance for MZ twins with one or more monochorionic markers was 60%.[123] These data again strongly suggest that it was sharing a placenta rather than genes which was most important. A relationship between chorion type and concordance of abnormal behaviour between MZ and DZ pairs has also been supported by several other studies,[124,125] but not all.[126]

Nutrition

Pasamanick first suggested that maternal malnutrition may lead to behavioural abnormalities in childhood.[127] Nutrition and oxygen delivery are often inextricably interlinked and impaired fetal and placental growth in both singletons and twins discordant for schizophrenia may be a function of both factors.[51,115,128] Nutritional inadequacy in one form or another is one of the largest single nongenetic contributors to mental retardation and aberrant neural development.[129] As discussed elsewhere in this book, the relative imbalance of the current western diet may well lead to persistent metabolic and cardiovascular derangements.[130] The modern desire to be thin, which affects as many as 1% of pregnancies through inappropriate maternal dieting,[131] may also play a contributory role. Reduced caloric intake, nutritional imbalance (e.g., increased carbohydrate and reduced protein intake) and micronutrient deficiencies such as folate, homocystein and vitamin D may all impair fetal brain development, reduce glial proliferation, increase apoptosis, and lead to glutamate, serotonin and GABA neurotransmitter abnormalities.[129,132-137] In MC twins, there is a reduction of some essential and nonessential amino acids in the growth-restricted twin compared to their co-twins, including glycine.[138] Altered glycine metabolism may be important because of the glycine modulatory site on the NMDA receptor, which inhibits glutamate function. There is evidence that reduced glutaminergic activity contributes to the symptoms of schizophrenia.[139]

The effects of nutritional deprivation before birth persist well into adulthood and are associated with behavioural changes,[132,140,141] such as alterations in sleep-wake cycles and arousal.[142] Similar disturbances in sleep continuity, and in the balance of slow wave and rapid eye movement sleep are a consistent feature of schizophrenia.[143] Recent data show that indicators of intrauterine and childhood undernutrition have a complex association with increased risk of later schizophrenia,[128] but that it is prenatal not childhood growth which is most important.[144] For example, there is a reverse J-shaped association between adjusted birth weight and schizophrenia, with mean hazard ratio of 7.0 for males of low birth weight (<2.5 kg) and 3.4 for those of high birth weight (>4.0 kg). The Dutch Winter Famine studies have demonstrated a strong link between malnutrition in mid-gestation and later schizophrenia (around a 2 fold increase), whereas late-gestation undernutrition was associated with affective disorder, with exposure in the third trimester having a greater effect than exposure in the second trimester.[64,65,104,145-147] Taken with the data on injury in twins, these data further suggest that gestational timing rather than the nature of the event is more important to later behavioural outcomes.

Infection

Fetal infection has been suggested to be a possible etiologic factor based on epidemiological findings that individuals with schizophrenia and affective disorders tend to be born in winter/spring when compared to the general population and that there is a strong association between maternal influenza and mental illness in offspring.[64,148] Although schizophrenia has been linked with multiple infectious agents that differ in their antigenicity, modes of transmission, and teratogenic potential, it is likely that they share some pathogenic mechanisms.[149] Experimentally, potential mechanisms of action include induction of pro-inflammatory cytokines,[149,150] endotoxin-induced fever,[151-153] and hypotension and cerebral hypoperfusion/hypoxia.[154,155] Infection may also sensitise the brain to subsequent hypoxic injury.[156] In twin studies, there is evidence that a shared placenta and amniotic sac increases the risk of both fetuses being exposed to infection (chorioamnionitis),[64,157] whereas a dichorionic placenta helps to limit the spread of infection.[158]

Clues from the Preterm Infant

Schizophrenia shows a typically remitting and relapsing course.[5] If the neurodevelopmental hypothesis is correct, then why should a neurological injury sustained in utero lead to such variable symptoms in adulthood?

One possible link is that glia continue to be produced and myelination continues to develop well into middle age in key corticolimbic relay areas.[159-161] Glia are not simply an important but passive matrix for the brain. In addition to their traditional roles in neuronal migration and inflammatory processes, glia are now known to provide trophic support to neurons, to regulate local neuronal metabolism and neurotransmission, and the formation of synapses (including pruning).[162,163] The early appearance of ventriculomegaly suggests that there has been a profound loss of glia in prenatal life, as is seen on in utero imaging,[164] such that there may be an inadequate number in adulthood to consistently support neuronal function. The loss of glia in patients suffering from schizophrenia and affective disorders is further evidence for pathological events in mid-gestation as this is the maximal time of increased glial proliferation and differentiation.[165] The impact of prenatal loss of glia is perhaps best illustrated by the neurodevelopmental delay, behavioural abnormalities and increased rate of mental illness in children who are born prematurely,[166-168] Pathologically, such infants demonstrate highly selective early white matter damage, which leads to long-term reductions in grey matter volume and complexity of neuronal structures.[167,169,170] The neurological sequelae of preterm birth (including epilepsy, cerebral palsy and attention deficit) associated with these perinatal white matter lesions seem to be a consequence of the post-injury grey matter transformations.[171,172] This sequence of events is strikingly similar to those in patients with schizophrenia. Disturbingly, there is now evidence to show that preterms are also at greater risk of developing mental illnesses such as schizophrenia in adulthood.[173]

Importantly, however, infants do not necessarily deliver even after a major insult in utero. There is now increasing evidence to show that neurodevelopmental delay in term babies is also associated with white matter loss and subsequently impaired neuronal development which apparently had its origins much earlier in fetal life.[174-176]

Other Neuropathological Features

There are a number of other features of the neuropathology of schizophrenia, which suggest a fetal insult around mid-gestation. For example, it is known that normal human brain symmetry is determined early in development, during the early second trimester of gestation. Studies have suggested that the left side of the brain is generally more severely affected in schizophrenia than the right,[177] and thus that some event occurred during this stage. Similarly, gyrification occurs largely between weeks 16 and 19 weeks of gestation, and sulcal-gyral abnormalities have been found in imaging MRI studies of schizophrenic patients.[178,179] Finally, as Clouston himself observed, schizophrenia is associated with an increased risk for other congenital and physical abnormalities, such as cranio-facial abnormalites like cleft palate, which have their origins in mid-gestation.[180-183]

Cerebral Housekeeping or Implementing "Plan B"

The development of the brain is a highly complex coordinated process that can be roughly divided into neurogenesis, neuronal migration, glial proliferation, and neuronal differentiation. These events occur as part of a specific timetable in discrete critical windows of time, which is presumed to be largely under genetic control.[168] This unfolding maturational program can be derailed by environmental events; cell proliferation, differentiation, and migration can be slowed or inhibited or cells killed outright. Importantly, because many events only occur at a particular "critical window of time",[132,184] even if the event causing this impairment is acute (transient hypoxia due to placental infarction for example), the impairment is irreparable and this has consequences for subsequent neural development. The architectural plan for brain development started in utero does not, of course, reach completion until early adulthood, when final connections are made in the prefrontal and temporal lobes, and corticolimbic pathways. These are all key regions where aberrant neuronal development may contribute to the behavioural dysfunction of schizophrenia.

As discussed in relation to premature birth, cell loss may continue long after the acute injury has finished. Cells, be they neurons or glia, require other glia and neurons to provide the necessary support and signals cues to survive.[185-187] This balance is exquisitely fine. During normal development substantial numbers of initially generated cells do not form appropriate connections or are in excess of requirements. These cell are removed by physiological apoptosis.[188] Critically, however, programmed cell death is also triggered when cells lose essential input from other cells, for example due to injury elsewhere in the brain. In such a pathological situation, upregulation of apoptosis is a normal part of the complex 'social' controls that ensure that individual cells behave for the good of the whole.[185] The brain will thus develop according to an alternate architectural plan—Plan B. Inevitably this leads to a smaller, less complex brain, but a functional one; in as much as function is defined by the ultimate human prime directive: the ability to reproduce.

A critical part of this process may not be simply abnormal neuronal connections, but loss of white matter cells, consistent with the association of mild ventriculomegaly with later impaired grey matter development and neurodevelopmental delay and behavioural difficulties. The hypothesis that loss of glial support is a major contributor to long-term outcome is consistent with the pathological profile of patients with schizophrenia of variable neuronal loss, but a consistent reduction in soma size and abnormal synaptic connections. Oligodenrocytes enhance the number of functional synapses that form between neurons, and regulate neuronal activity. Glia are also a primary source for the growth factors necessary to inhibit apoptosis.[189] Insulin-like growth factor (IGF-I) is a key mediator of normal brain development; regulating neural stem cell proliferation, differentiation, and maturation, as well as promoting myelination, neurite outgrowth and synaptogenesis.[190] In recent years it has been proposed that a derangement of the IGF axis may be involved in the aetiology of schizophrenia,[191] and that the excessive synaptic pruning which is a feature of the schizophrenic brain, is a function not of late (post-natal) neurodevelopmental events, but rather occur secondary to diminished trophic cues.

Is there clinical evidence for such increased, on-going apoptosis? Recent imaging data shows that children who go onto develop schizophrenia have accelerated loss of cortical grey matter compared to controls during adolescence.[192] This deficit enveloped increasing amounts of cortex throughout adolescence, starting in parietal regions, and then swept forward into sensory and motor regions. By 18 years of age this process had moved into the critical areas of the brain known to be key to schizophrenia; the dorsolateral prefrontal and temporal cortices - areas which initially were not affected. This aberrant development is also seen in MZ twins discordant for schizophrenia.[193] It is likely that this is an upregulation of the normal remodelling of the brain is in part mediated by an upregulation of physiological apoptosis.[186] Consistent with this there is some evidence that apoptotic processes are upregulated in the brain of schizophrenic patients at postmortem,[194] and that alterations in glutamate receptor activity seem to be important.[195]

Perspective

This chapter has examined the hypothesis that schizophrenia and other mental illnesses may have at least in part their origin in preceding fetal neurodevelopmental injury. Although the combined epidemiological, neuroanatomical, behavioural, and imaging evidence is highly suggestive, the data cannot yet definitively distinguish the roles of inherited predisposition and environmental triggers. Considerable work remains to properly understand the impact of timing and the nature of different adverse events in utero on the brain, and how these relate to the post-natal development of disease. Such knowledge would offer at the very least, improved detection of children at risk of later mental illness and thus the potential for earlier intervention. Regardless of the precise origins of the disease, there is now inconvertible proof that schizophrenia is an evolving disease that involves both significant premorbid developmental problems and progressive anatomical and cellular deterioration during childhood and adolescence well before the 'mental illness' appears fully in adulthood. Current data strongly suggest that the most likely mechanisms involve upregulation of physiological programmed cell death and a pathological imbalance in excitatory neurotransmission. This very long-term evolution offers the tantalising possibility that some intervention, whether pharmacological or behavioural might be able to arrest the progression of the disease before the florid symptoms appear, or even to favourably remodel the brain.

Acknowledgements

The authors would like to thank the Health Research Council of New Zealand for their support.

References

1. Beveridge A. Voices of the mad: Patients' letters from the Royal Edinburgh Asylum, 1873-1908. Psychol Med 1997; 27(4):899-908.
2. In: Harrison PJ, Roberts GW, eds. The Neuropathology of Schizophrenia. Progress and Interpretation. Oxford: Oxford University Press, 2000.
3. Plum F. Prospects for research on schizophrenia. III. Neurophysiology. Neuropathological findings. Neurosci Res Program Bull 1972; 10(4):384-388.
4. Shenton ME, Dickey CC, Frumin M et al. A review of MRI findings in schizophrenia. Schizophr Res 2001; 49(1-2):1-52.
5. Harrison PJ. The neuropathology of schizophrenia. A critical review of the data and their interpretation. Brain 1999; 122(Pt 4):593-624.
6. Pearlson GD. Neurobiology of schizophrenia. Ann Neurol 2000; 48(4):556-566.
7. Woods BT. Is schizophrenia a progressive neurodevelopmental disorder? Toward a unitary pathogenetic mechanism. Am J Psychiatry 1998; 155(12):1661-1670.
8. Benes FM. Emerging principles of altered neural circuitry in schizophrenia. Brain Res Brain Res Rev 2000; 31(2-3):251-269.
9. Miyamoto S, LaMantia AS, Duncan GE et al. Recent advances in the neurobiology of schizophrenia. Mol Interv 2003; 3(1):27-39.
10. Harrison PJ. The neuropathology of primary mood disorder. Brain 2002; 125(Pt 7):1428-1449.
11. Hirschhorn JN. Genetic epidemiology of type 1 diabetes. Pediatr Diabetes 2003; 4(2):87-100.
12. Owen MJ, Williams NM, O'Donovan MC. The molecular genetics of schizophrenia: New findings promise new insights. Mol Psychiatry 2004; 9(1):14-27.
13. Harrison PJ, Owen MJ. Genes for schizophrenia? Recent findings and their pathophysiological implications. Lancet 2003; 361(9355):417-419.
14. Allen NB, Lewinsohn PM, Seeley JR. Prenatal and perinatal influences on risk for psychopathology in childhood and adolescence. Dev Psychopathol 1998; 10(3):513-529.
15. Glasson EJ, Bower C, Petterson B et al. Perinatal factors and the development of autism: A population study. Arch Gen Psychiatry 2004; 61(6):618-627.
16. McGrath JJ, Feron FP, Burne TH et al. The neurodevelopmental hypothesis of schizophrenia: A review of recent developments. Ann Med 2003; 35(2):86-93.
17. Church SM, Cotter D, Bramon E et al. Does schizophrenia result from developmental or degenerative processes? J Neural Transm Suppl 2002; (63):129-147.
18. Walker J, Curtis V, Murray RM. Schizophrenia and bipolar disorder: Similarities in pathogenic mechanisms but differences in neurodevelopment. Int Clin Psychopharmacol 2002; 17(Suppl 3):S11-19.
19. Weinberger DR, McClure RK. Neurotoxicity, neuroplasticity, and magnetic resonance imaging morphometry: What is happening in the schizophrenic brain? Arch Gen Psychiatry 2002; 59(6):553-558.
20. Clouston TS. The Neuroses of Development. Edinburgh: Oliver and Boyd, 1891.
21. Clouston TS. Clinical Lectures on Mental Diseases. 3rd ed. London: Churchill, 1892.
22. Clouston TS. Clinical Lectures on Mental Diseases. 6th ed. London: Churchill, 1904.
23. O'Connell P, Woodruff PW, Wright I et al. Developmental insanity or dementia praecox: Was the wrong concept adopted? Schizophr Res 1997; 23(2):97-106.
24. Kraeplin E. Dementia praecox and paraphrenia. New York: Krieger, 1919.
25. de Haan L, Bakker JM. Overview of neuropathological theories of schizophrenia: From degeneration to progressive developmental disorder. Psychopathology 2004; 37(1):1-7.
26. Bleuler E. Dementia praecox or the group of schizophrenias. New York: International Universities Press, 1911.
27. Weinberger DR. Implications of normal brain development for the pathogenesis of schizophrenia. Arch Gen Psychiatry 1987; 44(7):660-669.
28. Murray RM, Lewis SW. Is schizophrenia a neurodevelopmental disorder? Br Med J Clin Res ed 1987; 295(6600):681-682.
29. Cannon M, Caspi A, Moffitt TE et al. Evidence for early-childhood, pan-developmental impairment specific to schizophreniform disorder: Results from a longitudinal birth cohort. Arch Gen Psychiatry 2002; 59(5):449-456.
30. Jones PB, Tarrant CJ. Developmental precursors and biological markers for schizophrenia and affective disorders: Specificity and public health implications. Eur Arch Psychiatry Clin Neurosci 2000; 250(6):286-291.
31. Remschmidt H. Early-onset schizophrenia as a progressive-deteriorating developmental disorder: Evidence from child psychiatry. J Neural Transm 2002; 109(1):101-117.

32. Isohanni M, Jones PB, Moilanen K et al. Early developmental milestones in adult schizophrenia and other psychoses. A 31-year follow-up of the Northern Finland 1966 Birth Cohort. Schizophr Res 2001; 52(1-2):1-19.
33. Jones P, Rodgers B, Murray R et al. Child development risk factors for adult schizophrenia in the British 1946 birth cohort. Lancet 1994; 344(8934):1398-1402.
34. Sporn AL, Addington AM, Gogtay N et al. Pervasive developmental disorder and childhood-onset schizophrenia: Comorbid disorder or a phenotypic variant of a very early onset illness? Biol Psychiatry 2004; 55(10):989-994.
35. Arnold SE. Cellular and molecular neuropathology of the parahippocampal region in schizophrenia. Ann NY Acad Sci 2000; 911:275-292.
36. Royston MC, Roberts GW. Schizophrenia. When neurons go astray. Curr Biol 1995; 5(4):342-344.
37. Jakob H, Beckmann H. Prenatal developmental disturbances in the limbic allocortex in schizophrenics. J Neural Transm 1986; 65(3-4):303-326.
38. Akbarian S, Kim JJ, Potkin SG et al. Maldistribution of interstitial neurons in prefrontal white matter of the brains of schizophrenic patients. Arch Gen Psychiatry 1996; 53(5):425-436.
39. Baumann B, Bogerts B. The pathomorphology of schizophrenia and mood disorders: Similarities and differences. Schizophr Res 1999; 39(2):141-148, discussion 162.
40. Benes FM, Kwok EW, Vincent SL et al. A reduction of nonpyramidal cells in sector CA2 of schizophrenics and manic depressives. Biol Psychiatry 1998; 44(2):88-97.
41. Lewis DA, Glantz LA, Pierri JN et al. Altered cortical glutamate neurotransmission in schizophrenia: Evidence from morphological studies of pyramidal neurons. Ann NY Acad Sci 2003; 1003:102-112.
42. Beasley CL, Cotter DR, Everall IP. Density and distribution of white matter neurons in schizophrenia, bipolar disorder and major depressive disorder: No evidence for abnormalities of neuronal migration. Mol Psychiatry 2002; 7(6):564-570.
43. Highley JR, Walker MA, McDonald B et al. Size of hippocampal pyramidal neurons in schizophrenia. Br J Psychiatry 2003; 183:414-417.
44. Akil M, Lewis DA. Cytoarchitecture of the entorhinal cortex in schizophrenia. Am J Psychiatry 1997; 154(7):1010-1012.
45. Hatten ME. Central nervous system neuronal migration. Annu Rev Neurosci 1999; 22:511-539.
46. Chan WY, Lorke DE, Tiu SC et al. Proliferation and apoptosis in the developing human neocortex. Anat Rec 2002; 267(4):261-276.
47. Wolkin A, Rusinek H, Vaid G et al. Structural magnetic resonance image averaging in schizophrenia. Am J Psychiatry 1998; 155(8):1064-1073.
48. Christensen J, Holcomb J, Garver DL. State-related changes in cerebral white matter may underlie psychosis exacerbation. Psychiatry Res 2004; 130(1):71-78.
49. Roessmann U, Gambetti P. Pathological reaction of astrocytes in perinatal brain injury. Immunohistochemical study. Acta Neuropathol (Berl) 1986; 70(3-4):302-307.
50. Stevens JR. Neuropathology of schizophrenia. Arch Gen Psychiatry 1982; 39(10):1131-1139.
51. Kunugi H, Urushibara T, Murray RM et al. Prenatal underdevelopment and schizophrenia: A case report of monozygotic twins. Psychiatry Clin Neurosci 2003; 57(3):271-274.
52. Uranova NA, Vostrikov VM, Orlovskaya DD et al. Oligodendroglial density in the prefrontal cortex in schizophrenia and mood disorders: A study from the Stanley Neuropathology Consortium. Schizophr Res 2004; 67(2-3):269-275.
53. Hof PR, Haroutunian V, Friedrich Jr VL et al. Loss and altered spatial distribution of oligodendrocytes in the superior frontal gyrus in schizophrenia. Biol Psychiatry 2003; 53(12):1075-1085.
54. Davis KL, Stewart DG, Friedman JI et al. White matter changes in schizophrenia: Evidence for myelin-related dysfunction. Arch Gen Psychiatry 2003; 60(5):443-456.
55. Uranova N, Orlovskaya D, Vikhreva O et al. Electron microscopy of oligodendroglia in severe mental illness. Brain Res Bull 2001; 55(5):597-610.
56. Hulshoff Pol HE, Schnack HG, Mandl RC et al. Focal white matter density changes in schizophrenia: Reduced inter-hemispheric connectivity. Neuroimage 2004; 21(1):27-35.
57. Rajkowska G, Halaris A, Selemon LD. Reductions in neuronal and glial density characterize the dorsolateral prefrontal cortex in bipolar disorder. Biol Psychiatry 2001; 49(9):741-752.
58. Rajkowska G, Miguel-Hidalgo JJ, Makkos Z et al. Layer-specific reductions in GFAP-reactive astroglia in the dorsolateral prefrontal cortex in schizophrenia. Schizophr Res 2002; 57(2-3):127-138.
59. Flynn SW, Lang DJ, Mackay AL et al. Abnormalities of myelination in schizophrenia detected in vivo with MRI, and post-mortem with analysis of oligodendrocyte proteins. Mol Psychiatry 2003; 8(9):811-820.
60. Chance SA, Highley JR, Esiri MM et al. Fiber content of the fornix in schizophrenia: Lack of evidence for a primary limbic encephalopathy. Am J Psychiatry 1999; 156(11):1720-1724.

61. Hakak Y, Walker JR, Li C et al. Genome-wide expression analysis reveals dysregulation of myelination-related genes in chronic schizophrenia. Proc Natl Acad Sci USA 2001; 98(8):4746-4751.
62. Hyde TM, Ziegler JC, Weinberger DR. Psychiatric disturbances in metachromatic leukodystrophy. Insights into the neurobiology of psychosis. Arch Neurol 1992; 49(4):401-406.
63. Laruelle M, Kegeles LS, Abi-Dargham A. Glutamate, dopamine, and schizophrenia: From pathophysiology to treatment. Ann NY Acad Sci 2003; 1003:138-158.
64. Brown AS, Susser ES. In utero infection and adult schizophrenia. Ment Retard Dev Disabil Res Rev 2002; 8(1):51-57.
65. Hulshoff Pol HE, Hoek HW, Susser E et al. Prenatal exposure to famine and brain morphology in schizophrenia. Am J Psychiatry 2000; 157(7):1170-1172.
66. Verdoux H, Sutter AL. Perinatal risk factors for schizophrenia: Diagnostic specificity and relationships with maternal psychopathology. Am J Med Genet 2002; 114(8):898-905.
67. Cannon M, Jones PB, Murray RM. Obstetric complications and schizophrenia: Historical and meta-analytic review. Am J Psychiatry 2002; 159(7):1080-1092.
68. Geddes JR, Lawrie SM. Obstetric complications and schizophrenia: A meta-analysis. Br J Psychiatry 1995; 167(6):786-793.
69. Buka SL, Tsuang MT, Lipsitt LP. Pregnancy/delivery complications and psychiatric diagnosis. A prospective study. Arch Gen Psychiatry 1993; 50(2):151-156.
70. Howes OD, McDonald C, Cannon M et al. Pathways to schizophrenia: The impact of environmental factors. Int J Neuropsychopharmacol 2004; 7(Suppl 1):S7-S13.
71. McNeil TF. Perinatal risk factors and schizophrenia: Selective review and methodological concerns. Epidemiol Rev 1995; 17(1):107-112.
72. Preti A, Cardascia L, Zen T et al. Risk for obstetric complications and schizophrenia. Psychiatry Res 2000; 96(2):127-139.
73. Bennedsen BE, Mortensen PB, Olesen AV et al. Preterm birth and intra-uterine growth retardation among children of women with schizophrenia. Br J Psychiatry 1999; 175:239-245.
74. Bennedsen BE, Mortensen PB, Olesen AV et al. Obstetric complications in women with schizophrenia. Schizophr Res 2001; 47(2-3):167-175.
75. Modrzewska K. The offspring of schizophrenic parents in a North Swedish isolate. Clin Genet 1980; 17(3):191-201.
76. Nilsson E, Lichtenstein P, Cnattingius S et al. Women with schizophrenia: Pregnancy outcome and infant death among their offspring. Schizophr Res 2002; 58(2-3):221-229.
77. Sacker A, Done DJ, Crow TJ. Obstetric complications in children born to parents with schizophrenia: A meta-analysis of case-control studies. Psychol Med 1996; 26(2):279-287.
78. Gunduz H, Woerner MG, Alvir JM et al. Obstetric complications in schizophrenia, schizoaffective disorder and normal comparison subjects. Schizophr Res 1999; 40(3):237-243.
79. Kinney DK, Yurgelun-Todd DA, Tohen M et al. Pre and perinatal complications and risk for bipolar disorder: A retrospective study. J Affect Disord 1998; 50(2-3):117-124.
80. Zornberg GL, Buka SL, Tsuang MT. The problem of obstetrical complications and schizophrenia. Schizophr Bull 2000; 26(2):249-256.
81. Rosso IM, Cannon TD, Huttunen T et al. Obstetric risk factors for early-onset schizophrenia in a Finnish birth cohort. Am J Psychiatry 2000; 157(5):801-807.
82. Moller HJ. Bipolar disorder and schizophrenia: Distinct illnesses or a continuum? J Clin Psychiatry 2003; 64(Suppl 6):23-27, discussion 28.
83. Rees S, Harding R. Brain development during fetal life: Influences of the intra-uterine environment. Neurosci Lett 2004; 361(1-3):111-114.
84. MacLennan A. A template for defining a causal relation between acute intrapartum events and cerebral palsy: International consensus statement. Bmj 1999; 319(7216):1054-1059.
85. George S, Gunn AJ, Westgate JA et al. Fetal heart rate variability and brainstem injury after asphyxia in preterm fetal sheep. Am J Physiol Regul Integr Comp Physiol 2004.
86. Bennet L, Rossenrode S, Gunning MI et al. The cardiovascular and cerebrovascular responses of the immature fetal sheep to acute umbilical cord occlusion. J Physiol 1999; 517(Pt 1):247-257.
87. Keunen H, Blanco CE, van Reempts JL et al. Absence of neuronal damage after umbilical cord occlusion of 10, 15, and 20 minutes in midgestation fetal sheep. Am J Obstet Gynecol 1997; 176(3):515-520.
88. Shelley H. Glycogen reserves and their changes at birth and in anoxia. Br med Bull 1961; 17(2):137-143.
89. Mott JC. The ability of young mammals to withstand total oxygen lack. Br Med Bull 1961; 17:144-148.
90. Gunn AJ, Quaedackers JS, Guan J et al. The premature fetus: Not as defenseless as we thought, but still paradoxically vulnerable? Dev Neurosci 2001; 23(3):175-179.

91. Grafe MR. The correlation of prenatal brain damage with placental pathology. J Neuropathol Exp Neurol 1994; 53(4):407-415.
92. Ment LR, Schwartz M, Makuch RW et al. Association of chronic sublethal hypoxia with ventriculomegaly in the developing rat brain. Brain Res Dev Brain Res 1998; 111(2):197-203.
93. Tashima L, Nakata M, Anno K et al. Prenatal influence of ischemia-hypoxia-induced intrauterine growth retardation on brain development and behavioral activity in rats. Biol Neonate 2001; 80(1):81-87.
94. Mallard EC, Rehn A, Rees S et al. Ventriculomegaly and reduced hippocampal volume following intrauterine growth-restriction: Implications for the aetiology of schizophrenia. Schizophr Res 1999; 40(1):11-21.
95. Dieni S, Rees S. Dendritic morphology is altered in hippocampal neurons following prenatal compromise. J Neurobiol 2003; 55(1):41-52.
96. Baud O, Daire JL, Dalmaz Y et al. Gestational hypoxia induces white matter damage in neonatal rats: A new model of periventricular leukomalacia. Brain Pathol 2004; 14(1):1-10.
97. Kohlhauser C, Mosgoller W, Hoger H et al. Myelination deficits in brain of rats following perinatal asphyxia. Life Sci 2000; 67(19):2355-2368.
98. Bennet L, Quaedackers JS, Guan J et al. Chronically evolving white matter and cortical cell loss after asphyxia in the mid-gestation sheep fetus are mediated by caspase-dependent apoptotic mechanisms. Pediatr Res 2003; 53(4):347A.
99. Cao Y, Gunn AJ, Bennet L et al. Insulin-like growth factor (IGF)-1 suppresses oligodendrocyte caspase-3 activation and increases glial proliferation after ischemia in near-term fetal sheep. J Cereb Blood Flow Metab 2003; 23(6):739-747.
100. Guan J, Bennet L, George S et al. Insulin-like growth factor-1 reduces postischemic white matter injury in fetal sheep. J Cereb Blood Flow Metab 2001; 21(5):493-502.
101. Ikeda T, Murata Y, Quilligan EJ et al. Physiologic and histologic changes in near-term fetal lambs exposed to asphyxia by partial umbilical cord occlusion. Am J Obstet Gynecol 1998; 178(1 Pt 1):24-32.
102. Boog G. Obstetrical complications and subsequent schizophrenia in adolescent and young adult offsprings: Is there a relationship? Eur J Obstet Gynecol Reprod Biol 2004; 114(2):130-136.
103. Cannon TD, van Erp TG, Rosso IM et al. Fetal hypoxia and structural brain abnormalities in schizophrenic patients, their siblings, and controls. Arch Gen Psychiatry 2002; 59(1):35-41.
104. Van Erp TG, Saleh PA, Rosso IM et al. Contributions of genetic risk and fetal hypoxia to hippocampal volume in patients with schizophrenia or schizoaffective disorder, their unaffected siblings, and healthy unrelated volunteers. Am J Psychiatry 2002; 159(9):1514-1520.
105. Torrey EF. Are we overestimating the genetic contribution to schizophrenia? Schizophr Bull 1992; 18(2):159-170.
106. Walker E, Downey G, Caspi A. Twin studies of psychopathology: Why do the concordance rates vary? Schizophr Res 1991; 5(3):211-221.
107. Pharoah PO. Errors in birth registrations and coding of twins and higher order multiples. Twin Res 2002; 5(4):270-272.
108. Machin G. Placentation in multiple births. Twin Res 2001; 4(3):150-155.
109. Suddath RL, Christison GW, Torrey EF et al. Anatomical abnormalities in the brains of monozygotic twins discordant for schizophrenia. N Engl J Med 1990; 322(12):789-794.
110. Pharoah PO. Neurological outcome in twins. Semin Neonatol 2002; 7(3):223-230.
111. Pharoah PO, Price TS, Plomin R. Cerebral palsy in twins: A national study. Arch Dis Child Fetal Neonatal Ed 2002; 87(2):F122-124.
112. Adegbite AL, Castille S, Ward S et al. Neuromorbidity in preterm twins in relation to chorionicity and discordant birth weight. Am J Obstet Gynecol 2004; 190(1):156-163.
113. Laplaza FJ, Root L, Tassanawipas A et al. Cerebral palsy in twins. Dev Med Child Neurol 1992; 34(12):1053-1063.
114. Glinianaia SV, Pharoah PO, Wright C et al. Fetal or infant death in twin pregnancy: Neurodevelopmental consequence for the survivor. Arch Dis Child Fetal Neonatal Ed 2002; 86(1):F9-15.
115. Smith GN, Flynn SW, McCarthy N et al. Low birthweight in schizophrenia: Prematurity or poor fetal growth? Schizophr Res 2001; 47(2-3):177-184.
116. Petterson B, Nelson KB, Watson L et al. Twins, triplets, and cerebral palsy in births in Western Australia in the 1980s. Bmj 1993; 307(6914):1239-1243.
117. Grether JK, Nelson KB, Cummins SK. Twinning and cerebral palsy: Experience in four northern California counties, births 1983 through 1985. Pediatrics 1993; 92(6):854-858.
118. Dube J, Dodds L, Armson BA. Does chorionicity or zygosity predict adverse perinatal outcomes in twins? Am J Obstet Gynecol 2002; 186(3):579-583.

119. Benirschke K. The biology of the twinning process: How placentation influences outcome. Semin Perinatol 1995; 19(5):342-350.
120. Lewi L, Van Schoubroeck D, Gratacos E et al. Monochorionic diamniotic twins: Complications and management options. Curr Opin Obstet Gynecol 2003; 15(2):177-194.
121. Bejar R, Vigliocco G, Gramajo H et al. Antenatal origin of neurologic damage in newborn infants. II. Multiple gestations. Am J Obstet Gynecol 1990; 162(5):1230-1236.
122. Larroche JC, Girard N, Narcy F et al. Abnormal cortical plate (polymicrogyria), heterotopias and brain damage in monozygous twins. Biol Neonate 1994; 65(6):343-352.
123. Davis JO, Phelps JA, Bracha HS. Prenatal development of monozygotic twins and concordance for schizophrenia. Schizophr Bull 1995; 21(3):357-366.
124. Sokol DK, Moore CA, Rose RJ et al. Intrapair differences in personality and cognitive ability among young monozygotic twins distinguished by chorion type. Behav Genet 1995; 25(5):457-466.
125. Riese ML. Effects of chorion type on neonatal temperament differences in monozygotic twin pairs. Behav Genet 1999; 29(2):87-94.
126. Wichers MC, Danckaerts M, Van Gestel S et al. Chorion type and twin similarity for child psychiatric symptoms. Arch Gen Psychiatry 2002; 59(6):562-564.
127. Pasamanick B, Rogers ME, Lilienfeld AM. Pregnancy experience and the development of behavior disorders in children. Am J Psychiatry 1956; 112(8):613-618.
128. Wahlbeck K, Forsen T, Osmond C et al. Association of schizophrenia with low maternal body mass index, small size at birth, and thinness during childhood. Arch Gen Psychiatry 2001; 58(1):48-52.
129. Morgane PJ, Mokler DJ, Galler JR. Effects of prenatal protein malnutrition on the hippocampal formation. Neurosci Biobehav Rev 2002; 26(4):471-483.
130. Ozanne SE, Fernandez-Twinn D, Hales CN. Fetal growth and adult diseases. Semin Perinatol 2004; 28(1):81-87.
131. Turton P, Hughes P, Bolton H et al. Incidence and demographic correlates of eating disorder symptoms in a pregnant population. Int J Eat Disord 1999; 26(4):448-452.
132. Morgane PJ, Austin-LaFrance R, Bronzino J et al. Prenatal malnutrition and development of the brain. Neurosci Biobehav Rev 1993; 17(1):91-128.
133. Steiger JL, Alexander MJ, Galler JR et al. Effects of prenatal malnutrition on GABAA receptor alpha1, alpha3 and beta2 mRNA levels. Neuroreport 2003; 14(13):1731-1735.
134. Carmichael SL, Shaw GM, Schaffer DM et al. Dieting behaviors and risk of neural tube defects. Am J Epidemiol 2003; 158(12):1127-1131.
135. Mattson MP, Shea TB. Folate and homocysteine metabolism in neural plasticity and neurodegenerative disorders. Trends Neurosci 2003; 26(3):137-146.
136. McGrath J. Hypothesis: Is low prenatal vitamin D a risk-modifying factor for schizophrenia? Schizophr Res 1999; 40(3):173-177.
137. Craciunescu CN, Brown EC, Mar MH et al. Folic acid deficiency during late gestation decreases progenitor cell proliferation and increases apoptosis in fetal mouse brain. J Nutr 2004; 134(1):162-166.
138. Bajoria R, Sooranna SR, Ward S et al. Placental transport rather than maternal concentration of amino acids regulates fetal growth in monochorionic twins: Implications for fetal origin hypothesis. Am J Obstet Gynecol 2001; 185(5):1239-1246.
139. Coyle JT, Tsai G, Goff D. Converging evidence of NMDA receptor hypofunction in the pathophysiology of schizophrenia. Ann NY Acad Sci 2003; 1003:318-327.
140. Almeida SS, Tonkiss J, Galler JR. Prenatal protein malnutrition affects the social interactions of juvenile rats. Physiol Behav 1996; 60(1):197-201.
141. Almeida SS, Tonkiss J, Galler JR. Prenatal protein malnutrition affects exploratory behavior of female rats in the elevated plus-maze test. Physiol Behav 1996; 60(2):675-680.
142. Datta S, Patterson EH, Vincitore M et al. Prenatal protein malnourished rats show changes in sleep/wake behavior as adults. J Sleep Res 2000; 9(1):71-79.
143. Boivin DB. Influence of sleep-wake and circadian rhythm disturbances in psychiatric disorders. J Psychiatry Neurosci 2000; 25(5):446-458.
144. Gunnell D, Rasmussen F, Fouskakis D et al. Patterns of fetal and childhood growth and the development of psychosis in young males: A cohort study. Am J Epidemiol 2003; 158(4):291-300.
145. Hoek HW, Susser E, Buck KA et al. Schizoid personality disorder after prenatal exposure to famine. Am J Psychiatry 1996; 153(12):1637-1639.
146. Susser E, Hoek HW, Brown A. Neurodevelopmental disorders after prenatal famine: The story of the Dutch Famine Study. Am J Epidemiol 1998; 147(3):213-216.
147. Brown AS, van Os J, Driessens C et al. Further evidence of relation between prenatal famine and major affective disorder. Am J Psychiatry 2000; 157(2):190-195.

148. Davies G, Welham J, Chant D et al. A systematic review and meta-analysis of Northern Hemisphere season of birth studies in schizophrenia. Schizophr Bull 2003; 29(3):587-593.
149. Brown AS, Hooton J, Schaefer CA et al. Elevated maternal interleukin-8 levels and risk of schizophrenia in adult offspring. Am J Psychiatry 2004; 161(5):889-895.
150. Gilmore JH, Fredrik Jarskog L, Vadlamudi S et al. Prenatal infection and risk for schizophrenia: IL-1beta, IL-6, and TNFalpha inhibit cortical neuron dendrite development. Neuropsychopharmacology 2004.
151. Baena RC, Busto R, Dietrich WD et al. Hyperthermia delayed by 24 hours aggravates neuronal damage in rat hippocampus following global ischemia. Neurology 1997; 48(3):768-773.
152. Kim Y, Busto R, Dietrich WD et al. Delayed postischemic hyperthermia in awake rats worsens the histopathological outcome of transient focal cerebral ischemia. Stroke 1996; 27(12):2274-2280, discussion 2281.
153. Laptook AR, Corbett RJ. The effects of temperature on hypoxic-ischemic brain injury. Clin Perinatol 2002; 29(4):623-649, vi.
154. Dalitz P, Harding R, Rees SM et al. Prolonged reductions in placental blood flow and cerebral oxygen delivery in preterm fetal sheep exposed to endotoxin: Possible factors in white matter injury after acute infection. J Soc Gynecol Investig 2003; 10(5):283-290.
155. Duncan JR, Cock ML, Scheerlinck JP et al. White matter injury after repeated endotoxin exposure in the preterm ovine fetus. Pediatr Res 2002; 52(6):941-949.
156. Mallard C, Welin AK, Peebles D et al. White matter injury following systemic endotoxemia or asphyxia in the fetal sheep. Neurochem Res 2003; 28(2):215-223.
157. Davis JO, Phelps JA. Twins with schizophrenia: Genes or germs? Schizophr Bull 1995; 21(1):13-18.
158. Phung DT, Blickstein I, Goldman RD et al. The northwestern twin chorionicity study: I. discordant inflammatory findings that are related to chorionicity in presenting versus nonpresenting twins. Am J Obstet Gynecol 2002; 186(5):1041-1045.
159. Bartzokis G, Beckson M, Lu PH et al. Age-related changes in frontal and temporal lobe volumes in men: A magnetic resonance imaging study. Arch Gen Psychiatry 2001; 58(5):461-465.
160. Bartzokis G. Schizophrenia: Breakdown in the well-regulated lifelong process of brain development and maturation. Neuropsychopharmacology 2002; 27(4):672-683.
161. Benes FM, Turtle M, Khan Y et al. Myelination of a key relay zone in the hippocampal formation occurs in the human brain during childhood, adolescence, and adulthood. Arch Gen Psychiatry 1994; 51(6):477-484.
162. Barres BA, Barde Y. Neuronal and glial cell biology. Curr Opin Neurobiol 2000; 10(5):642-648.
163. Broadie K. Axon pruning: An active role for glial cells. Curr Biol 2004; 14(8):R302-304.
164. Gilmore JH, van Tol J, Kliewer MA et al. Mild ventriculomegaly detected in utero with ultrasound: Clinical associations and implications for schizophrenia. Schizophr Res 1998; 33(3):133-140.
165. Kinney HC, Back SA. Human oligodendroglial development: Relationship to periventricular leukomalacia. Semin Pediatr Neurol 1998; 5(3):180-189.
166. Perlman JM. White matter injury in the preterm infant: An important determination of abnormal neurodevelopment outcome. Early Hum Dev 1998; 53(2):99-120.
167. Inder TE, Volpe JJ. Mechanisms of perinatal brain injury. Semin Neonatol 2000; 5(1):3-16.
168. Peterson BS. Brain imaging studies of the anatomical and functional consequences of preterm birth for human brain development. Ann NY Acad Sci 2003; 1008:219-237.
169. Peterson BS, Vohr B, Staib LH et al. Regional brain volume abnormalities and long-term cognitive outcome in preterm infants. Jama 2000; 284(15):1939-1947.
170. Ajayi-Obe M, Saeed N, Cowan FM et al. Reduced development of cerebral cortex in extremely preterm infants. Lancet 2000; 356(9236):1162-1163.
171. Marin-Padilla M. Developmental neuropathology and impact of perinatal brain damage. II: White matter lesions of the neocortex. J Neuropathol Exp Neurol 1997; 56(3):219-235.
172. Ment LR, Vohr B, Allan W et al. The etiology and outcome of cerebral ventriculomegaly at term in very low birth weight preterm infants. Pediatrics 1999; 104(2 Pt 1):243-248.
173. Ichiki M, Kunugi H, Takei N et al. Intra-uterine physical growth in schizophrenia: Evidence confirming excess of premature birth. Psychol Med 2000; 30(3):597-604.
174. Wang LW, Huang CC, Yeh TF. Major brain lesions detected on sonographic screening of apparently normal term neonates. Neuroradiology 2004; 46(5):368-373.
175. Filippi CG, Ulug AM, Deck MD et al. Developmental delay in children: Assessment with proton MR spectroscopy. AJNR Am J Neuroradiol 2002; 23(5):882-888.
176. Harbord MG, Finn JP, Hall-Craggs MA et al. Myelination patterns on magnetic resonance of children with developmental delay. Dev Med Child Neurol 1990; 32(4):295-303.
177. James AC, Crow TJ, Renowden S et al. Is the course of brain development in schizophrenia delayed? Evidence from onsets in adolescence. Schizophr Res 1999; 40(1):1-10.

178. White T, Andreasen NC, Nopoulos P et al. Gyrification abnormalities in childhood- and adolescent-onset schizophrenia. Biol Psychiatry 2003; 54(4):418-426.
179. Yucel M, Stuart GW, Maruff P et al. Paracingulate morphologic differences in males with established schizophrenia: A magnetic resonance imaging morphometric study. Biol Psychiatry 2002; 52(1):15-23.
180. Waddington JL, Lane A, Scully P et al. Early cerebro-craniofacial dysmorphogenesis in schizophrenia: A lifetime trajectory model from neurodevelopmental basis to 'neuroprogressive' process. J Psychiatr Res 1999; 33(6):477-489.
181. Gourion D, Goldberger C, Bourdel MC et al. Minor physical anomalies in patients with schizophrenia and their parents: Prevalence and pattern of craniofacial abnormalities. Psychiatry Res 2004; 125(1):21-28.
182. Trixler M, Tenyi T, Csabi G et al. Minor physical anomalies in schizophrenia and bipolar affective disorder. Schizophr Res 2001; 52(3):195-201.
183. Tenyi T, Trixler M, Csabi G et al. Minor physical anomalies in nonfamilial unipolar recurrent major depression. J Affect Disord 2004; 79(1-3):259-262.
184. Rice D, Barone Jr S. Critical periods of vulnerability for the developing nervous system: Evidence from humans and animal models. Environ Health Perspect 2000; 108(Suppl 3):511-533.
185. Goldberg JL, Barres BA. The relationship between neuronal survival and regeneration. Annu Rev Neurosci 2000; 23:579-612.
186. Johnston MV. Clinical disorders of brain plasticity. Brain Dev 2004; 26(2):73-80.
187. Zahir N, Weaver VM. Death in the third dimension: Apoptosis regulation and tissue architecture. Curr Opin Genet Dev 2004; 14(1):71-80.
188. Davies AM. Regulation of neuronal survival and death by extracellular signals during development. EMBO J 2003; 22(11):2537-2545.
189. Raff MC, Barres BA, Burne JF et al. Programmed cell death and the control of cell survival. Philos Trans R Soc Lond B Biol Sci 1994; 345(1313):265-268.
190. Guan J, Bennet L, Gluckman PD et al. Insulin-like growth factor-1 and post-ischemic brain injury. Prog Neurobiol 2003; 70(6):443-462.
191. Gunnell D, Holly JM. Do insulin-like growth factors underlie associations of birth complications, fetal and preadult growth with schizophrenia? Schizophr Res 2004; 67(2-3):309-311.
192. Thompson PM, Vidal C, Giedd JN et al. Mapping adolescent brain change reveals dynamic wave of accelerated gray matter loss in very early-onset schizophrenia. Proc Natl Acad Sci USA 2001; 98(20):11650-11655.
193. Cannon TD, Thompson PM, van Erp TG et al. Cortex mapping reveals regionally specific patterns of genetic and disease-specific gray-matter deficits in twins discordant for schizophrenia. Proc Natl Acad Sci USA 2002; 99(5):3228-3233.
194. Jarskog LF, Selinger ES, Lieberman JA et al. Apoptotic proteins in the temporal cortex in schizophrenia: High Bax/Bcl-2 ratio without caspase-3 activation. Am J Psychiatry 2004; 161(1):109-115.
195. Harris LW, Sharp T, Gartlon J et al. Long-term behavioural, molecular and morphological effects of neonatal NMDA receptor antagonism. Eur J Neurosci 2003; 18(6):1706-1710.

CHAPTER 18

Hypoxia, Fetal Growth and Developmental Origins of Health and Disease

Dino A. Giussani*

Abstract

The compelling evidence linking small size at birth with later cardiovascular disease, obtained from epidemiological studies of human populations of more than a dozen countries,[1] has clearly renewed and amplified a clinical and scientific interest into the determinants of fetal growth, birth weight and the development of cardiovascular function and dysfunction before and after birth. As early as the 1950s Penrose[2] highlighted that an important determinant of birth weight was the quality of the intrauterine environment, being twice as great a determinant of the rate of fetal growth than the maternal or fetal genotype. Studies of birth weights of relatives[2] together with strong evidence from animal cross-breeding experiments[3,4] have clearly supported this contention. One of the great qualifiers of the fetal environment is the maternal nutritional status during pregnancy. As such, the reciprocal association between low birth weight and increased risk of high blood pressure in adulthood, as first described by Barker,[1] has literally exploded a new field of research investigating the effects of maternofetal nutrition on fetal growth, birth weight and subsequent cardiovascular disease. However, the fetus nourishes itself also with oxygen, and in contrast to the international effort which is assessing the effects of maternofetal under-nutrition on early development, the effects of maternofetal under-oxygenation on fetal growth, birth weight and subsequent increased risk of disease have been little addressed. Here, evidence is presented, which supports the concept that fetal hypoxia alone may provide a candidate prenatal stimulus contributing to fetal growth restriction and the developmental origins of cardiovascular health and disease.

The Fetal Cardiovascular Defence to Short- and Long-Term Hypoxia

Hypoxia is one of the major challenges that the fetus may face during gestation. The immediate fetal defence to hypoxia is largely dependent on its cardiovascular system. During acute hypoxia the fetal strategy is to make best use of the available oxygen delivery. Hence, in response to acute hypoxia, a redistribution of the fetal cardiac output occurs, which shunts blood flow away from peripheral towards essential circulations in order to protect hypoxia-sensitive organs like the fetal brain[5,6] (Fig. 1). Should the duration of the hypoxic challenge become prolonged, the initial homeostatic cardiovascular defence becomes enhanced. In response to chronic hypoxia, there is persistent redistribution of blood flow towards essential circulations, secondary to chronic elevations in peripheral vascular tone.[7-9] Whether this sustained homeostatic vascular response in the fetus becomes vestigial and a maladaptation, akin to the erythrocytotic[10] or pulmonary vasoconstrictor[11] responses to hypoxia in highland neonates and adults, which if transient are beneficial but if persistent lead to pathology, is unclear at present. However, it is

*Dino A. Giussani—Department of Physiology, University of Cambridge, Downing Street, Cambridge CB2 3EG, U.K. Email: dag26@cam.ac.uk

Early Life Origins of Health and Disease, edited by E. Marelyn Wintour and Julie A. Owens.

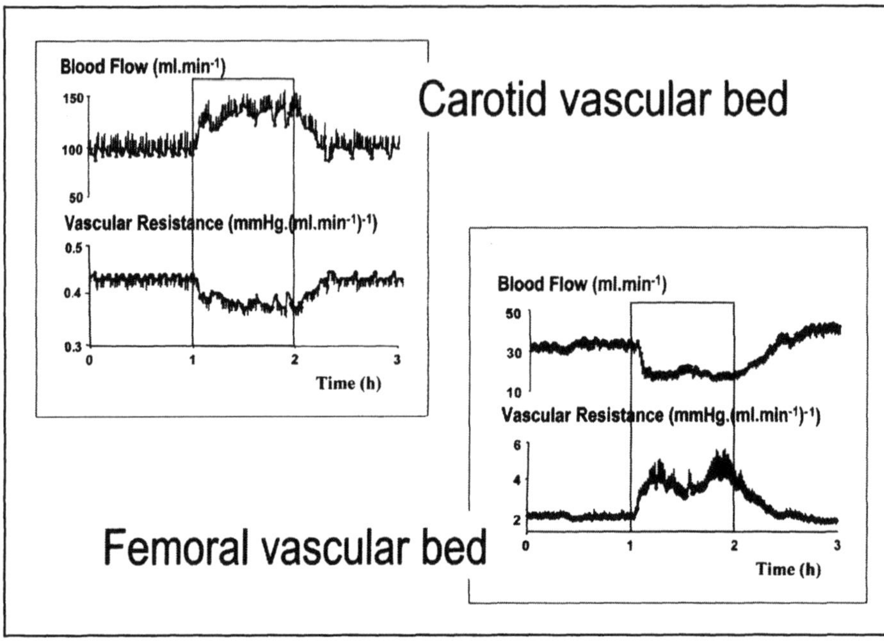

Figure 1. The fetal vascular defence to acute hypoxia. Values represent the minute mean ± S.E.M for blood flow and vascular resistance in the carotid and femoral circulations in 8 fetal sheep between 125-130 days of gestation. Note that the carotid vascular bed undergoes vasodilatation during acute hypoxaemia (box) as indexed by a fall in carotid vascular resistance and an increase in carotid blood flow. In marked contrast, the femoral vascular bed undergoes vasoconstriction, aiding the redistribution of blood flow away from peripheral circulations.

suggested that plausible biological trade-offs of fetal persistent peripheral vasoconstriction are asymmetric growth retardation and increased cardiac afterload, proposing that the classical phenotypic association between intrauterine growth retardation with cardiovascular dysfunction in adult life may originate from the same developmental stimulus—fetal hypoxia.

Hypoxia and Fetal Growth Retardation

Although several studies in animals have shown that chronic hypoxia during pregnancy can lead to slow, disproportionate fetal growth,[12,13] whether the effects are due to sustained under-oxygenation or partial under-nutrition is uncertain as chronic hypoxia also reduces maternal food intake.[13] In human populations, materno-fetal hypoxia occurs most commonly during the hypobaric hypoxia of pregnancy at high altitude. In support of data gathered from animal experiments, several investigators have also reported reduced birth weight and asymmetric growth retardation in human babies with increasing altitude.[14-16] However, because most high altitude populations are also impoverished, the extent to which this reduction in fetal growth is governed by maternal nutritional status or the hypoxia of high altitude, again, remains uncertain. To assess the partial contributions of fetal under-oxygenation and under-nutrition in the control of fetal growth, we have recently adopted a two-prong approach addressing questions in a specific human population and in a specific experimental animal model.

Epidemiological studies of human populations were carried out in Bolivia as this country is geographically and socio-economically unique. Bolivia lies in the heart of South America and it is split by the Andean cordillera into areas of very high altitude to the west of the country (4000 m) and sea level areas as the east of the country spans into the Brazilian Amazon. Facilitating

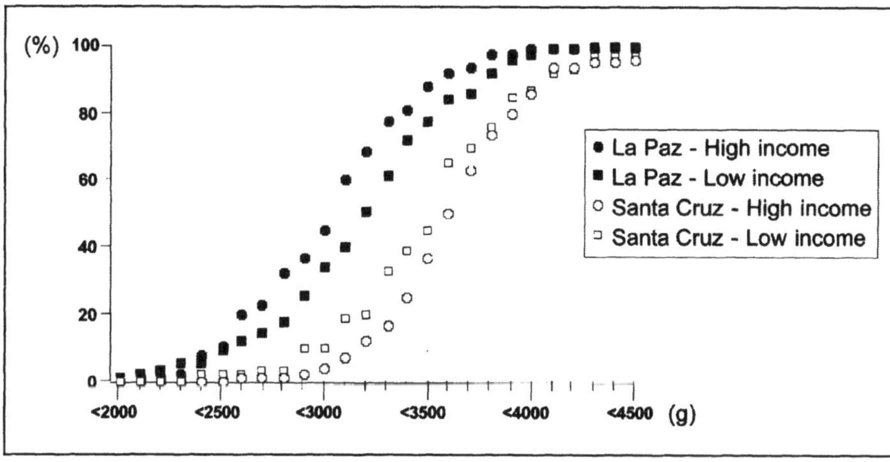

Figure 2. High altitude versus socio-economic stastus on birth weight. The curves represent the cumulative frequency distribution across all birth weights in term babies born from mothers of opposing economic status in La Paz (4000 m) and in Santa Cruz (sea level), in Bolivia. Modified from Giussani et al. Pediatric Research 2001; 49:490-494.[16]

the study design, the two largest cities, and therefore the most populated with approximately 2 million inhabitants each, are La Paz (4000 m) and Santa Cruz (400 m). Bolivia is also socio-economically unique as both La Paz and Santa Cruz are made up of striking economically-divergent populations.[17] In third world countries, and especially in Bolivia, there is an unsurprising strong relationship between socio-economic status and nutritional status.[18] Joining strengths with Barker, a recent study investigated whether the intrauterine growth retardation observed in the high altitude regions of Bolivia was primarily due to intrauterine hypoxia or due to the maternal socio-economic nutritional status.[16] Birth weight records were obtained from term pregnancies in La Paz and Santa Cruz, especially from obstetric hospitals selectively attended by wealthy or impoverished mothers. Plots of the cumulative frequency distribution across all birth weights gathered revealed a pronounced shift to the left in the curve of babies from high altitude than from low altitude, despite similarly high maternal economic status (Fig. 2). Interestingly, when lowland babies born from mothers with high or low economic status were compared, a shift to the left in birth weight occurred in low versus high income groups, however this shift was not as pronounced as the effect on birth weight of high altitude hypoxia alone. Additional data also showed that highland babies from poor families did not have the greatest leftward shift of the relationship as one would have expected. Rather, counter-intuitively, these babies were actually heavier than highland babies born from families with a high socio-economic status. The apparent conundrum is easily explained by assessing the ancestry of the families. In our study, the low socio-economic group of La Paz contained a high percentage (92%) of women from Amerindian origin with Aymara indian paternal and maternal surnames.[16] In contrast, the high socio-economic group of La Paz contained a high European admixture. These findings are reminiscent of the observations of Hass et al[19] and Moore[20] who suggested that fetal growth retardation at altitude is correlated to the duration of high altitude residence, independent of maternal nutrition: the longest resident population experiencing the least decline and the shortest residence groups demonstrating the most reduction in birth weight. Accordingly, reductions in birth weight at elevations greater than 3000m above sea level are greatest in Colorado, intermediate in Andeans and least in Tibetans.[20]

The second prong of our approach exploited the chick embryo as an animal model. In contrast to all mammals, in avian species the effects of hypoxia on the fetus can be assessed

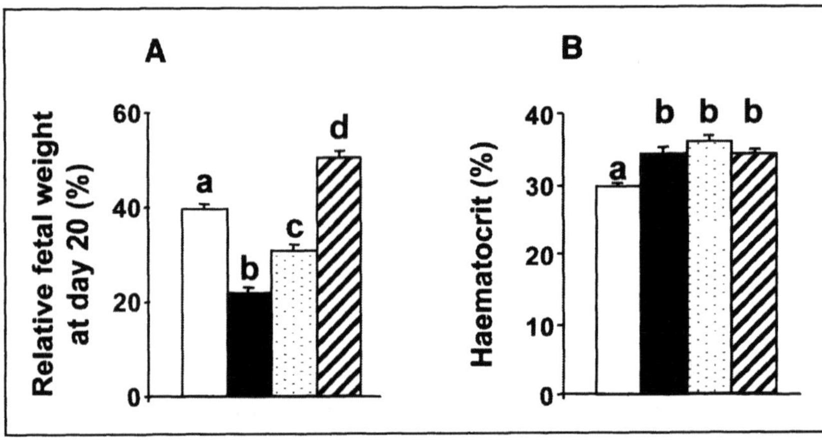

Figure 3. The role of oxygen in the growth of the chick embryo. Values are mean ±S.E.M. for the fetal weight at the end of the incubation period expressed as a percentage of the initial egg mass (A) and the fetal haematocrit at the end of the incubation period (B). Groups are sea level chick embryos incubated either at sea level (SLSL, open bar, n = 31) or high altitude (SLHA, closed bar, n = 19) and high altitude embryos incubated at high altitude (HAHA, stippled bar, n = 33) or sea level (HASL, hatched bar, n = 30). Different letters are significantly different by one way ANOVA + Student Newman Keuls Test ($P < 0.05$).

directly, without additional effects of hypoxia on the mother and the placenta, and without confounding problems associated with reductions in maternal food intake. The study reasoned that if oxygen alone has a real role in the direct control of fetal growth, then fertilised eggs from hens native to sea level should show growth restriction when incubated at high altitude, and fertilised eggs from hens native to high altitude, which usually show growth restriction, should at least recover their growth when incubated at sea level.[21] The data of the study showed that incubation of sea level embryos at high altitude led to a 45% growth restriction, but incubation at high altitude of embryos from hens native to high altitude only led to 22% growth restriction (Fig. 3). The embryonic growth restriction of incubations at high altitude was asymmetric as brain weight was preserved at the expense of body length. Another component of this study showed that when fertilised eggs laid by hens native to high altitude were incubated at sea level, the resulting embryos not only recovered their growth, but they grew heavier than sea level controls. The haematocrit data reveal that this group of embryos retained an increased oxygen carrying capacity despite incubation at sea level (Fig. 3). This suggests that embryos from hens native to high altitude incubated at sea level had a greater oxygen content than sea level controls, further supporting a role for oxygen in the control of fetal growth. The mechanism via which elevated haematocrit levels are maintained in the absence of a hypoxic stimulus is unknown, but the data may reflect an adaptive response, transmitted by the mother to the oocyte prior to egg laying, predictive of fetal development in a hypoxic environment. Another example of a predictive adaptive response[22] is that of the meadow vole, in which the photoperiodic history of the dam prior to conception, rather than the perinatal thermal environment, can better determine the offspring's coat thickness at birth.[23] Alternatively, the maintained elevated haematocrit in HASL embryos in the present study may highlight that genetic control of factors determining oxygen carrying capacity is regulated very early on in the developmental process of the oocyte by the available oxygen concentration at that time.

Hypoxia and Developmental Origins of Cardiovascular Disease

An increasing number of experimental studies are beginning to support the argument that developmental hypoxia can give rise to cardiovascular dysfunction before and after birth.

Studies in fetal sheep have shown that chronic hypoxia, secondary to pregnancy at high altitude, sustained anaemia or chronic fetal placental embolization, led to suppressed cardiac function and contractility, and resulted in cardiac hypertrophy (see ref. 24). Studies in chick and rat embryos have shown that the exposure to sustained hypoxia during development, which leads to asymmetric growth retardation, is accompanied by ventricular and aortic hypertrophy.[24,25] Three elegant studies have now shown that fetal hypoxia alone may have persisting consequences for cardiac and vascular health in the offspring at adulthood. Li et al[26] reported that exposure of pregnant rats to hypoxia from day 15 to 21 of gestation produced offspring with increased cardiac susceptibility to ischaemic-reperfusion injury at 6 months of age. Williams et al[27] reported that exposure of pregnant rats to hypoxia from day 15 to 21 of gestation produced offspring with impaired NO-dependent vasodilatation in the mesenteric circulation at 4 months of age. Ruijtenbeek et al[28] reported that exposure of chick embryos to hypoxia from day 6 to 19 of the incubation period produced offspring with exaggerated responses to peri-arterial sympathetic nerve stimulation and down regulated NO-dependent dilator function in the femoral vascular endothelium at 15-16 weeks of adulthood.

In humans, it remains to be elucidated whether hypoxia-induced reductions in birth weight, as a result of placental insufficiency or of pregnancy at high altitude are associated with increased health risks after birth. Clearly, it is close to impossible to isolate the partial effects of fetal hypoxia and fetal undernutrition in promoting intrauterine growth retardation in human pregnancies complicated with placental insufficiency or preeclampsia. In contrast, investigation of the contribution of fetal hypoxia alone in promoting effects on fetal growth and programming health risks before and after birth can be achieved in human pregnancy at high altitude, carefully controlled for maternal nutritional status, making this a powerful model. Preliminary data[29] suggest that the rates of infant mortality, within the first year of newborn life, are positively correlated with altitude, increasing at a rate of *ca.* 8 deaths per 1000 m increase, and that this relationship is independent of the maternal socio-economic and nutritional status. However, the negligible number of studies in which basal arterial blood pressure was measured in adult residents rather than climbers at high altitude, report conflicting results suggesting either a higher incidence of hypertension in the inhabitants of high altitude regions of Saudi Arabia[30] or lower resting blood pressure in Peruvian highlanders.[31] These inconsistencies may be related to the high altitude residence ancestry of the individuals being studied. Thus, while hypoxia during development may increase the risk of hyertension at adulthood in people who originated from sea level regions, highland natives may have developed a protection, masking the effect. This point reemphasises the clear need for future epidemiological studies of human populations at altitude to relate the effects of prenatal hypoxia with postnatal cardiovascular function, separately, in lowland and highland natives. It is also likely that the deleterious programming effects of prenatal hypoxia on cardiovascular function in later life may express themselves, not during basal conditions, but only once the cardiovascular system is stressed. In this context, it is extremely interesting to highlight one study in India reporting a much higher incidence of ischaemic stroke in 20-50 year old men resident at high altitude.[32]

In summary, our observations in human babies and experiments in the chick embryo at high altitude strongly support the hypothesis that fetal oxygen, independent of genetic and maternal nutritional factors, is an important regulator of fetal growth. Accumulating evidence suggests that prenatal hypoxia alone can also be a potent stimulus triggering a developmental origin of cardiovascular dysfunction in the offspring. In both human and avian species, prolonged high altitude residence ancestry can develop a protection against unwanted biological trade-offs, such as the effects of hypoxia on fetal growth. Whether this protection spans into the postnatal and adult periods, minimising the risk of developing cardiovascular disease, remains to be elucidated. The mechanism of this protection is clearly of paramount scientific interest and important clinical application, not only in pregnancy at high altitude but in sea level pregnancies complicated with reduced oxygen delivery to the fetus, such as during placental insufficiency and preeclampsia.

References

1. Barker DJP. Mothers, babies, and disease in later life. Edinburgh: Churchill Livingstone, 1998.
2. Penrose LS. Some recent trends in human genetics. Cardiologia 1954; 6(Suppl):521-529.
3. Walton A, Hammond J. Proc R Soc Lond B Biol Sci 1938; 125:311-335.
4. Giussani DA, Forhead AJ, Gardner DS et al. Postnatal cardiovascular function after manipulation of fetal growth by embryo transfer in the horse. J Physiol 2003; 547(1):67-76.
5. Cohn HE, Sacks EJ, Heymann MA et al. Cardiovascular responses to hypoxemia and acidemia in fetal lambs. Am J Obstet Gynecol 1974; 120:817-824.
6. Giussani DA, Spencer JAD, Moore PJ et al. Afferent and efferent components of the cardiovascular reflex responses to acute hypoxia in term fetal sheep. J Physiol 1993; 461:431-449.
7. Giussani DA, Riquelme RA, Moraga FA et al. Chemoreflex and endocrine components of the cardiovascular response to acute hypoxaemia in the llama fetus. Am J Physiol 1996; 271:R73-R83.
8. Mulder AL, van Golde JC, Prinzen FW et al. Cardiac output distribution in response to hypoxia in the chick embryo in the second half of the incubation time. J Physiol 1998; 508:281-287.
9. Gardner DS, Fletcher AJW, Swann M et al. A novel method for controlled, long term compression of the umbilical cord in fetal sheep. J Physiol 2001; 535(1):217-29.
10. Monge CC, Arregui A, León-Velarde F. Pathophysiology and epidemiology of chronic mountain sickness. Int J Sports Med 1992; 13(Suppl 1):S79-81.
11. Maggiorini M, León-Velarde F. High-altitude pulmonary hypertension: A pathophysiological entity to different diseases. Eur Respir J 2003; 22(6):1019-25.
12. Chang JHT, Rutledge JC, Stoops D et al. Hypobaric hypoxia-induced intrauterine growth retardation. Biol Neonate 1984; 46:10-13.
13. De Grauw TJ, Myers R, Scott WJ. Fetal growth in rats from different levels of hypoxia. Biol Neonate 1986; 49:85-89.
14. Gonzales GF, Guerra-Gracia R. Características hormonales y antropométricas del embarazo y del recien nacido en la altura. In: Gonzales GF, ed. Reproducción humana en la Altura. Lima Peru: Consejo Nacional de Ciencia y Tecnología, 1993:125-141.
15. Zamudio S, Droma T, Norkyel KY et al. Protection from intrauterine growth retardation in Tibetans at high altitude. Am J Phys Anthropol 1993; 91:215-224.
16. Giussani DA, Phillips PS, Anstee S et al. Effects of altitude vs. economic status on birth weight and body shape at birth. Ped Res 2001; 49(4):490-494.
17. Mapa de Pobreza. Una guía para la acción social. Ministerio de Desarrollo Humano, República de Bolivia. 2nd ed. 1995.
18. Post GB, Lujan C, San-Miguel JL et al. The nutritional intake of Bolivian boys. The relation between altitude and socioeconomic status. Int J Sports Med 1994; 15(Suppl 2):S100-S105.
19. Haas JD, Frongillo EF, Stepcik C et al. Altitude, ethnic and sex differences in birthweight and length in Bolivia. Hum Biol 1980; 52:459-477.
20. Moore LG. Maternal O_2 transport and fetal growth in Colorado, Peru and Tibet high-altitude residents. Am J Hum Biol 1990; 2:627-637.
21. Salinas CE, Villena M, Blanco CE et al. The role of oxygen in fetal growth. J Soc Gynecol Investig 2003a; 10(2 Suppl):305A.
22. Gluckman PD, Hanson MA. Living with the past: Evolution, development, and patterns of disease. Science 2004; 305:1733-1736.
23. Lee TM, Zucker I. Vole infant development influenced perinatally by maternal photoperiodic history. Am J Physiol 1988; 255:R831-R838.
24. Bae S, Xiao Y, Li G et al. Effect of maternal chronic hypoxic exposure during gestation on apoptosis in fetal rat heart. Am J Physiol 2003; 285(3):H983-90.
25. Salinas CE, Villena M, Blanco CE et al. Protection against hypoxia-induced cardiomegaly in chick embryos from hens native to high altitude. J Soc Gynecol Investig 2003b; 10(2 Suppl):108A.
26. Li G, Xiao Y, Estrella JL et al. Effect of fetal hypoxia on heart susceptibility to ischemia and reperfusion injury in the adult rat. J Soc Gynecol Investig 2003; 10(5):265-74.
27. Williams SJ, Hemmings DG, McMillen IC et al. Maternal hypoxia during late gestation in rats impairs endothelium-dependent relaxation in mesenteric arteries from four month old male offspring. J Soc Gynecol Investig 2004; 10(2 Suppl):184A.
28. Ruijtenbeek K, Kessels CG, Janssen BJ et al. Chronic moderate hypoxia during in ovo development alters arterial reactivity in chickens. Pflugers Archives 2003; 447(2):158-67.
29. Giussani DA. High altitude and infant mortality: A study of 99 provinces in Bolivia. J Soc Gynecol Investig 2003; 10(2 Suppl):308A.
30. Khalid ME, Ali ME, Ahmed EK et al. Pattern of blood pressures among high and low altitude residents of southern Saudi Arabia. J Hum Hypertens 1994; 8(10):765-9.
31. Ruiz L, Peñaloza D. Altitude and hypertension. Mayo Clin Proc 1977; 52(7):442-5.
32. Jha SK, Anand AC, Sharma V et al. Stroke at high altitude: Indian experience. High Alt Med Biol 2002; 3(1):21-27.

Index

F

G

H

I

K

S

T

U

V

Z

MIX
Papier aus verantwortungsvollen Quellen
Paper from responsible sources
FSC® C105338

If you have any concerns about our products,
you can contact us on
ProductSafety@springernature.com

In case Publisher is established outside the EU,
the EU authorized representative is:
Springer Nature Customer Service Center GmbH
Europaplatz 3, 69115 Heidelberg, Germany

Printed by Libri Plureos GmbH
in Hamburg, Germany